Springer Series in Pharmaceutical Statistics

Recent advances in statistics have led to new concepts and solutions in different areas of pharmaceutical research and development. "Springer Series in Pharmaceutical Statistics" focuses on developments in pharmaceutical statistics and their practical applications in the industry. The main target groups are researchers in the pharmaceutical industry, regulatory agencies and members of the academic community working in the field of pharmaceutical statistics. In order to encourage exchanges between experts in the pharma industry working on the same problems from different perspectives, an additional goal of the series is to provide reference material for non-statisticians. The volumes will include the results of recent research conducted by the authors and adhere to high scholarly standards. Volumes focusing on software implementation (e.g. in SAS or R) are especially welcome. The book series covers different aspects of pharmaceutical research, such as drug discovery, development and production.

More information about this series at http://www.springer.com/series/15122

Meinhard Kieser

Methods and Applications of Sample Size Calculation and Recalculation in Clinical Trials

 Springer

Meinhard Kieser
Institute of Medical Biometry
and Informatics
University of Heidelberg
Heidelberg, Baden-Württemberg, Germany

ISSN 2366-8695   ISSN 2366-8709   (electronic)
Springer Series in Pharmaceutical Statistics
ISBN 978-3-030-49530-5   ISBN 978-3-030-49528-2   (eBook)
https://doi.org/10.1007/978-3-030-49528-2

Mathematics Subject Classification: 62P10

This Springer imprint is published by the registered company Springer Nature Switzerland AG
The registered company address is: Gewerbestrasse 11, 6330 Cham, Switzerland

To Yvonne

Preface

Correct calculation of the required sample size is essential for medical research projects. In clinical trials, a gain in knowledge can only be achieved if the sample size of the study is sufficiently high. At the same time, the sample size should not be larger than necessary due to ethical and economic reasons. Therefore, sample size calculation is one of the most frequent and important topics of biostatistical consulting as well as a major issue when assessing clinical trial protocols in ethics committees.

This book provides methods for sample size calculation and recalculation for a wide range of scenarios encountered when planning clinical trials. To facilitate a deeper understanding of the procedures, the derivations of the methods are presented within a common general framework. Furthermore, application of the methods is illustrated by real clinical trial examples and source code written in the software R implementing the presented approaches is provided. In addition, numerous hints are given to published contributions on the topic. By this, the book aims at giving an up-to-date overview on sample size calculation and recalculation addressing both application-oriented practitioners and biostatisticians who are also interested in the theoretical background of the methods they apply.

Numerous colleagues have contributed to the success of this book. First of all, I would like to thank Andrea Wendel for her extremely patient and careful writing of the manuscript. I gratefully acknowledge the very valuable contributions made by Dr. Marietta Kirchner and Dr. Katrin Jensen that considerably improved the presentation and content. Dr. Laura Benner and Dr. Svenja Schüler kindly prepared the R code implementing the presented sample size methods, performed computations, and reviewed various chapters. Large parts of the book are based on courses I held within the Master Program "Medical Biometry/Biostatistics" conducted by the Institute of Medical Biometry and Informatics (IMBI) at the University of Heidelberg and within summer schools on the topic. The book profited from discussions with the students as well as from the thorough review of draft versions of the book by my colleagues at the IMBI Lukas Baumann, Rouven Behnisch, Samuel Kilian, Johannes Krisam, Jan Meis, Maximilian Pilz, and Stella Erdmann, who also contributed R code, performed computations, and prepared figures. In addition, I am grateful to Springer Publishers and the Editors of the Springer Series in Pharmaceutical Statistics for giving me the opportunity to publish this book. A special thank goes to Dr. Eva Hiripi from Springer for her important advice and support. Definitely, I would not have been able to prepare the work without the continuing encouragement by my wife Yvonne. Her unique ability to inspire and motivate and to ensure at the same time welcome distractions created perfect conditions for writing this book.

Heidelberg, Germany   Meinhard Kieser
April 2020

Contents

Part I
Basics

Chapter 1
Introduction

1.1 Background and Overview

In various aspects, medical research projects are similar to expeditions. Starting point is the desire to move into new areas and to explore open questions. To be successful, in both cases an interdisciplinary team is required where all members have excellent expertise, formidable commitment, and exceptional power of endurance. And there is another aspect expeditions and medical research projects have in common: If the enterprise is completed, full attention focuses on the results. All those considerations made in the planning stage and the sophisticated preparations required for starting the venture have stepped in the background or have even been forgotten. This matches to the fact that in the context of medical research projects the contributions of biostatistics are frequently associated with the data analysis only. In fact, this is an essential task of biostatistics which, however, can unfold its full potential only if an adequate planning has been conducted before. In fact, mistakes made during the planning phase can often no more repaired later on, not even with the best available analysis method. In this context, appropriate calculation of the required sample size plays a key role. If the sample size is too small, the chance of reaching the study aim and of obtaining a reliable answer to the research question is smaller than desired. If the sample size is chosen higher than required, too much patients are unnecessarily included and undergo study-related interventions, and the research project lasts longer and needs more resources. Some studies may even not be feasible in case of a too high sample size and the underlying research question thus has to remain unanswered. Therefore, choice of both a too small and a too high sample size is to be criticized for ethical, scientific, and economic reasons alike.

Notwithstanding its paramount importance, sample size calculation is frequently inadequately performed and reported. During the last decades, the proportion of clinical trial publications reporting sample size calculations has increased (Chen et al. 2013). However, comprehensive assessments of the quality of reporting sample size calculations for clinical trials revealed that there are still substantial deficits. This

© Springer Nature Switzerland AG 2020
M. Kieser, *Methods and Applications of Sample Size Calculation and Recalculation
in Clinical Trials*, Springer Series in Pharmaceutical Statistics,
https://doi.org/10.1007/978-3-030-49528-2_1

applies to articles published in general or indication-specific medical journals [see, for example, Lee and Tse (2017) and Charles et al. (2009) or Tam et al. (2020), Speich (2019), Copsey et al. (2018), Abdulatif et al. (2015), McKeown et al. (2015), and Bariani et al. (2012), respectively] as well as to research protocols submitted to ethics committees (Rauch et al. 2020; Clark et al. 2013). These investigations unanimously found that sample size calculation is often erroneous, based on implausible or unjustified assumptions, and reported incomplete such that the results can not be replicated. From these findings, it can be concluded that there is a high need for improvement. This book tries to make a contribution to this.

In biostatistical consulting, the question concerning the required sample size is one of the most frequent ones. Naturally, a universally valid answer is not possible. In contrast, the method to be applied for sample size calculation depends on the research question, the study design, and the scale level of the primary outcome(s). An all-encompassing overview on the multitude of methods will not be possible within this book. However, a wide area of application fields is covered including, for example, superiority, non-inferiority, and equivalence trials, trials with two groups and multi-armed trials, and both trials assessing efficacy and safety.

Basic principles of statistical testing and sample size calculation are presented in Chap. 2 of Part I of the book. Part II comprises methods for designs with fixed sample size (Chaps. 3–16). Some portions of this material were already presented in the book *Sample Size Calculation in Medical Research* [in German] (Kieser 2018). In fixed sample size designs, the required sample size is calculated in the planning stage and then the trial is conducted including this number of patients. However, during the course of the study, new information may become available from interim data or from external sources, for example from the results of trials which were completed in the meantime. The updated knowledge may cause doubts on the validity of assumptions made upfront. Part III of the book (Chaps. 17–31) presents methods that enable using such information to modify the initially specified sample size mid-course. Two different approaches are considered. Firstly, blinded sample size recalculation procedures to be applied in internal pilot study designs are described. Secondly, methods for sample size reassessment in adaptive designs are presented that are based on data arising from unblinded interim analyses. Biostatistical methodology can only be applied in practice if related software is available. The sample size calculation methods presented in this book have been implemented in the freely available software R. The source code of these programs or links to packages in which such programs are implemented as well as exemplary calls of functions are provided in the Appendix.

The basic setting considered in this book is that of prospective clinical trials comparing two or more interventions. In most such trials, patients are allocated equally to the treatment groups. However, there are situations where unequal allocation may be advantageous, for example to increase recruitment rates by raising the appeal of study participation, to reduce the impact of a learning curve in complex interventions, or to reduce costs. Moreover, as we will see later in the book, settings exist where it is beneficial even from a statistical point of view to apply unequal allocation. Therefore, the methods for sample size calculation presented in this book are throughout given for arbitrary ratios of the group-wise sample sizes.

Before providing some general points to consider when calculating the required sample size, two introductory examples are presented illustrating some of the typical tasks and challenges biostatisticians are confronted with when planning clinical trials. These examples will be taken up in Part II of the book to demonstrate application of methods for sample size calculation.

1.2 Examples

1.2.1 The ChroPac Trial

The incidence of chronic pancreatitis is increasing in industrialized countries. Typical symptoms are upper abdominal pain, nausea and vomiting, weight loss, and possibly diabetes type 1. Several surgical treatment options exist for chronic pancreatitis. Established procedures are duodenum-preserving pancreatic head resection (DPPHR) and partial pancreatoduodenectomy. Although the results of several small randomized trials indicated superiority of DPPHR over partial pancreatoduodenectomy with respect to short-term outcomes, there remained considerable uncertainty concerning the robustness of these results. Therefore, the ChroPac trial (Diener et al. 2010, 2017a) was performed to compare DPPHR with partial pancreatoduodenectomy in a large-scaled, randomized study. Primary endpoint was the mean quality of life within 24 months after surgery, measured after 6, 12, and 24 months by the physical functioning scale of the European Organization for Research and Treatment of Cancer (EORTC) QLQ-C30 questionnaire (Aaronson et al. 1993). Furthermore, a number of secondary outcomes were documented such as other scales of the EORTC QLQ-C30 questionnaire, mortality, perioperative measures (duration of surgery, intraoperative blood loss), general and pancreas-associated postoperative morbidity, and safety. After inclusion of the patients in the trial, they were randomly allocated in a balanced way to the two groups. In order to exclude conscious or unconscious impact on the outcomes due to awareness of the treatment group membership, both patient and outcome assessment were performed masked to group assignment. As the primary aim of the ChroPac trial was demonstration of a difference between the surgical procedures with respect to the functional scale of the EORTC QLQ-C30 questionnaire within 24 months after surgery, sample size calculation is based on this variable. The values for this scale range between 0 and 100 and high values indicate good functionality. A difference of 10 points is considered as clinically relevant (King 1996; Osoba et al. 1998; Maringwa et al. 2011), and in previous trials values of about 20 points were observed for the standard deviation. It was assumed that about 15% of the randomized patients may withdraw before achieving the final study visit. If the mean physical functioning score of the EORTC QLQ-C30 questionnaire is equal for the two surgical procedures to be compared, the probability that the statistical test falsely suggests a difference should be not higher than 0.05. If, however, there is a difference to the extent specified above, it should be detected with a probability of at least 0.80. This was the starting basis when planning the trial, and the question was which sample size was required.

1.2.2 The Parkinson Trial

Morbus Parkinson is one of the most frequent diseases of the central nervous system. Characteristic symptoms are shaking, rigidity, slowness of movement, and difficulty with walking. The motor symptoms of Morbus Parkinson are due to reduced production of dopamine, and this leads the way to a therapeutic approach. Although a compensation of the dopamine deficit will not lead to cure, the symptoms may be alleviated. The clinical trial performed by Jankovic et al. (2007) investigated efficacy and safety of the nonergoline dopamine agonist rotigotine in patients with early Parkinson disease. Here, the active ingredient is delivered by a transdermal system applied to the patient's skin once a day independent of mealtimes which is advantageous as compared to oral application. This treatment was compared with application of a transdermal system without active ingredient (placebo). Neither the patients nor the attending physicians were aware of the actual treatment. The patients were randomly allocated to the two groups, whereby twice as much patients should receive the active treatment. By this, the study participants had a higher chance to receive the investigational drug. Furthermore, additional information on efficacy and safety of the novel treatment was gathered. The primary assessment of efficacy was based on the sum of scores from the "activities of daily living" and "motor examination" components of the Unified Parkinson Disease Rating Scale (UPDRS) (Fahn et al. 1987). In the analysis, the responder rates should be compared between the active and the placebo group. A response to therapy was defined as a decrease of 20% or more in the primary outcome variable at the end of the 27-weeks treatment phase compared to baseline. In the planning stage, a response rate of at most 30% in the placebo group and of at least 50% in the active treatment group was assumed. For the scenario of responder rates of 30% versus 50%, the probability of achieving a statistically significant result should be 0.80. If there is no difference in responder rates between the two groups, the probability for falsely declaring the active treatment as superior should be at most 0.025. Again, the question was which sample size is required for this trial. Furthermore, the question may arise whether there is a price to be paid for unequal sample size allocation in terms of additional sample size and if so, how large it is. Moreover, one may wonder whether alternative choices of the primary outcome may lead to smaller sample sizes. Potential candidates may, for example, be the original non-dichotomized sum of scores of the "activities of daily living" and "motor examination" subscales of the UPDRS, a categorized version where this sum score is split up to more than two classes, or the time until occurrence of response.

In the following chapters, methods for determining the required sample size will be presented for the above-described and many other settings. Their knowledge and understanding enable to perform appropriate sample size calculations in a multitude of application scenarios. Application is illustrated by means of real clinical trial examples. Before, the subsequent section presents considerations that should generally be made whenever sample sizes are calculated.

1.3 General Considerations When Calculating Sample Sizes

This section presents issues which are generally to be considered and quantities which have to be specified in any sample size calculation. Information on these points is not only required to be able to determine the sample size. They also need to be reported in study protocols and publications to make the considerations underlying the choice of the sample size transparent and reproducible. Various guidance documents exist giving recommendations for reporting sample size calculations in clinical trial protocols as, for example, the ICH E9 Guideline *Statistical Principles for Clinical Trials* (ICH 1998) or the Standard Protocol Items: Recommendations for Interventional Trials (SPIRIT) Statement (Chan et al. 2013). Furthermore, the Consolidated Standards for Reporting Trials (CONSORT) Statement (Schulz et al. 2010) was developed to improve reporting in clinical trial publications and includes items on sample size calculation. The description of the conception and methodology underlying sample size calculation has to comprise information on the following key elements.

Trial objective. Most clinical trials compare interventions with the aim of demonstrating differences between the interventions under investigation or superiority of one of them over the other. However, there are also scenarios where non-inferiority or equivalence, respectively, of a new treatment to an established standard is to be shown (see Chaps. 8–10). In these cases, additionally to the items given below, the non-inferiority or equivalence margin, respectively, has to be stated and justified.

Trial design. The parallel group design is by far the most commonly applied approach in clinical trials. This is the reason why we focus in the following on parallel group comparisons. Here, each patient is allocated to receive exactly one of the investigated treatments during the study. An example of an alternative design is the crossover design, where each patient receives several treatments consecutively separated by washout periods. This design may be more efficient than the parallel group design. However, application is restricted to very specific situations as, for example, when investigating treatments in chronic and stable diseases.

Primary outcome, on which sample size calculation is based. In confirmatory analysis (see Sect. 2.1), the primary research question of a clinical trial is assessed. It is advantageous if the related working hypothesis can be formulated in a single test problem by means of a single outcome. Then, the specified significance level (see below) has not to be adjusted. However, there are also situations where the various facets of a disease or a therapeutic effect, respectively, cannot be described by a single endpoint (see Sect. 11.3). In any case, the outcome(s) on which sample size calculation is based has/have to be specified, and it/they should be identical to that/those analyzed in confirmatory analysis.

Applied statistical test. The required sample size essentially depends on the statistical test applied in the analysis. Therefore, sample size calculation should be based on it. The test should be chosen such that it is adequate for the scale level of the outcome and that it is most efficient in the design situation at hand.

Significance level. The significance level defines the probability of erroneously rejecting the null-hypothesis, i.e. of a type I error. In medical research and especially in clinical trials, a two-sided significance level of $\alpha = 0.05$ has been established as the standard approach. In a regulatory environment, application of half of the two-sided level is recommended if a one-sided test is performed, i.e. a one-sided level of 0.025 (ICH E9 1998; CPMP 2000). Levels differing from these values should be justified. For example, clinical trials in rare diseases are often restricted in the number of patients that can be recruited in an acceptable timeframe. To achieve sufficient power with the available sample size, the significance level may be increased (with corresponding consequences for the probability of false-positive study results). If more than one hypothesis is tested in confirmatory analysis, an appropriate multiple test procedure has to be applied to control the (multiple) type I error rate. The consequences for the significance level to be used for sample size calculation have to be explained and taken into account (see Chap. 11).

Statistical power. The statistical power is the probability of correctly rejecting the null-hypothesis, i.e. of making no type II error. Commonly used values for the power are $1 - \beta = 0.80$ or 0.90. The reason why one generally accepts a higher probability for a type II error β as compared to the common type I error rate of 0.05 is solely the required (and feasible) sample size. Any increase in type II error rate increases the chance of failing to detect a true alternative. In most clinical trials, such a wrong decision means that efficacy is not demonstrated although the treatment under investigation is in fact efficacious. In cases where a planned study can presumably not be repeated, a value of 0.95 for the power is sometimes applied because then a type II error is particularly fatal. Use of power values below 0.80 have to be justified. In these cases, the question arises whether it is ethically and economically acceptable that a gain in knowledge can be expected with moderate probability only in view of the additional study-specific measures applied to the patients and the resources spent. In clinical trials in rare diseases, for example, one may argue that the required number of patients can only be recruited if the value for the power is decreased below the commonly used figures. Applying such an approach may then be the only option for evaluating a novel therapy at all.

Assumed clinically relevant difference. Performing a clinical trial with the aim of demonstrating a difference between various interventions is acceptable from an ethical and economical viewpoint only if the assumed difference is clinically relevant for the patient population. In this context, the minimally clinically relevant difference denotes the smallest difference patients regard as beneficial and which (under the assumption of no essential adverse effects and moderate costs) gives rise to a change of the treatment strategy (Jaeschke et al. 1989). When calculating the sample size, this value is the smallest difference that makes sense to be used. However, if there is sufficient evidence that the actual difference is larger than the clinically relevant one, then it seems reasonable to base sample size calculation on a larger value. The notation "expected difference with clinical relevance" includes both these aspects. Applying strictly the minimally clinically relevant difference leads to the maximum sample size in the respective situation. However, assuming a larger difference results in a lower power than aspired in cases where the actual difference is smaller than projected but still clinically relevant. For some outcomes and medical indications, established values for the (minimally) clinically relevant difference exist. If no such values exist for the planned trial, profound clinical expertise is required when defining the assumed difference, and its clinical relevance has to be justified. The Difference Elicitation in Trials (DELTA) project developed guidance on the specification and reporting of the target difference in sample size calculation (Hislop et al. 2014; Cook et al. 2014, 2015, 2018; Sones et al. 2018).

Assumed value of nuisance parameters. Nuisance parameters denote quantities that are not involved in the formulation of the hypotheses translating the research question in statistical terms but which have nevertheless impact on the power of the test and thus on the required sample size. For example, the variance plays such a role for the two-group comparison with normally distributed outcomes (see Chap. 3). Values for nuisance parameters to be inserted in sample size calculation can be obtained, if available, from pilot studies or meta-analyses collecting and summarizing the results of previous trials. However, commonly these values vary between studies and, therefore, a certain degree of uncertainty cannot be avoided in the planning stage. In order to be able to judge the impact on the sample size, it is advisable to include sensitivity analyses in sample size calculation by varying the value of the nuisance parameter(s) in a plausible range. Two attractive alternatives to this approach are applying Bayesian ideas (Sect. 3.5) or using the design with internal pilot study (Chaps. 18–26). In clinical trials performed in the internal pilot study design, the initial assumptions about the value of the nuisance parameter are reviewed during the course of the trial and the sample size can be adjusted accordingly, if required.

Sample size allocation ratio. In clinical trials, the study participants are usually allocated to the interventions in a balanced way, i.e., the sample sizes per group are equal. In many cases as, for example, for the two-group test for difference with normally distributed outcomes (see Chap. 3), this approach is statistically optimal as, all other quantities fixed, it leads to the smallest total sample size. However, there are also situations where it may be considered to allocate more patients to some group(s) [see the reviews on the use of unequal allocation in clinical trials and the underlying rationale by Dumville et al. (2006) and Peckham et al. (2015)]. For example, if a novel therapy is compared to placebo or an established control treatment and if more patients are allocated to the experimental treatment, more information on safety is gained for the new intervention than in case of balanced allocation. Furthermore, it may be more appealing for potential study participants if he/she is allocated with a higher probability to an otherwise not available therapy than to the control. Even in situations where balanced allocation is optimal, the loss in efficiency as compared to the optimal allocation is usually moderate as long as the allocation ratio is not too extreme (see, for example, Fig. 3.3). In some settings, unequal allocation with more patients assigned to the active group may be even advantageous from a statistical viewpoint (see, for example, Sect. 9.3). If unequal allocation is an option, it is advisable to perform the sample size calculation for various allocation ratios to investigate the impact on the sample size.

Assumed rate of withdrawals. It is more the exception than a rule in clinical trials, that all measurements of all study participants can be captured as specified in the protocol. Reasons for deviations from the planned schedule are, for example, early termination of patients due to a marked improvement (or worsening) of the health condition, adverse effects (related to the study interventions or not), or causes that are not connected to the trial. The so-called "intention-to-treat principle" [see, for example, Fisher et al. (1990)] requires that all patients that enter the study have to be included in the analysis. If data is not complete, this can be achieved by applying appropriate methods for imputing missing values. However, incomplete data inevitably leads to a loss of information and thus power. This should already be taken into account in sample size calculation. The simplest way consists in adjusting the calculated sample size upwards by the assumed withdrawal rate. Implicitly, this approach assumes that the withdrawals do not provide any information at all with regard to the primary outcome(s). Therefore, this adjustment generally results in an overcompensation of the information loss, but it assures that the aspired power is not undershot. For example, in the ChroPac trial (Sect. 1.2.1), sample size calculation provided a required total of 172 patients. In the planning stage, about 15% of patients were expected to drop out before regular study end. Therefore, it was decided to randomize a total of 200 patients (Diener et al. 2010). A more refined method for adjusting the sample size due to missing data was proposed by Cook and Zea (2019) for two-group comparisons with binary data. Alternatively, simulation studies may be performed implementing the expected pattern and rate of drop-outs as well as the specified analysis method (see Chap. 16). By this, sample size determination can be tailored to the specific anticipated scenario.

Applied software. In order to be able to reproduce the sample size calculation, the applied software has to be specified. When selecting an appropriate software for sample size calculation, the following aspects should be taken into account:

- the available spectrum of methods should fit to the requirements of the user
- the implemented algorithms should be documented and validated
- the usability should be acceptable
- the price should be appropriate for the user.

Importance and weighting of the above-mentioned aspects depend on the working environment of the user. There exist a multitude of statistical tests and related sample size calculation methods and there is a high diversity in terminology with respect to the designation of the various procedures. Therefore, the methods implemented in the software should in any case be exactly described together with literature references that unequivocally define the underlying methodology.

Chapter 2
Statistical Test and Sample Size Calculation

2.1 The Main Principle of Statistical Testing

When analyzing clinical trials, two objectives are to be distinguished. On the one hand, the results obtained for the study participants have to be described adequately. This is the task of the so-called *descriptive analysis*. Here, the empirical distribution of the captured data is summarized by meaningful measures which is complemented by informative graphics. An example of an informative graphical display of the results is the so-called box-and-whisker plot. It summarizes the essential features of the distribution of data with at least ordinal scale level in a concise way. Figure 2.1 shows the box-and-whisker plots for the primary outcome of the ChroPac trial (see Sect. 1.2.1), the mean of the three post-surgical measurements of the functional scale of the EORTC QLQ-C30 questionnaire, for the two surgical procedures under investigation.

The upper and lower boundary of the box marks the 25% and 75% percentile, respectively (i.e., 25% or 75% of the observed data are smaller or equal of this value; as a consequence, the middle 50% of the data lie within the box), the middle horizontal line in the box identifies the median (i.e., 50% of the observed data are smaller or equal to this value), and "×" denotes the mean. The "whiskers" at the end of the box identify the distance from the boundaries of the box to the smallest or largest observed value, respectively, which is at most 1.5 times the box length away from the respective end of the box. The stars mark outliers that are located outside this distance.

The description of the results for the study participants is an important component of data analysis. However, beyond that it is of special interest what the study results imply for the population of all patients whose characteristics correspond to those of the study participants (i.e. for which the study participants constitute a representative sample). This question is answered by the so-called *confirmatory analysis* (or inferential analysis). Here, a pair of hypotheses is formulated for parameters characterizing the distribution of the outcome in the population, the null-hypothesis H_0 and

© Springer Nature Switzerland AG 2020
M. Kieser, *Methods and Applications of Sample Size Calculation and Recalculation in Clinical Trials*, Springer Series in Pharmaceutical Statistics,
https://doi.org/10.1007/978-3-030-49528-2_2

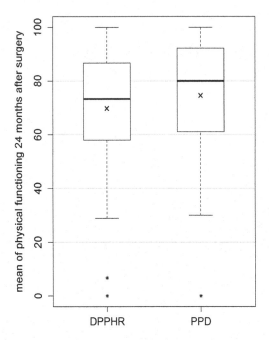

Fig. 2.1 Box-and-whisker plots for the primary outcome of the ChroPac trial "mean of physical functioning score of the EORTC QLQ-C30 questionnaire within 24 months after surgery" (see Sect. 1.2.1; "DPPHR" = duodenum-preserving pancreatic head resection; "PPD" = partial pancreatoduodenectomy). Figure modified according to Diener et al. (2017b)

the alternative hypothesis H_1. The latter describes the study aim, and null- and alternative hypothesis together constitute the complete parameter space. For the ChroPac trial, the null-hypothesis states that there is no difference in the population means with respect to the average of the functional scale of the EORTC QLQ-C30 questionnaire during the 24 months following surgery, the alternative hypothesis claims that the population means are different. Confirmatory analysis is based on a statistical test. Statistical tests define a decision-making procedure for rejecting or accepting H_0 which assures that the probability of erroneously rejecting the null-hypothesis— so-called type I error—is not larger than a specified value α, the significance level. Figure 2.2 illustrates the relationship between study participants ("sample") and population as well as the role of descriptive and confirmatory analysis.

The decision-making procedure defined by a statistical test is based on a so-called test statistic T, whose value is calculated from the study results, and a rejection region R_α. The null-hypothesis is rejected, if the test statistic falls in the rejection region, whereby R_α is chosen such that the probability for this event is not larger than α if the null-hypothesis is true, i.e. $\Pr_{H_0}(T \in R_\alpha) \leq \alpha$. Besides a type I error, another incorrect decision can be made when applying a statistical test, namely falsely accepting H_0; this is denoted as type II error with related probability β. In the ChroPac trial, a type II error occurs if there is a difference in means between the two

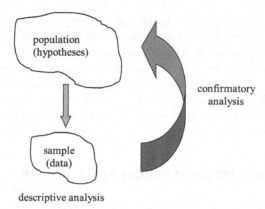

Fig. 2.2 Schematic representation of the roles of descriptive and confirmatory analysis

surgical procedures with respect to the average functional quality of life after surgery for the population of patients with chronic pancreatitis but the study results lead to a value of the test statistic that does not fall in the rejection region R_α, i.e. H_0 cannot be rejected. The probability $1 - \beta$, which is complementary to the probability of a type II error, is denoted as power. Its value depends, amongst others, on the chosen sample size. The other way round, it is the aim of sample size calculation to assure that the sample size is sufficiently large such that the power achieves a specific value. Figure 2.3 shows the possible test decisions and the related probabilities.

The probabilities α and β are also denoted as consumer or sponsor risk, respectively. The meaning of this notation can be illustrated by considering the Parkinson trial (see Sect. 1.2.2). Here, the study aim, as defined by the alternative hypothesis, is to demonstrate superiority of the active treatment with respect to the response rate. A type I error occurs if superiority of the active treatment is claimed although its efficacy is at most as large as that of placebo. This is to the disadvantage of future patients, i.e. "consumers", who then apply an inefficacious (or even harmful) therapy. A type II error is made if an existing superiority of the active therapy is not detected in the confirmatory analysis due to a test decision for accepting H_0. It is especially in the interest of the study team, i.e. the "sponsors", to quantify and limit this risk.

		test decision for	
		H_0	H_1
actually true	H_0	✓ $(1-\alpha)$	type I error (α)
	H_1	type II error (β)	✓ $(1-\beta)$

Fig. 2.3 Test decisions and related probabilities

This is one of the major tasks during the planning stage of a clinical trial and it is accomplished by an appropriate determination of the sample size.

We have already pointed out in Chap. 1 that it is desirable both from an ethical and an economical viewpoint to choose the sample size neither too small nor too large. Therefore, the choice of an adequate sample size is likewise important for the study initiators, the study participants, and the future patients. In the following section, the principle of sample size calculation will be described in a general context.

2.2   The Main Principle of Sample Size Calculation

The task of sample size calculation is to assure a specified power of $1 - \beta$ for the statistical test that will be applied in the analysis at level α and for a defined parameter constellation assumed for the alternative hypothesis. Therefore, the basic elements of sample size calculation are given by a test problem H_0 versus H_1 with a test statistic T and a related rejection region R_α such that the null-hypothesis is rejected if $T \in R_\alpha$. The level α is controlled (and exhausted) by the test if the condition

$$\Pr_{H_0}(T \in R_\alpha) = \alpha \qquad (2.1)$$

holds true. Here and in the following we denote by H_1 the parameter set complementary to H_0 and by H_A a specific constellation in H_1 which is defined by specifying particular parameter values. For a specific alternative hypothesis H_A, the power is given by $\Pr_{H_A}(T \in R_\alpha)$. The desired power $1 - \beta$ is achieved, if the following equation holds true:

$$\Pr_{H_A}(T \in R_\alpha) = 1 - \beta. \qquad (2.2)$$

The task of sample size calculation is to determine the sample size such that (2.2) is fulfilled for a statistical test ensuring the level equation (2.1).

We now consider the more specific situation of a continuous test statistic T and the case that H_0 is rejected for "sufficiently large" values of T; the latter situation holds true for numerous fields of application, for example for all one-sided test problems. The rejection region has then the shape $R_\alpha =]c_\alpha, \infty[, c_\alpha \in \mathbb{R}$. For this scenario, Fig. 2.4 illustrates exemplarily the density functions of the distribution of T under H_0 and H_A, which are denoted by f_0 and f_A, as well as the critical value c_α and the related probabilities of a type I and type II error.

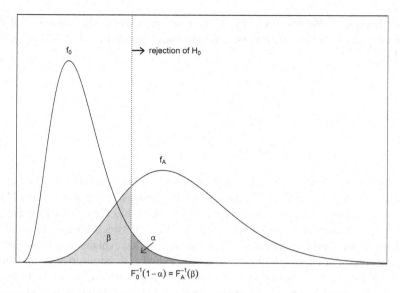

$$F_0^{-1}(1-\alpha) = F_A^{-1}(\beta)$$

Fig. 2.4 Density function of the test statistic under the null-hypothesis (f_0) and the specified alternative hypothesis (f_A) together with related probability of a type I (α) and a type II (β) error

If the distribution function of the test statistic under H_0 or H_A, respectively is denoted by F_0 and F_A, it can easily be seen from Fig. 2.4 that the equations (2.1) and (2.2) are equivalent

$$F_0(c_\alpha) = 1 - \alpha \tag{2.3}$$

and

$$F_A(c_\alpha) = \beta. \tag{2.4}$$

Eliminating c_α leads to the equation

$$F_0^{-1}(1 - \alpha) = F_A^{-1}(\beta) \tag{2.5}$$

or equivalently

$$F_A\big(F_0^{-1}(1 - \alpha)\big) = \beta. \tag{2.6}$$

The distribution function F_A (and usually also F_0) depends on the sample size. It is the task of sample size calculation to solve Eq. (2.5)—or equivalently (2.6)—for n. As the sample size can only reach integer values, there exists generally no sample size for which the right side of (2.5) is exactly equal to the $(1 - \alpha)$-quantile of the null-distribution of T or, equivalently, for which the value of the left side of Eq. (2.6) is exactly equal to the specified type II error rate β, respectively. Therefore, if the exact sample size is to be determined, we are searching for the smallest integer for which

$$F_0^{-1}(1 - \alpha) \leq F_A^{-1}(\beta) \tag{2.7}$$

or equivalently

$$F_A\big(F_0^{-1}(1 - \alpha)\big) \leq \beta. \tag{2.8}$$

In the following, this sample size will be denoted as the solution of (2.5) or (2.6), respectively. When determining approximate sample sizes, solution of (2.5) or (2.6), respectively, may result in a non-integer value which has then to be rounded up (taking into account the sample size allocation ratio between groups) in order to assure the desired power (see Sect. 3.3 for a more detailed explanation of this issue).

For a particular clinical trial, the values for α and β to be inserted in the equations (2.5) and (2.6) or inequalities (2.7) and (2.8), respectively, are fixed. Furthermore, the distribution function F_0 and the critical value c_α are to be known; otherwise one could not perform a level-α test. The only remaining tasks for sample size calculation are therefore determination of the distribution function F_A under a specified alternative hypothesis H_A and solution of (2.7) or (2.8), respectively, for the required sample size.

An important special case of the general setting described above is the situation that the test statistic follows under H_0 the standard normal distribution and under H_A a normal distribution with variance also equal to 1 but which is, compared to the standard normal distribution, shifted by $nc > 0$. Then, $F_0^{-1}(1 - \alpha) = z_{1-\alpha}$ and $F_A^{-1}(\beta) = nc - z_{1-\beta}$, where z_γ is the γ-quantile of the standard normal distribution. Therefore, (2.5) results in

$$z_{1-\alpha} = nc - z_{1-\beta} \tag{2.9}$$

or equivalently

$$nc = z_{1-\alpha} + z_{1-\beta}. \tag{2.10}$$

For a specified alternative, the extent of the shift depends on the sample size n and, hence, n is included in the expression for nc. The required sample size can thus simply be obtained by solving (2.10) for n.

As we will see in the following chapters, application of the general principles described above leads to sample size calculation methods for a wide range of design settings.

Part II
Sample Size Calculation

Chapter 3
Comparison of Two Groups for Normally Distributed Outcomes and Test for Difference or Superiority

3.1 Background and Notation

Outcomes that are (at least approximately) normally distributed are frequent in clinical trials. Furthermore, this chapter is for another reason of special importance. Many statistical tests apply test statistics for decision-making that follow (at least approximately) the normal distribution. This is the reason why for various settings the derivation of sample size calculation methods as well as the basic structure of sample size formulas are very similar to those presented below.

We consider in the following comparisons between a control (C) and an experimental (E) group with respect to a normally distributed outcome. The trial aim is to demonstrate a difference between the two groups or to show superiority of E to C. The group-wise sample sizes are denoted by n_C and n_E, respectively, the total sample size by $n = n_C + n_E$, and the sample size allocation ratio by $r = n_E/n_C$.

3.2 z-Test

With the z-test for two independent samples, the expectations of a normally distributed outcome with known variance can be compared between two groups. The underlying statistical model is given by

$$X_{Cj} \sim N(\mu_C, \sigma^2), \quad j = 1, \ldots, n_C \tag{3.1a}$$

$$X_{Ej} \sim N(\mu_E, \sigma^2), \quad j = 1, \ldots, n_E. \tag{3.1b}$$

Here, X_{Cj} and X_{Ej} denote the mutually independent random variables describing the outcomes under control and experimental treatment, respectively, $N(\mu, \sigma^2)$ is the

© Springer Nature Switzerland AG 2020
M. Kieser, *Methods and Applications of Sample Size Calculation and Recalculation in Clinical Trials*, Springer Series in Pharmaceutical Statistics,
https://doi.org/10.1007/978-3-030-49528-2_3

normal distribution with expectation μ and variance σ^2, where σ^2 is assumed to be known, and n_C and n_E are the group-wise sample sizes. The equality of the variance for both groups is not necessarily required but was only chosen due to technical simplicity of the presentation. In practice, the situation of known values for the variance can hardly be found. However, the derivation of the method for sample size calculation is quite simple for this scenario and is therefore well-suited for a first step into entering the world of sample size calculation. Furthermore, we will see that the results are rather similar to those obtained when dropping this assumption (see Sect. 3.3). Moreover, as already mentioned above, the structure of the resulting sample size formula is identical to those for many other fields of application.

We start assuming that the study aim is to demonstrate a difference between group C and group E. The related two-sided test problem is given by

$$H_0: \mu_E - \mu_C = 0$$
$$\text{versus} \tag{3.2}$$
$$H_1: \mu_E - \mu_C \neq 0.$$

From (3.1a) and (3.1b), the z-test statistic together with its distribution can be derived:

$$Z = \sqrt{\frac{n_C \cdot n_E}{n_C + n_E}} \cdot \frac{\bar{X}_E - \bar{X}_C}{\sigma} \sim N(nc, 1), \tag{3.3}$$

where $\bar{X}_C = \frac{1}{n_C} \cdot \sum_{j=1}^{n_C} X_{Cj}$ and $\bar{X}_E = \frac{1}{n_E} \cdot \sum_{j=1}^{n_E} X_{Ej}$ denote the means in group C and E, and $nc = \sqrt{\frac{n_C \cdot n_E}{n_C + n_E}} \cdot \frac{\mu_E - \mu_C}{\sigma} = \sqrt{\frac{r}{(1+r)^2} \cdot n} \cdot \frac{\mu_E - \mu_C}{\sigma}$ the non-centrality parameter. Thus, under the null-hypothesis Z follows the standard normal distribution, i.e. $Z \sim N(0, 1)$. If the γ-quantile of the standard normal distribution is denoted by z_γ and a realization of the random variable Z by z, the following decision rule defines a two-sided statistical test with type I error rate α for test problem (3.2): "Reject H_0, if $|z| > z_{1-\alpha/2}$, accept H_0 otherwise".

Figure 3.1 shows the density functions of the z-test statistic under the null-hypothesis and for a specified alternative $\Delta_A = (\mu_E - \mu_C)_A > 0$.

For alternative hypotheses $\Delta_A > 0$ which are relevant in practical applications, the probability that the test statistic falls in the region $z < -z_{1-\alpha/2}$ is very close to zero. Therefore, for sample size calculation, the situation described in Sect. 2.2 applies (with "α" replaced by "$\alpha/2$"), and the following equations are to be resolved for sample size calculation: $F_0^{-1}(1 - \alpha/2) = z_{1-\alpha/2}$ and $F_A^{-1}(\beta) = nc - z_{1-\beta}$. As already expressed by Eq. (2.10), the basic equation (2.5) therefore results in the current situation in

$$nc = z_{1-\alpha} + z_{1-\beta}. \tag{3.4}$$

If the ratio of the sample sizes in the two groups is denoted by $r = n_E/n_C$, (3.4) can be written as

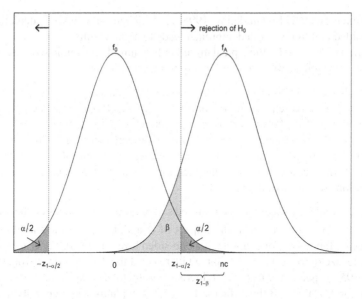

Fig. 3.1 Density function of the test statistic of the z-test under the null-hypothesis (f_0) and for a specified alternative hypothesis (f_A) together with related probability of a type I error (α, two-sided, or $\alpha/2$, one-sided) and a type II error (β)

$$\sqrt{\frac{r}{(1+r)^2}} \cdot n \cdot \frac{\Delta_A}{\sigma} = z_{1-\alpha/2} + z_{1-\beta},$$

which results in the following formula for the total sample size

$$n = n_C + n_E = \frac{(1+r)^2}{r} \cdot \left(z_{1-\alpha/2} + z_{1-\beta}\right)^2 \cdot \left(\frac{\sigma}{\Delta_A}\right)^2 \qquad (3.5)$$

with group-wise sample sizes $n_C = \left(\frac{1}{1+r}\right) \cdot n$ and $n_E = \left(\frac{r}{1+r}\right) \cdot n$.

If the study aim is demonstrating superiority of group E over group C instead of showing a difference between the groups, and if higher values of the outcome correspond to more favorable results (as is, for example, the case for the ChroPac trial introduced in Sect. 1.2.1), the one-sided test problem is given by

$$H_0: \mu_E - \mu_C \le 0$$
$$\text{versus} \qquad (3.6)$$
$$H_1: \mu_E - \mu_C > 0.$$

For a one-sided test at level α, sample size calculation can be performed by applying Eq. (3.5) and just replacing "$\alpha/2$" by "α". However, in the regulatory context it is recommended to conduct one-sided tests at half of the level used for two-sided testing (see, for example, ICH E9 Guideline, 1998, p. 25). Then, sample size calculation for the one-sided and the two-sided approach leads to equal results.

Equation (3.5) includes the typical ingredients required for sample size calculation and illustrates their qualitative relationship to the sample size:

- probability of a type I error α: n increases with decreasing α
- power $1 - \beta$: n increases with increasing $1 - \beta$ (i.e. with decreasing β)
- assumed difference between groups Δ_A: n increases with decreasing Δ_A
- variability (here: variance σ^2): n increases with increasing variability
- sample size allocation ratio r: For many settings, including the one considered here, n is the smallest for $r = 1$ and increases the more r deviates from balanced allocation (see, for example Fig. 3.3).

If Δ_A, σ^2, and r are specified, the required sample size just depends on the quantity $\left(z_{1-\alpha/2} + z_{1-\beta}\right)^2$. For the z-test, the impact of α and $1 - \beta$ on the sample size is illustrated in Table 3.1. For example, if a one-sided level of $\alpha = 0.025$ is applied and the sample size ratio r is kept constant, an increase of the desired power from 0.80 to 0.90 or 0.95, respectively, leads to an increase of the total sample size by about one third (factor 1.34) or two third (factor 1.66). This explains why generally a higher type II error rate as compared to the type I error rate is accepted and why sample size calculation is usually performed for power values of 0.80 or 0.90: Sample sizes resulting from higher power values can commonly only hardly, if at all, be realized within an acceptable time frame. General considerations on the choice of α and $1 - \beta$ can be found in Sect. 1.3.

Table 3.1 Impact of significance level $\alpha/2$, one-sided, and power $1 - \beta$ on the total sample size for the z-test. The displayed factor is the multiple of the total sample size required for $\alpha/2 = 0.025$ and $1 - \beta = 0.80$

Significance level $\alpha/2$	Power $1 - \beta$	Factor
0.005	0.80	1.49
	0.90	1.90
	0.95	2.27
	0.99	3.06
0.025	0.80	1.0
	0.90	1.34
	0.95	1.66
	0.99	2.34
0.05	0.80	0.79
	0.90	1.09
	0.95	1.38
	0.99	2.01

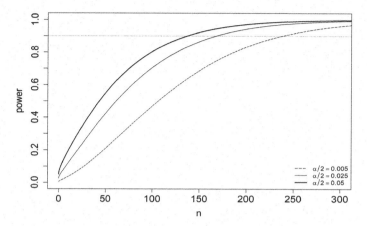

Fig. 3.2 Power of the z-test at one-sided significance level $\alpha/2$ depending on the total sample size $(\Delta_A = 10, \sigma = 20, r = n_E/n_C = 1$; see Example 3.1)

Example 3.1 Sample size calculation for the **ChroPac trial** (see Sect. 1.2.1) was based on the assumptions $\Delta_A = 10$ and $\sigma = 20$. A power of $1 - \beta = 0.90$ was desired for the specified alternative and the total sample size should be balanced between the groups, i.e. $r = 1$. Assuming for illustrative purposes that the variance is known and that thus the z-test can be applied in the analysis, a total sample size of $2 \times 85 = 170$ results for a one-sided significance level of $\alpha/2 = 0.025$. Figure 3.2 shows the power depending on the total sample size for various one-sided significance levels $\alpha/2$ and the planning situation of the ChroPac trial. The antagonistic effects of the probabilities of a type I error and a type II error are evident: Keeping both quantities "small" can only be achieved at the price of a large sample size.

In Eq. (3.5), the total sample size can be seen as a function $n(r)$ of the allocation ratio r. Calculating the first and second derivation with respect to r leads to the result that the total sample size is minimized for balanced groups, i.e. for $r = 1$. Thus, equal allocation of the sample size to the two groups is statistically optimal in this situation. The ratio $\frac{n(r)-n(1)}{n(1)} = \frac{(r-1)^2}{4r}$ defines the percentage increase of the total sample size compared to balanced allocation if the values for $1 - \beta$, Δ_A, and σ^2 are kept fixed and if the sample size is split up in the ratio $r = n_E/n_C$ between the two groups (Fig. 3.3).

The increase is moderate if the sample sizes of the groups do not differ too much and amounts, for example, to 4.2% for $r = 3/2$ and 12.5% for $r = 2$. However, the additionally required sample size sharply increases for more extreme allocation ratios and equals 33.3% for $r = 3$ and 80.0% for $r = 5$.

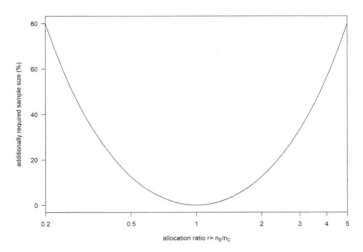

Fig. 3.3 Percentage increase of the total sample size for the z-test to achieve the same power for a sample size allocation ratio $r = n_E/n_C$ as for balanced allocation, i.e. for $r = 1$

3.3 t-Test

The two-sample t-test is based on the same statistical model (3.1a) and (3.1b) as the z-test, but now the variance σ^2 is unknown. The test statistic is given by

$$T = \sqrt{\frac{n_C \cdot n_E}{n_C + n_E}} \cdot \frac{\bar{X}_E - \bar{X}_C}{S} \tag{3.7}$$

with the pooled standard deviation

$$S = \sqrt{\frac{(n_C - 1) \cdot S_C^2 + (n_E - 1) \cdot S_E^2}{n_C + n_E - 2}} \tag{3.8a}$$

and estimated variances in group C and E, respectively:

$$S_C^2 = \frac{1}{n_C - 1} \cdot \sum_{j=1}^{n_C} (X_{Cj} - \bar{X}_C)^2, \quad S_E^2 = \frac{1}{n_E - 1} \cdot \sum_{j=1}^{n_E} (X_{Ej} - \bar{X}_E)^2. \tag{3.8b}$$

Under H_0: $\mu_E - \mu_C = 0$, T follows the central t-distribution with $n_C + n_E - 2$ degrees of freedom. Denoting the related $(1 - \alpha/2)$-quantile by $t_{1-\alpha/2,n_C+n_E-2}$, the following decision rule defines a two-sided test with type I error rate α for the test problem (3.2): "Reject H_0 if $|t| > t_{1-\alpha/2,n_C+n_E-2}$, accept H_0 otherwise". For a one-sided test at type I error rate $\alpha/2$ for the test problem (3.6), the decision rule is "Reject H_0 if $t > t_{1-\alpha/2,n_C+n_E-2}$, accept H_0 otherwise". Under an alternative hypothesis H_A specified by $\Delta_A = (\mu_E - \mu_C)_A \neq 0$, T follows the non-central t-distribution with

the same number of degrees of freedom and the same non-centrality parameter as
for the *z*-test, i.e.

$$nc = \sqrt{\frac{r}{(1+r)^2} \cdot n \cdot \left(\frac{\Delta_A}{\sigma}\right)} \qquad (3.9)$$

with $r = n_E/n_C$. The exact sample size can be obtained by iteratively searching
for the smallest integer n fulfilling the inequality (2.7) or (2.8), respectively. If the
distribution functions of the central and non-central *t*-distribution with $n-2$ degrees
of freedom are denoted by $F_{T;n-2,0}$ and $F_{T;n-2,nc}$, respectively, the required sample
size to achieve a power $1 - \beta$ for a two-sided test at level α or a one-sided test at
level $\alpha/2$, respectively, is thus given by the smallest integer n fulfilling

$$F_{T;n-2,0}^{-1}(1 - \alpha/2) \leq F_{T;n-2,nc}^{-1}(\beta). \qquad (3.10)$$

The sample size obtained from (3.5) for the *z*-test is a good starting value for the
searching procedure which is never larger than the solution of (3.10). As for the
z-test, balanced sample size allocation, i.e. $r = 1$, leads to the minimum total sample
size and is thus optimal.

Numerous approximation formulas exist which allow calculation of the (approximate) sample size for the *t*-test without iterative computations and without requiring
application of the non-central *t*-distribution. Due to the broad availability of statistical software, these approximations are nowadays actually superflous. However, to
illustrate the quantitative relationship between the sample sizes required for the *z*- and
the *t*-test, we provide the approximation formulas proposed by Guenther (1981) for
balanced groups and by Schouten (1999) for the general case of arbitrary allocation
ratios:

$$n_{GS,C} = \left(\frac{1+r}{r}\right) \cdot \left(z_{1-\alpha/2} + z_{1-\beta}\right)^2 \cdot \left(\frac{\sigma}{\Delta_A}\right)^2 + \frac{\left(z_{1-\alpha/2}\right)^2}{2 \cdot (1+r)} \qquad (3.11a)$$

$$n_{GS,E} = r \cdot n_{GS,C}. \qquad (3.11b)$$

From these formulas follows that the approximate total sample size required for the
t-test is by just $\left(z_{1-\alpha/2}\right)^2/2$ larger than the exact one needed for the *z*-test:

$$n_{GS} = n_{GS,C} + n_{GS,E} = n_{z\text{-test}} + \frac{\left(z_{1-\alpha/2}\right)^2}{2}. \qquad (3.11c)$$

For the common two-sided significance level $\alpha = 0.05$ we have $\left(z_{1-\alpha/2}\right)^2/2 =$
1.92. Table 3.2 shows the total sample sizes resulting from the various calculation
methods for a wide range of standardized group differences Δ_A/σ, sample size
allocation ratios $r = n_E/n_C = 1$ and 2, two-sided type I error rate $\alpha = 0.05$ (or

Table 3.2 Total sample size for the z-test (exact) and the t-test (exact and according to the approximation formula of Guenther-Schouten) ($\alpha/2 = 0.025$, one-sided, $1 - \beta = 0.80$, $r = n_E/n_C = 1\,(2)$)

Δ_A/σ	z-test	t-test	
		Exact	Guenther-Schouten approximation
0.1	3140 (3534)	3142 (3534)	3142 (3534)
0.2	786 (885)	788 (885)	788 (885)
0.3	350 (393)	352 (396)	352 (396)
0.4	198 (222)	200 (225)	200 (225)
0.5	126 (144)	128 (144)	128 (144)
0.6	88 (99)	90 (102)	90 (102)
0.7	66 (75)	68 (75)	66 (75)
0.8	50 (57)	52 (60)	52 (60)
0.9	40 (45)	42 (48)	42 (48)
1.0	32 (36)	34 (39)	34 (39)
1.25	22 (24)	24 (27)	24 (27)
1.5	14 (18)	18 (18)	16 (18)

one-sided $\alpha/2 = 0.025$) and power $1 - \beta = 0.80$. It turns out that the Guenther-Schouten formula provides a quite accurate approximation to the exact sample size. Furthermore, the difference between the sample sizes for the z- and the t-test are for most practical applications negligible. Nevertheless, as a matter of principle, sample size calculation should always be based on the best available method, i.e. that assuring maximum accuracy, which is in the current case exact calculation.

It should be noted that different strategies exist for rounding the value obtained from a closed sample size formula to an integer value. In Table 3.2, the values obtained from Eq. (3.5) for the z-test are rounded up to a multiple of 2 for balanced allocation ($r = 1$), and to a multiple of 3 for $r = 2$ to maintain the specified allocation ratio. In most of the numerical examples given in the book, r is an integer and the group-wise sample sizes are determined in this way if not stated otherwise. By this, the ratio of the sample sizes is exactly equal to the specified allocation ratio.

A different strategy, which is generally applicable for any arbitrary $r > 0$, is to round the sample size up to the next integer for each group separately. In detail, the unrounded n is used to calculate $n_C = \left(\frac{1}{1+r}\right) \cdot n$ and $n_E = \left(\frac{r}{1+r}\right) \cdot n$, and the returned values are both rounded up to the next integer. The resulting allocation ratio might then be slightly different to the specified allocation ratio. For example, for $\Delta_A/\sigma = 0.5$ and $r = 2$, the sample size formula (3.5) for the z-test yields the value 141.28. If the aim is to exactly maintain the given allocation ratio, a sample size of 144 ($= 48 + 96$), which is the next higher integer divisible by 3, is used. The alternative strategy would lead to group-wise sample sizes of $n_C = 1/3 \cdot 141.28 = 47.09 \approx 48$ and $n_E = 2/3 \cdot 141.28 = 94.19 \approx 95$ resulting in a total sample size of 143 with an actual r of 1.98. The latter strategy is used throughout in the R codes provided in the Appendix that implement the sample size calculation methods.

If the sample size is determined iteratively [as for the *t*-test according to (3.10)], in each step the particular total sample size n is split up to the group-wise sample sizes according to $n_C = \left(\frac{1}{1+r}\right) \cdot n$ and $n_E = \left(\frac{r}{1+r}\right) \cdot n$, the resulting sample sizes n_C and n_E are rounded up to the next integer, and the achieved power is calculated for the associated sample size and allocation ratio. This strategy is implemented in the provided R codes for iterative sample size determination.

Example 3.2 In the **ChroPac trial** (see Sect. 1.2.1), sample size calculation was performed for evaluating the primary outcome with a two-sided *t*-test at level $\alpha = 0.05$. For an assumed clinically relevant difference between groups of $\Delta_A = 10$ and a standard deviation of $\sigma = 20$ (i.e. $\Delta_A / \sigma = 0.5$), a power of $1 - \beta = 0.90$ should be assured. For balanced groups, the exact sample size amounts to $n = 172$. The same sample size results from the approximation formula of Guenther-Schouten, while the formula for the z-test provides a sample size of 170. If an allocation ratio of 2:1 or 3:1, respectively, is chosen, all three calculation methods result in total sample sizes of 192 or 228.

Remarks

1. When applying the sample size formulas (3.5), (3.10), or (3.11a)–(3.11c), respectively, the quantity inserted for the population variance σ^2 is commonly regarded as a fixed value. However, in practice the value for σ^2 is frequently obtained from a pilot study or a published trial and is thus a random variable. As a consequence, both sample size and power are also random variables. If a single sample is available for sample size calculation, Kieser and Wassmer (1996) showed that the probability of achieving the planned power is less than 0.50 when using the z-test sample size formula (3.5) and when replacing the population variance by the sample variance. Furthermore, the following result was derived: If the upper boundary of the one-sided $(1 - \gamma)$ confidence interval for σ^2 is inserted in formula (3.5), the probability to achieve the desired power is equal to $1 - \gamma$. When inserting in formula (3.5) the upper boundary of the one-sided 80% (90%) confidence interval for σ^2 instead of the sample variance obtained from a pilot study, this results in an inflation of the sample size by a factor of 1.34, 1.21, 1.14, and 1.09 (1.55, 1.34, 1.22, 1.15) for total pilot study sample sizes of 25, 50, 100, or 200 patients. For a clinical trial application of this approach see Palmer et al. (2012). Chen et al. (2013) considered the situation that more than one sample is available at the planning stage. Their simulation studies indicated that using the weighted average of the sample variances may result in an insufficient probability of achieving the aspired power. Recommendations are given how to choose the value to be inserted in place of the population variance depending on the number of available samples.
2. If the *t*-test is applied in the analysis, the question arises how strict the assumption of normal distribution of the data has to be fulfilled. The latter aspect can be addressed by the central limit theorem. It states that the distribution of the mean

of independent identically distributed random variables with finite variance is approximately normally distributed for large sample sizes (see, for example, Durrett 2010, p. 124ff); this property is denoted as "asymptotic normality". For the derivation of the distributional properties of the t-test statistic, it is not required for the nominator that the outcomes itself are normally distributed but it is sufficient that the difference of means, i.e. $\bar{X}_E - \bar{X}_C$, follows the normal distribution. However, according to the central limit theorem, if the sample size is "sufficiently large" this holds, at least approximately, also true in case of non-normally distributed data. Figure 3.4 shows as an example the density of the central chi-square distribution with three degrees of freedom and the density of means calculated from ten observations drawn from this distribution. It can be seen that, although the original distribution is skewed, the distribution of the mean is close to a normal distribution even for this small sample size. This

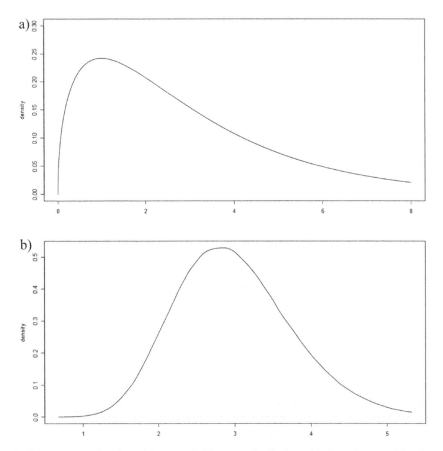

Fig. 3.4 **a** Density function of the central chi-square distribution with three degrees of freedom; **b** density function of the mean calculated from ten observations drawn from the central chi-square distribution with three degrees of freedom

explains the robustness of the *t*-test against deviations from the normal distribution assumption (see, for example, Posten 1978). If the distribution of the original observations is not "too distorted", the approximation is satisfactory already for comparatively small sample sizes. For the sample size of the ChroPac trial, the normal distribution assumption is unproblematic. Moreover, extensive simulation studies have been performed demonstrating considerable robustness of the *t*-test under heterogeneity of variances [see, for example, Posten et al. (1982) and for a summary of the robustness properties of the two-sample *t*-test Posten (1984)].

3.4 Analysis of Covariance

If there is a remarkable impact of one or several covariates on a normally distributed outcome and if the relationship can be described by a linear model, it is advisable to apply an analysis of covariance (ANCOVA). By this, a biased estimation of the treatment group difference can be prevented and the variance of the estimator can be reduced. For these reasons, application of adjusted analyses that take into account influential covariate(s) is also recommended by regulatory guidelines [see, for example, EMA (2015) and FDA (2019)].

We consider here the case of a single random covariate; for the situation of multiple random covariates see Remark 1 at the bottom of this section. The outcome X and the covariate W are assumed to be bivariate normally distributed in each group $i = C, E$, with variance σ^2 and σ_W^2, respectively, and correlation ρ. We suppose the statistical model

$$X_{Cj} = \tau_C + \beta \cdot \left(W_{Cj} - \bar{W}\right) + \varepsilon_{Cj}, \quad j = 1, \ldots, n_C \qquad (3.12a)$$

$$X_{Ej} = \tau_E + \beta \cdot \left(W_{Ej} - \bar{W}\right) + \varepsilon_{Ej}, \quad j = 1, \ldots, n_E, \qquad (3.12b)$$

where X_{ij} and W_{ij} denote the value of the outcome variable and the covariate, respectively, for patient j in group i, $\bar{W} = \frac{1}{n_C + n_E} \cdot \sum_{i=C,E} \sum_{j=1}^{n_i} W_{ij}$ is the overall mean of the covariate for the pooled sample of both groups, β is the slope (assumed to be equal in both groups), and ε_{ij} are independent normally distributed errors with expectation 0 and constant variance. The difference $\tau_E - \tau_C = E(X_E | W) - E(X_C | W)$ denotes the treatment group difference adjusted for the covariate W, for which the one- and two-sided test problem, respectively, is formulated analogously to (3.2) and (3.6):

$$H_0: \tau_E - \tau_C = 0$$
$$\text{versus}$$
$$H_1: \tau_E - \tau_C \neq 0$$

and

$$H_0: \tau_E - \tau_C \le 0$$
$$\text{versus}$$
$$H_1: \tau_E - \tau_C > 0.$$

From the above model equations and the data, the least squares estimate $\hat{\beta}$ can be obtained resulting in the following estimates of τ_C and τ_E

$$\hat{\tau}_C = \bar{X}_C - \hat{\beta} \cdot (\bar{W}_C - \bar{W})$$

$$\hat{\tau}_E = \bar{X}_E - \hat{\beta} \cdot (\bar{W}_E - \bar{W}),$$

where $\bar{X}_C$ and $\bar{X}_E$ and $\bar{W}_C$ and $\bar{W}_E$, respectively, are the means of the outcome and the covariate in the respective group. Analogously to the t-test, the ANCOVA test statistic is given by

$$T_{ANCOVA} = \frac{\hat{\tau}_E - \hat{\tau}_C}{\sqrt{\widehat{Var}(\hat{\tau}_E - \hat{\tau}_C)}}, \tag{3.13}$$

where the estimator of the variance of the adjusted treatment group difference estimate is given by

$$\widehat{Var}(\hat{\tau}_E - \hat{\tau}_C) = \frac{n_C + n_E - 2}{n_C + n_E - 3} \cdot (1 - \hat{\rho}^2) \cdot S^2 \cdot \left(\frac{1}{n_C} + \frac{1}{n_E} + \frac{(\bar{W}_E - \bar{W}_C)^2}{(n_C + n_E - 2) \cdot S_W^2} \right) \tag{3.14}$$

with the pooled variances [see (3.8a) and (3.8b) in Sect. 3.3] S^2 and S_W^2 for the outcome and the covariate, respectively, and the estimated correlation $\hat{\rho}$. Under the null-hypothesis, T_{ANCOVA} follows the central t-distribution with $n_C + n_E - 3$ degrees of freedom. Under an alternative hypothesis H_A specified by $\Delta_A = (\tau_E - \tau_C)_A \ne 0$, T_{ANCOVA} follows the non-central t-distribution with the same number of degrees of freedom and non-centrality parameter $nc = \sqrt{\frac{r}{(1+r)^2} \cdot n} \cdot \frac{\Delta_A}{\sqrt{1-\rho^2} \cdot \sigma}$, where $r = n_E/n_C$. This follows from Shan and Ma (2014) who considered the test statistic T_{ANCOVA}^2. Note that the non-centrality parameter for the ANCOVA differs from that of the t-test just by the factor $(1 - \rho^2)$ in the denominator and that both coincide for $\rho = 0$. As for the t-test, the method for exact calculation can be derived from the basic inequality (2.7) by inserting the respective distribution functions under the null- and the alternative hypothesis:

$$F_{T;n-3,0}^{-1}(1 - \alpha/2) \le F_{T;n-3,nc}^{-1}(\beta), \tag{3.15}$$

where $F_{T;n-3,0}$ and $F_{T;n-3,nc}$, respectively, denote the distribution functions of the central and non-central t-distribution with $n - 3$ degrees of freedom. Iteratively searching for the smallest integer n fulfilling (3.15) provides the exact required sample size. Again, as for the z- and the t-test, balanced allocation (i.e. $r = 1$) is optimal in the sense that it leads to the smallest total sample size.

The following considerations result in closed approximation formulas that provide insight in the quantitative relationship between the sample sizes required for ANCOVA, t-test, and z-test. For sample sizes common in clinical trials, $(n_C + n_E - 2)/(n_C + n_E - 3) \approx 1$ holds true. Furthermore, if the patients are randomly allocated to the groups, the covariates are balanced between the investigated treatments. Therefore, $(\bar{W}_E - \bar{W}_C)^2/((n_C + n_E - 2) \cdot S_W^2)$ is "small" and can be neglected as compared to the other terms in (3.14). Applying these approximations, the approximate expression for T_{ANCOVA} differs from the test statistic of the t-test just by the factor $(1 - \hat{\rho}^2)$ in the variance term constituting the denominator. Hence, for "large" sample sizes, it holds approximately true that $T_{ANCOVA} \sim N(0, 1)$ under the null-hypothesis and that $T_{ANCOVA} \sim N(nc, 1)$ under a specified alternative H_A. Therefore, we can apply Eq. (2.10) which results in

$$\sqrt{\frac{r}{(1+r)^2} \cdot n} \cdot \left(\frac{\Delta_A}{\sqrt{(1 - \rho^2)} \cdot \sigma} \right) = z_{1-\alpha/2} + z_{1-\beta}. \tag{3.16}$$

Solving (3.16) for n leads to the following approximation formula for the total sample size required for ANCOVA which was proposed by Frison and Pocock (1999):

$$n_{FP} = \frac{(1+r)^2}{r} \cdot \left(z_{1-\alpha/2} + z_{1-\beta} \right)^2 \cdot \frac{(1 - \rho^2) \cdot \sigma^2}{\Delta_A^2}. \tag{3.17}$$

We have already seen in Sect. 3.3 that the Guenther-Schouten correction provides adequate sample sizes when approximating the t-distribution by the normal distribution. In analogy, applying this idea results in the Guenther-Schouten-corrected sample size for the ANCOVA approach:

$$n = n_{FP} + \frac{\left(z_{1-\alpha/2} \right)^2}{2}. \tag{3.18}$$

The group-wise sample sizes are then obtained as $n_C = \left(\frac{1}{1+r} \right) \cdot n$ and $n_E = \left(\frac{r}{1+r} \right) \cdot n$, respectively. The sample sizes provided by (3.18) are quite accurate. For example, considering a one-sided level $\alpha/2 = 0.025$, power values of $1 - \beta = 0.80$ and 0.90, standardized effects $\frac{(\tau_C - \tau_E)_A}{\sigma} = 0.1, 0.2, \ldots, 1.5$, correlations $\rho = 0.1, 0.2, \ldots, 0.9$, and $r = 1, 2, 3$, covers total sample sizes between 8 and 4164. For by far the most of these scenarios, the approximate sample size obtained from (3.18) and the exact one resulting from (3.15) are equal. In the few instances where they differ, the approximate total sample size is by the smallest amount possible lower than the exact one, i.e. by 2 for $r = 1$, by 3 for $r = 2$, and by 4 for $r = 3$. Nevertheless, the exact method should be used as it assures that the desired power is in fact achieved under the assumptions made in a specific setting.

The sample size formulas (3.17) and (3.18) show that the sample size can be considerable decreased when applying an ANCOVA instead of a t-test. For example, if the correlation between outcome and covariate amounts to $\rho = 0.5$, the saving is 25%, and for $\rho = 0.7$ it is nearly 50%. Correlations between outcome and covariates of this magnitude are not uncommon. For example, baseline values measured before start of an intervention are frequently strongly correlated with the outcome value during or at the end of therapy. In clinical trials where a baseline value of the outcome is measured, application of a baseline-adjusted ANCOVA is recommended by regulatory guidelines (EMA 2015, p. 7). As demonstrated above, this has the potential to lead to a substantially smaller sample size as compared to those for an unadjusted analysis.

Example 3.3 In the **ChroPac trial** (see Sect. 1.2.1), the primary analysis was specified as an ANCOVA adjusting for the functional scale of the EORTC QLQ-C30 questionnaire measured preoperatively. To take this evaluation approach into account in sample size calculation, an assumption would have to be made concerning the correlation between the value of the endpoint before surgery and the mean of the measurements 6, 12, and 24 weeks after. However, in the planning stage of the trial no data was available from which information could have been derived for this quantity. Therefore, the sample size was determined "conservatively" for the t-test (i.e. for $\rho = 0$) resulting in a total sample size of 172. Under the assumption $\rho = 0.4$ (0.5, 0.6, 0.7) and the same parameter specifications otherwise ($\alpha = 0.05$, two-sided, $1 - \beta = 0.90$, $(\tau_E - \tau_C)_A = 10$, and $\sigma = 20$), the required exact total sample size for the ANCOVA calculated from (3.15) equals to the approximate one obtained from (3.18) and amounts to 144 (130, 110, 88). By the way, the estimated correlation obtained from the analysis of the ChroPac trial amounts to $\hat{\rho} = 0.56$. If this information had been available in the planning stage, sample size calculation would have benefited considerably.

Remarks

1. For the case of multiple random covariates, exact sample size calculation for ANCOVA was provided by Shieh (2017). Zimmermann et al. (2020) presented various approximation formulas for this situation and investigated their accuracy. Again, information on the correlation between the outcome and the covariates is needed for determining the sample size.
2. Tang (2018) derived exact sample size calculation methods for ANCOVA with multiple covariates taking into account stratification factors. Furthermore, he showed by simulations (which included also unstratified ANCOVA) that the power and sample size formulas are very accurate also for non-normally distributed covariates, even if the sample size is small.
3. In Sect. 3.3, Remark 2, it was mentioned that the t-test is quite robust in the presence of violations of the underlying assumptions. Similar results hold true for ANCOVA in case that the assumptions of normality and homogeneity of regression slopes are violated [see, for example, Sullivan and D'Agostino (1996), (2002), and (2003) and references given there].

3.5 Bayesian Approach

3.5.1 Background

In this book, we restrict to the framework of frequentist inference with corresponding hypothesis testing approach. The Bayesian methods for sample size calculation considered in this chapter stick to this setting and are therefore in the literature sometimes denoted as "hybrid classical and Bayesian" techniques (Spiegelhalter et al. 2004). The "traditional" power concept that we have exclusively considered up to now is a conditional one in that a power of $1 - \beta$ for rejecting the null-hypothesis H_0 is conditional on the assumed specific alternative H_A (and potentially on the value of further parameters not included in H_A). It is therefore in the literature sometimes denoted as conditional frequentist power (Brutti et al. 2014). Using a Bayesian approach allows to include prior uncertainty on the value of parameters that have an impact on the required sample size. In the subsequent section, this procedure will be illustrated for the comparison of two groups with normally distributed outcomes and the situation that there is some uncertainty about the treatment effect (what is usually the case).

3.5.2 Methods

As in Sect. 3.2, we assume two groups with normally distributed outcomes $X_{ij} \sim N(\mu_i, \sigma^2)$, $i = C, E, j = 1, \ldots, n_i$, with equal and known variance σ^2. The z-test statistic Z given by (3.3) is applied to assess the one-sided test problem H_0: $\mu_E - \mu_C \leq 0$ versus H_1: $\mu_E - \mu_C > 0$. For an alternative H_A specified by a value $\Delta_A = (\mu_E - \mu_C)_A > 0$ and for total sample size n and sample size allocation ratio $r = n_E/n_C$, the "traditional" power TP to reject H_0 at one-sided level $\alpha/2$ is then given by

$$TP(\Delta_A) = \Pr(Z > z_{1-\alpha/2}|\Delta_A) = 1 - \Phi\left(z_{1-\alpha/2} - \sqrt{\frac{r}{(1+r)^2} \cdot n} \cdot \left(\frac{\Delta_A}{\sigma}\right)\right),$$

(3.19)

where Φ denotes the distribution function of the standard normal distribution.

The value Δ_A is chosen as a difference that we would not like to miss. Therefore, it includes both the aspects of clinical relevance and of expected magnitude of the effect (see Sect. 1.3). However, there is usually some uncertainty about the true effect. We assume that in the planning phase the knowledge on the treatment effect can be summarized by a prior distribution with density $f(\Delta)$. An appropriate prior distribution may, for example, be derived by formal prior elicitation [see, for example, Dallow et al. (2018) and references given there]. Then, the expected power EP averages the traditional power over the prior distribution of the treatment effect:

$$EP = \int_{-\infty}^{\infty} \Pr\left(Z > z_{1-\alpha/2}|\Delta\right) \cdot f(\Delta)d\Delta$$

$$= 1 - \int_{-\infty}^{\infty} \Phi\left(z_{1-\alpha/2} - \sqrt{\frac{r}{(1+r)^2} \cdot n} \cdot \left(\frac{\Delta}{\sigma}\right)\right) \cdot f(\Delta)d\Delta \tag{3.20}$$

Equation (3.20) provides the unconditional probability to reject H_0 in the frequentist analysis. This quantity is alternatively denoted as average power (Spiegelhalter et al. 1994) and is also used in other and more general contexts under the names Bayesian predictive power (Spiegelhalter et al. 1986), assurance (O'Hagan et al. 2005), or probability of success (Wang et al. 2013). Note that this concept includes the "traditional" power: When there is no uncertainty in the planning phase or when existing uncertainty is ignored, a point-mass is chosen as prior density and the expression for the expected power equals to the conditional frequentist power. Note that the so-called conditional expected power is also considered in the literature (Brown et al. 1987; Carleglio et al. 2015). This quantity is defined as the expected power given that $\Delta > 0$, i.e. conditional on that the experimental treatment is superior to the control. Usually, the difference between expected power and conditional expected power is negligible, and the former concept is used in the following due to technical simplicity.

Let us for illustrative purposes consider the situation of a normal prior distribution, i.e. $\Delta \sim N\left(\Delta_A, \tau^2\right)$. The value to be chosen for τ^2 can, for example, be obtained from historical data, meta-analyses of previous trials, or from expert opinion surveys. Alternatively, if there exists a guess of the prior probability that the true treatment effect is negative, the standard deviation of the normal prior that leads to this value can be obtained by $\tau = -\Delta_A/\Phi^{-1}(\Pr(\Delta < 0))$. For example, for the **ChroPac trial** (see Sect. 1.2.1) with $\Delta_A = 10$, τ amounts to 3.24, 4.30, and 6.08 for $\Pr(\Delta < 0) = 0.001, 0.01$, and 0.05, respectively. Figure 3.5 shows the densities of the related normal priors for these scenarios.

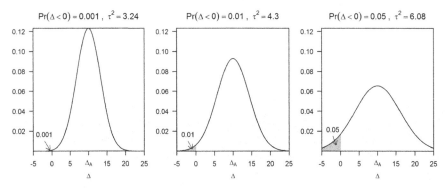

Fig. 3.5 Densities of the normal prior distributions $N\left(\Delta_A, \tau^2\right)$ with $\Delta_A = 10$ and $\tau^2 = \left[\Delta_A/\Phi^{-1}(\Pr(\Delta < 0))\right]^2$ for prior probabilities of a negative treatment effect $\Pr(\Delta < 0) = 0.001$, 0.01, and 0.05

Having fixed the prior distribution, the expected power can be calculated. For the normal prior $N(\Delta_A, \tau^2)$, the expected power is given by

$$EP = 1 - \Phi\left(\sqrt{\frac{\frac{(1+r)^2}{r \cdot n} \cdot \sigma^2}{\frac{(1+r)^2}{r \cdot n} \cdot \sigma^2 + \tau^2}} \cdot \left[z_{1-\alpha/2} - \sqrt{\frac{r}{(1+r)^2} \cdot n} \cdot \left(\frac{\Delta_A}{\sigma}\right)\right]\right)$$

$$= 1 - \Phi\left(\frac{z_{1-\alpha/2} \cdot \vartheta - \Delta_A}{\sqrt{\vartheta^2 + \tau^2}}\right), \tag{3.21}$$

with $\vartheta^2 = \frac{(1+r)^2}{r \cdot n} \cdot \sigma^2$. Note that with this notation $TP = 1 - \Phi\left(\frac{z_{1-\alpha/2} \cdot \vartheta - \Delta_A}{\vartheta}\right)$, which demonstrates that EP and TP just differ by the term τ^2 in the denominator.

Of course, for $\tau \to 0$ the expected power approaches the traditional power. Furthermore, it holds true that $z_{1-\alpha/2} - \sqrt{\frac{r}{(1+r)^2} \cdot n} \cdot \left(\frac{\Delta_A}{\sigma}\right) < 0$ if and only if $TP(\Delta_A) > 0.5$; analogously, $z_{1-\alpha/2} - \sqrt{\frac{r}{(1+r)^2} \cdot n} \cdot \left(\frac{\Delta_A}{\sigma}\right) > 0$ or $z_{1-\alpha/2} - \sqrt{\frac{r}{(1+r)^2} \cdot n} \cdot \left(\frac{\Delta_A}{\sigma}\right) = 0$, respectively, hold true if and only if $TP(\Delta_A) < 0.5$ or $TP(\Delta_A) = 0.5$. Therefore, for the relevant scenario $TP(\Delta_A) > 0.5$, the expected power obtained by averaging the traditional power over a normal prior with mean Δ_A is smaller than the traditional power. For $TP(\Delta_A) < 0.5$ $(TP(\Delta_A) = 0.5)$, it holds true that $EP < TP(\Delta_A)$ $(EP = TP(\Delta_A))$. The other way round, the sample size necessary to assure an expected power $1 - \beta$ is higher or smaller, respectively, than that required for a traditional power of $1 - \beta$ for $TP(\Delta_A) > 0.5$ or $TP(\Delta_A) < 0.5$. The sample size can be obtained by determining the smallest integer for which EP given by Eq. (3.21) is at least $1 - \beta$. The following example illustrates that taking into account the uncertainty about the treatment effect may have a substantial impact on the required sample size or power, respectively.

Example 3.4 The **ChroPac trial** (see Sect. 1.2.1) assumed a clinically relevant difference of $\Delta_A = 10$ and a standard deviation of $\sigma = 20$ for the primary endpoint. To assure a traditional power of $1 - \beta = 0.90$ for a one-sided z-test at level $\alpha = 0.025$ with balanced groups $(r = 1)$, a total sample size of 170 is required (see Sect. 3.2, Example 3.1). Figure 3.6 shows for a total balanced sample size of $n = 170$ the traditional power and the expected power (for normal priors $N(\Delta_A, \tau^2)$ with $\tau^2 = \left[\Delta_A / \Phi^{-1}(\Pr(\Delta < 0))\right]^2$ for values of $\Pr(\Delta < 0)$ of 0.001, 0.01, and 0.05) in dependence of Δ_A.

Table 3.3a shows the expected power for some scenarios with normal prior distribution; the sample size of 170 patients is required when assuming the treatment effect to be fixed at $\Delta_A = 10$. Table 3.3b gives the required total sample size to achieve an expected power of 0.90. It can be seen that taking the uncertainty about the effect into account may have considerable consequences. The impact increases with increasing extent of vagueness corresponding to larger values of τ^2.

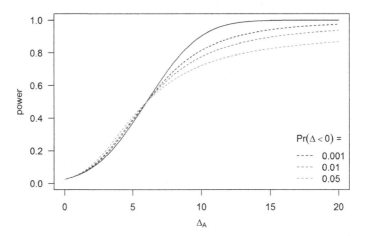

Fig. 3.6 Traditional power (solid line) and expected power (dashed lines) for a total balanced sample size $n = 170$, $\sigma = 20$ and the one-sided z-test at level $\alpha/2 = 0.025$ depending on the specified alternative Δ_A (see Example 3.4). For the expected power, normal priors $N(\Delta_A, \tau^2)$ with $\tau^2 = [\Delta_A/\Phi^{-1}(\mathrm{Pr}(\Delta < 0))]^2$ are used, where $\mathrm{Pr}(\Delta < 0)$ is the prior probability of a negative treatment effect

Table 3.3 (a) Expected power EP for total balanced sample size of 170 and (b) total balanced sample size n required to achieve an expected power of 0.90 for the z-test at one-sided level $\alpha/2 = 0.025$ under the assumptions $\Delta_A = 10$ and $\sigma = 20$ for normal priors $N(\Delta_A, \tau^2)$ with $\tau^2 = [\Delta_A/\Phi^{-1}(\mathrm{Pr}(\Delta < 0))]^2$

(a) $n = 170$			(b) $EP = 0.90$		
$\mathrm{Pr}(\Delta < 0)$	τ^2	EP	$\mathrm{Pr}(\Delta < 0)$	τ^2	n
0.0	0.00	0.90	0.0	0.00	170
0.001	10.47	0.814	0.001	10.47	268
0.01	18.48	0.775	0.01	18.48	400
0.05	36.96	0.721	0.05	36.96	1408

Remarks

1. Rufibach et al. (2016) investigated the characteristics of the expected power for alternative prior distributions.
2. Ciarleglio et al. (2016) extended the approach described above for normally distributed endpoints by specifying additionally a separate prior distribution for σ thus taking also the uncertainty with respect to this quantity into account. Ciarleglio and Arendt (2017) presented a procedure for two-group comparisons with respect to binary outcomes that includes prior distributions for the two rates.

Chapter 4
Comparison of Two Groups for Continuous and Ordered Categorical Outcomes and Test for Difference or Superiority

4.1 Background and Notation

The statistical tests considered in Chap. 3 rely (at least asymptotically) on the assumption of normal distribution. In practice, normality can rarely be assured. In case of non-normally distributed outcomes, it crucially depends on the underlying distributions (which are usually unknown) for which sample sizes the central limit theorem ensures that the distribution of the mean is sufficiently close to the normal distribution (see Remark 2 in Sect. 3.3). Therefore, there often remains some uncertainty whether or not the t-test strictly controls the defined type I error rate in the situation at hand. The Wilcoxon-Mann-Whitney test (Mann and Whitney 1947) assures control of the specified type I error rate for the related test problem under the following more general statistical model.

Let $X_{C1}, \ldots, X_{Cn_C}$ and $X_{E1}, \ldots, X_{En_E}$, respectively, be random samples of the independent random variables X_C and X_E with distribution functions F_C and F_E. This model includes both continuous and discrete metric with at least ordinal scale level. The general two-sided test problem is given by

$$H_0: F_C(t) = F_E(t) \text{ for all } t$$
$$\text{versus} \tag{4.1}$$
$$H_1: F_C(t) \neq F_E(t) \text{ for at least one } t.$$

In this setting (but also in other contexts; see, for example, Acion et al. 2006; Kieser et al. 2013), differences between two groups can be quantified by the so-called relative effect: $\pi = \Pr(X_E > X_C) + 0.5 \cdot \Pr(X_E = X_C)$. This quantity, which is in the literature also denoted as probabilistic index, can be interpreted as the probability that X_E has a tendency to larger values than X_C. If it can be assumed that the distribution functions F_C and F_E do not intersect, X_E is called stochastically larger than X_C if $\pi > 0.5$. For continuous data, $\Pr(X_C = X_E) = 0$ and therefore $\pi = \Pr(X_E > X_C)$. Furthermore, $\pi = 0.5$ if the null-hypothesis holds true. In applications, one is usually

© Springer Nature Switzerland AG 2020

M. Kieser, *Methods and Applications of Sample Size Calculation and Recalculation in Clinical Trials*, Springer Series in Pharmaceutical Statistics,
https://doi.org/10.1007/978-3-030-49528-2_4

not interested to demonstrate any arbitrary difference between the distribution functions of E and C. In contrast, it is mostly aspired to show a tendency to smaller or larger values, respectively, for one of the groups. This corresponds to expressing the alternative in terms of the relative effect by specifying a value $\pi_A \neq 0.5$. We assume in the following that higher values of the outcome indicate more favorable effects. Then, one-sided alternatives indicating superiority of E versus C can be specified by a value $\pi_A > 0.5$.

4.2 Continuous Outcomes

For continuous data, the test statistic of the Wilcoxon-Mann-Whitney test (Mann and Whitney 1947) for the above test problem (4.1) is the number of pairs (X_{Ci}, X_{Ej}), $i = 1, \ldots, n_C$, $j = 1, \ldots, n_E$, with $X_{Ej} > X_{Ci}$ which is given by $\sum_{i=1}^{n_C} \sum_{j=1}^{n_E} \delta(X_{Ej} - X_{Ci})$, where $\delta(z) = \begin{cases} 1 \text{ if } z > 0 \\ 0 \text{ if } z \leq 0 \end{cases}$. This test statistic provides an unbiased estimator $\hat{\pi}$ of π by $\hat{\pi} = \frac{1}{n_C \cdot n_E} \cdot \sum_{i=1}^{n_C} \sum_{j=1}^{n_E} \delta(x_{Ej} - x_{Ci})$, i.e. $E(\hat{\pi}) = \pi$, where $E(\hat{\pi})$ denotes the expectation of $\hat{\pi}$. Furthermore, under the null-hypothesis, the expectation of $\hat{\pi}$ is given by $E_{H_0}(\hat{\pi}) = 0.5$ and the variance by $\mathrm{Var}_{H_0}(\hat{\pi}) = \sigma_0^2 = (n_C + n_E + 1)/(12 \cdot n_C \cdot n_E)$. Under the null-hypothesis, the standardized test statistic $U = \frac{\hat{\pi} - E_{H_0}\hat{\pi}}{\sqrt{\mathrm{Var}_{H_0}\hat{\pi}}} = \frac{\sum_{i=1}^{n_C} \sum_{j=1}^{n_E} \delta(x_{Ej} - x_{Ci})/(n_C \cdot n_E) - 0.5}{\sigma_0}$ therefore follows asymptotically the standard normal distribution. Hence, for large sample sizes the following decision rule defines a statistical test with type I error rate α for the test problem (4.1): "Reject H_0 if $|u| > z_{1-\alpha/2}$, accept H_0 otherwise." For sample size calculation, the distribution of U under a specified alternative H_A has to be known. However, the variance σ_A^2 of $\hat{\pi}$ under a specific alternative depends on the distributions F_C and F_E which—this is the starting point of applying a test that does not rely on distributional assumptions—are usually unknown. Noether (1987) conjectured that for alternatives H_A that do not deviate "too much" from H_0, the variance of $\hat{\pi}$ can be approximated by σ_0^2. Under this assumption, it holds approximately true under an alternative specified by a value $\pi_A \neq 0.5$ that $U \sim N((\pi_A - 0.5)/\sigma_0, 1)$. To derive the sample size formula, we can now apply the basic equation (2.5) or more directly Eq. (2.10), i.e. $nc = z_{1-\alpha} + z_{1-\beta}$, that holds true for test statistics that follow the standard normal distribution under the null-hypothesis and a shifted normal distribution with variance 1 under the alternative. In the current situation, the non-centrality parameter is given by

$$nc = \frac{(\pi_A - 0.5)}{\sigma_0}$$
$$= \frac{(\pi_A - 0.5)}{\sqrt{\left(\frac{n_C + n_E + 1}{12 \cdot n_C \cdot n_E}\right)}}. \tag{4.2}$$

Employing $n_E = r \cdot n_C$ as well as the approximation $n_C + n_E + 1 \approx n_C + n_E = n$, it follows $nc \approx (\pi_A - 0.5) \cdot \sqrt{\frac{r}{(1+r)^2} \cdot 12 \cdot n}$. Therefore, from (2.10) we obtain the following sample size formula for the asymptotic Wilcoxon-Mann-Whitney test at two-sided level α or one-sided level $\alpha/2$, respectively, which was proposed by Noether (1987):

$$n = \frac{(1+r)^2}{r} \cdot \frac{\left(z_{1-\alpha/2} + z_{1-\beta}\right)^2}{12 \cdot (\pi_A - 0.5)^2} \tag{4.3}$$

with $r = n_E/n_C$, i.e. $n_C = \left(\frac{1}{1+r}\right) \cdot n$ and $n_E = \left(\frac{r}{1+r}\right) \cdot n$.

It should be pointed out that two simplifications were made when deriving the above formulas. Replacing $n + 1$ by n in the expression for σ_0 is uncritical for most practical solutions. However, the consequences of replacing σ_A^2 by σ_0^2 are not so obvious. For balanced groups, Vollandt and Horn (1997) derived an upper bound for the variance of the test statistic which holds true for any alternative and which can, alternatively to the Noether method, be used for a conservative sample size determination. Investigations of these two approaches dealing with the approximation of σ_A^2 by σ_0^2 showed that the Noether formula (4.3) produces quite accurate sample sizes (Vollandt and Horn 1997; Chakraborti et al. 2006). Alternatively, the exact variance σ_A^2 of $\hat{\pi}$ under a specific alternative can be derived (Gibbons and Chakraborti 2003; Wang et al. 2003) and used for exact sample size calculation (Wang et al. 2003; Chakraborti et al. 2006). It has to be noted that this approach is not only more complicated than the Noether method but that additionally assumptions concerning the underlying distribution functions F_C and F_E have to be made for the computation of σ_A^2.

For sample size calculation based on the Noether formula, a value π_A for the assumed alternative has to be specified. If higher values of the outcome indicate more benefit, the effect measure $\Pr(X_E > X_C)$ gives the probability that an outcome under therapy E is better than an outcome under C. This measure can be interpreted easily and has a number of further favorable features [see, for example, Kieser et al. (2013) and references given there]. However, it may be difficult to specify a clinically relevant difference in terms of this effect measure as such a value can commonly only be quantified by clinicians on the original scale of the outcome. To resolve this issue, two approaches can be pursued. The first method consists of assuming specific distributions F_C and F_E for the alternative hypothesis and calculating π_A from them. Example 4.1 illustrates this approach for the normal distribution. This technique for finding a suitable value for π_A is, however, in some kind going round in circles as it assumes the underlying distribution to be known while unknown distributions were the starting point for choosing the Wilcoxon-Mann-Whitney test for the analysis. The second method proposed by Chakraborti et al. (2006) requires availability of individual patient data from a pilot study from which the distribution is estimated. The basic idea is to construct from the pilot data two samples which are shifted by a pre-specified clinically relevant difference measured on the original scale of the outcome. Then, an estimate of π_A is obtained from the empirical distributions of these samples which is inserted in the sample size formula.

Example 4.1 For normally distributed outcomes, $\Pr(X_E > X_C) = \Phi\left(\frac{\Delta}{\sqrt{2}\cdot\sigma}\right)$ holds true, where $\Delta = \mu_E - \mu_C$ and σ^2 denotes the common variance of X_C and X_E, respectively. Therefore, for normally distributed outcomes, the assumptions made in the **ChroPac trial** (see Sect. 1.2.1), i.e. a clinically relevant difference of $\Delta_A = 10$ and $\sigma = 20$, translate to $\pi_A = \Phi\left(\frac{10}{\sqrt{2}\cdot20}\right) = 0.638$. Performing a two-sided Wilcoxon-Mann-Whitney test at level $\alpha = 0.05$ and aspiring a power of $1-\beta = 0.90$, a total of $184 = 2 \times 92$ patients are required for balanced groups, and $207 = 138+69$ in case of 2:1 allocation (i.e., $r = 2$) according to (4.3). The related sample sizes for the t-test are 172 and 192, respectively. The smaller sample sizes required for the t-test reflect the fact that this test is more efficient for normally distributed endpoints than the Wilcoxon-Mann-Whitney test.

Remarks

1. Zhao et al. (2006) derived an extension of Noether's formula for the case that the stratified Wilcoxon-Mann-Whitney test is applied thus accounting for stratification factors. This test is in the literature also denoted as van Elteren test (van Elteren 1960).
2. In clinical trials where patients in a critical state are included, death may occur before a measurement of the primary endpoint can be observed. Matsouka and Betensky (2015) considered the Wilcoxon-Mann-Whitney test in this situation when worst rank scores are assigned to patients with missing outcome due to death. They derived related sample size formulas and demonstrated their accuracy in simulation studies.

4.3 Ordered Categorical Outcomes

For many endpoints commonly used in clinical trials, the values of the outcome can be ordered, but differences between the values cannot be interpreted. If the observations of such ordinal data are discrete, they are denoted as ordered categorical data. Important examples are the severity of a disease or the extent of change of the patients' condition under therapy. For two-group comparisons, the results and the underlying probabilities can be displayed in a table as shown in Table 4.1.

4.3.1 Assumption-Free Approach

In this section, no assumption is made with respect to the probabilities p_{Ci} and p_{Ei}, $i = 1, \ldots, M$. Inserting the group-wise probabilities to observe data in the respective categories, the relative effect π can be expressed as

Table 4.1 Sample sizes (probabilities) of a $2 \times M$ table displaying ordered categorical data

	Category				Total
	Y_1	Y_2	$\cdots$	Y_M	
Group C	n_{C1} (p_{C1})	n_{C2} (p_{C2})	$\cdots$	n_{CM} (p_{CM})	n_C (1.0)
Group E	n_{E1} (p_{E1})	n_{E2} (p_{E2})	$\cdots$	n_{EM} (p_{EM})	n_E (1.0)
Total	$n_{C1} + n_{E1}$	$n_{C2} + n_{E2}$	$\cdots$	$n_{CM} + n_{EM}$	$n = n_C + n_E$

$$\pi = \sum_{i=2}^{M} \left(p_{Ei} \cdot \sum_{j=1}^{i-1} p_{Cj} \right) + 0.5 \cdot \sum_{i=1}^{M} p_{Ci} \cdot p_{Ei}. \tag{4.4}$$

An unbiased estimator of π is given by

$$\widehat{\pi} = \frac{1}{n_C \cdot n_E} \cdot \sum_{i=1}^{n_C} \sum_{j=1}^{n_E} \delta\left(X_{Ej} - X_{Ci}\right) \tag{4.5}$$

with $\delta(z) = \begin{cases} 1 & \text{if } z > 0 \\ 0.5 & \text{if } z = 0 \\ 0 & \text{if } z < 0. \end{cases}$

As for continuous data, the sum of pairs with $x_{Ei} > x_{Ci}$ is involved, but now also observations with the same value for the outcome ("ties") have to be taken into account.

The variance of $\widehat{\pi}$ under the null-hypothesis can be approximated by (Zhao et al. 2008)

$$\mathrm{Var}_{H_0}(\widehat{\pi}) = \sigma_0^2 \approx \frac{(1+r)^2}{12 \cdot r \cdot n} \cdot \left(1 - \sum_{i=1}^{M} \left(\frac{p_{Ci} + r \cdot p_{Ei}}{1+r} \right)^3 \right). \tag{4.6}$$

An asymptotic Wilcoxon-Mann-Whitney test for ordered categorical data as well as the related sample size calculation formula can now be derived by applying the same principles as in the previous Sect. 4.2. Note, however, that in contrast to the situation of continuous data the variance of $\widehat{\pi}$ under H_0 depends on the (commonly unknown) distribution functions F_C and F_E. Therefore, estimates for p_{Ci} and p_{Ei} have to be inserted in (4.6) resulting in an estimator of $\mathrm{Var}_{H_0}(\widehat{\pi})$. Using this quantity provides the test statistic $U = \dfrac{\widehat{\pi} - 0.5}{\sqrt{\mathrm{Var}_{H_0}(\widehat{\pi})}}$. Under the null-hypothesis, U follows asymptotically the standard normal distribution. For sample size calculation, the variance of $\widehat{\pi}$ for a specified alternative π_A is, as in Sect. 4.2, approximated by the variance under H_0 to obtain the approximate distribution of U under the alternative: $U \sim N((\pi_A - 0.5)/\sigma_0, 1)$. The sample size formula can then be derived from basic equation (2.5) or, more directly, from Eq. (2.10) that holds true for the situation

Table 4.2 Probabilities for meeting a specific number of error management criteria assumed for sample size calculation in Example 4.2

	Number of criteria						
	0	1	2	3	4	5	6
Control	0.01	0.03	0.05	0.12	0.55	0.22	0.02
FraTrix	0.00	0.01	0.02	0.03	0.31	0.43	0.20

that the test statistic is normally distributed with variance 1 both under the null- and the alternative hypothesis. Here, the non-centrality parameter is given by $nc = (\pi_A - 0.5)/\sigma_0$ and therefore Eq. (2.10) reads in the current situation $\frac{(\pi_A - 0.5)}{\sigma_0} = z_{1-\alpha/2} + z_{1-\beta}$. Thus, denoting the anticipated rates for category Y_i, $i = 1, \ldots, M$, in group C and E, respectively, by $p_{Ci,A}$ and $p_{Ei,A}$ and the overall rate for category Y_i by $\overline{p}_i = \left(\frac{r \cdot p_{Ci,A} + p_{Ei,A}}{1+r}\right)$, the following sample size formula provided by Zhao et al. (2008) results:

$$n = \frac{(1+r)^2}{r} \cdot \frac{\left(z_{1-\alpha/2} + z_{1-\beta}\right)^2}{12} \cdot \frac{\left(1 - \sum_{i=1}^{M} \overline{p}_i^3\right)}{(\pi_A - 0.5)^2} \tag{4.7}$$

with $r = n_E/n_C$. The group-wise sample sizes are given by $n_C = \left(\frac{1}{1+r}\right) \cdot n$ and $n_E = \left(\frac{r}{1+r}\right) \cdot n$.

Example 4.2 A high-level **safety culture in general practices** is essential for the quality of health care given to patients. The Frankfurt Patient Safety Matrix (FraTrix) aims at improving safety culture based on self-assessment of the practice team and potential subsequent implementation of specific measures (Müller et al. 2012). In an open, randomized, controlled trial, general practices were included to investigate whether four team sessions that used FraTrix lead to better safety culture as compared to the control intervention where the teams attended a short seminar on patient safety and error management (Hoffmann et al. 2014). Sample size calculation was based on the European Practice Assessment (EPA) indicator for error management (Engels 2005) which consists of six criteria. Based on the results of previous investigations, the probabilities for meeting a specific number of these criteria in the control and the FraTrix group, respectively, were assumed as displayed in Table 4.2.

Based on Eq. (4.4), the probabilities shown in Table 4.2 result in a value $\pi_A = 0.729$. The trial applied balanced randomization, a two-sided Wilcoxon-Mann-Whitney test at level $\alpha = 0.05$ was planned to be applied in the analysis, and the desired power was $1 - \beta = 0.80$. For these specifications, a total of $46 = 2 \times 23$ practices are required according to formula (4.7). If a 2:1 allocation had been applied between E and C, i.e. $r = 2$, the total number of practices would have to be increased to $51 = 34 + 17$. If a power of 0.90 should be achieved, the resulting number of practices is $60 = 2 \times 30$ or $66 = 44 + 22$ for 1:1 or 2:1 allocation, respectively.

Remarks

1. Sample size formula (4.7) is a special case of that presented by Zhao et al. (2008) for the more general situation of the stratified Wilcoxon-Mann-Whitney test (i.e. the van Elteren test; van Elteren 1960) for ordered categorical outcome variables.
2. The situation that death-censored observations exist was already mentioned in Sect. 4.2, Remark 2, for continuous data. Related methods for sample size calculation in case of ordinal data were also provided by Matsouaka and Betensky (2015).

4.3.2 Assuming Proportional Odds

Sometimes, a specific relationship between the probabilities of the group-wise categories can be assumed. A popular model is the so-called proportional odds model described by McCullagh (1980). Here, it is assumed that the odds ratios are equal for all possible two-by-two tables resulting from collapsing adjacent columns of the contingency table displayed in Table 4.1. Formally, this can be described by the log-odds-ratios

$$\theta_i = \ln\left(\frac{Q_{Ei} \cdot (1 - Q_{Ci})}{Q_{Ci} \cdot (1 - Q_{Ei})}\right), \quad i = 1, \ldots, M - 1,$$

where $Q_{Ci} = \sum_{j=1}^{i} p_{Cj}$ and $Q_{Ei} = \sum_{j=1}^{i} p_{Ej}, i = 1, \ldots, M - 1$. The proportional odds model assumes $\theta_1 = \theta_2 = \ldots = \theta_{M-1} = \theta$. With the arguments given below, Whitehead (1993) derived an asymptotic test and a related method to approximate the power or sample size, respectively, for alternatives fulfilling the proportional odds model, i.e. $\theta_1 = \theta_2 = \cdots = \theta_{M-1} = \theta_A \neq 0$. Here, the test problem $H_0: \theta = 0$ versus $H_1: \theta \neq 0$ is assessed by the test statistic $S = \frac{1}{n_C + n_E + 1} \cdot \sum_{i=1}^{M} (U_{Ei} - L_{Ei}) \cdot n_{Ci}$, where $U_{Ei} = \sum_{j=i+1}^{M} n_{Ej}, i = 1, \ldots, M - 1, L_{Ei} = \sum_{j=1}^{i-1} n_{Ej}, i = 2, \ldots, M$, and $U_{EM} = L_{E1} = 0$. For small θ and large sample size, S is approximately normally distributed. Under H_0, the expectation of S is 0 and the variance is approximated by

$$\hat{\sigma}_0^2 = \frac{r}{(1+r)^2} \cdot \frac{(n_C + n_E)}{3} \cdot \left(\frac{n_C + n_E}{n_C + n_E + 1}\right)^2 \cdot \left(1 - \sum_{i=1}^{M}\left(\frac{n_{Ci} + n_{Ei}}{n}\right)^3\right).$$

Hence, under the null-hypothesis, the test statistic $U = \frac{S}{\hat{\sigma}_0}$ follows approximately the standard normal distribution. Under a specified alternative θ_A, the expectation of S is approximated by $\theta_A \cdot \hat{\sigma}_0^2$ and the variance by $\hat{\sigma}_0^2$. Therefore, for large sample size, U follows under the alternative hypothesis approximately the normal distribution $N(\theta_A \cdot \sigma_0, 1)$ with

$$\sigma_0^2 = \frac{r}{(1+r)^2} \cdot \frac{(n_C+n_E)}{3} \cdot \left(\frac{n_C+n_E}{n_C+n_E+1}\right)^2 \cdot \left(1 - \sum_{i=1}^{M}\left(\frac{p_{Ci}+r\cdot p_{Ei}}{1+r}\right)^3\right)$$

$$\approx \frac{r}{(1+r)^2} \cdot \frac{n}{3} \cdot \left(1 - \sum_{i=1}^{M}\left(\frac{p_{Ci}+r\cdot p_{Ei}}{1+r}\right)^3\right).$$

Thus, we can again apply Eq. (2.10), i.e. $nc = z_{1-\alpha} + z_{1-\beta}$, that holds true for test statistics that follow normal distributions $N(0, 1)$ and $N(nc, 1)$ under the null- and the alternative hypothesis, respectively. Here, the non-centrality parameter is given by $nc = \theta_A \cdot \sigma_0$. Denoting as in Sect. 4.3.1 the anticipated rates for category Y_i, $i = 1, \ldots, M$, in group C and E, respectively, by $p_{Ci,A}$ and $p_{Ei,A}$ and the overall rate for category Y_i, by $\overline{p}_i = \left(\frac{r \cdot p_{Ci,A}+p_{Ei,A}}{1+r}\right)$, this results in the sample size formula

$$n = \frac{(1+r)^2}{r} \cdot \frac{3 \cdot \left(z_{1-\alpha/2}+z_{1-\beta}\right)^2}{\theta_A^2 \cdot \left(1 - \sum_{i=1}^{M}\overline{p}_i^3\right)} \tag{4.8}$$

with $r = n_E/n_C$, i.e. $n_C = \left(\frac{1}{1+r}\right)\cdot n$ and $n_E = \left(\frac{r}{1+r}\right)\cdot n$.

Example 4.3 An ischemic stroke happens when an artery to the brain is blocked. Then, blood and with it oxygen and nutrients are prevented from reaching the cells and tissues of the brain which may lead to serious neurological disability or death. When the **ESCAPE trial** was initiated, the relatively small rate of early reperfusion in patients with large-vessel occlusion was the major reason for the limited efficacy of the standard of care. The ESCAPE trial investigated whether stroke patients with small infarct core, proximal intracranial arterial occlusion, and moderate-to-good collateral circulation would benefit from rapid endovascular treatment (Goyal et al. 2015). ESCAPE was an academic-investigator-initiated, multicenter, randomized, open-label controlled trial with blinded outcome evaluation (so-called PROBE design) and balanced randomization comparing standard care alone (control) to standard care plus rapid endovascular treatment (experimental). Primary outcome was the modified Rankin scale (mRS) (Rankin 1957) 90 days after randomization. The mRS is a seven-level ordered categorical scale evaluating the degree of patient's disability ranging from $0 =$ no symptoms to $6 =$ dead. For sample size calculation, the scores of 5 (bedbound with severe disability) and 6 (dead) were combined. Based on previous trials, the probabilities displayed in Table 4.3 were supposed for the control group (Demchuck et al. 2015). Furthermore, the proportional odds model

Table 4.3 Probabilities for mRS categories assumed for sample size calculation in Example 4.3

	mRS					
	0	1	2	3	4	5–6
Control	0.10	0.20	0.10	0.15	0.20	0.25
Experimental	0.167	0.269	0.110	0.142	0.156	0.156

was assumed to hold true with a common log odds ratio of $\theta_A = \ln(1.8)$. Using the equation $Q_{Ei} = \frac{Q_{Ci}}{Q_{Ci}+(1-Q_{Ci})\cdot\exp(-\theta_A)}$, $i = 1,\ldots,6$, the resulting probabilities in the experimental group under the alternative hypothesis can be calculated (see Table 4.3).

A power of 0.90 was aspired when applying a two-sided test at level $\alpha = 0.05$. Under these specifications, Eq. (4.8) provides a total sample size of $378 = 2 \times 189$ for balanced randomization. If a randomization ratio of 2:1 between E and C had been applied, $426 = 284 + 142$ patients would have been required.

Besides the original mRS, dichotomized versions of this scale are frequently used as primary endpoint in stroke trials. In fact, a review by Nunn et al. (2016) showed that the majority of the identified randomized clinical trials in ischemic stroke with ordinal primary outcome used a dichotomized analysis although this approach comes along with a number of disadvantages as compared to using the original scale (see, for example, Bath et al. 2012). When applying a dichotomized endpoint, the cutpoint is commonly chosen such that it differentiates best between "good" and "bad" outcome for the investigated treatment and the case mix at hand. In the following chapter, it will be shown how to calculate the required sample size for such a binary endpoint.

Remarks

1. When deriving sample size formula (4.8), various approximations were applied. Kolassa (1995) showed that Eq. (4.8) may lead to inaccurate sample sizes if the alternative θ_A deviates heavily from 0 or if the group-wise sample sizes are small. Applying alternative approximations, Kolassa (1995) derived a method that results in more precise power calculations in these situations.

2. The asymptotic test to which sample size formula (4.8) relates may not control the specified type I error rate in case of small sample size (Kolassa 1995). In these situations, an exact test may be applied instead guaranteeing that the type I error rate is preserved at the nominal level. Hilton and Mehta (1993) proposed an algorithm to perform exact power and sample size calculations for this approach. Lee et al. (2002) showed that the sample sizes provided by Whitehead's formula (4.8) are close to those obtained by the much more complex exact calculations when the sample size is moderate to large. However, when the sample size is small—as is commonly the case when applying exact tests—exact calculations of power and sample size are recommended when using the exact test for the analysis.

Chapter 5
Comparison of Two Groups for Binary Outcomes and Test for Difference or Superiority

5.1 Background and Notation

In medical research, the result of an intervention is frequently measured by a binary endpoint as, for example therapy success, therapy response, or improvement of symptoms by a pre-defined extent. When comparing two treatments in clinical trials with binary outcome, this is done based on the observed rates. Various measures for expressing treatment group differences are used the most popular ones being the risk difference, the risk ratio (also denoted as relative risk), and the odds ratio. If the probability to observe the event of interest in the experimental (E) or control group (C), respectively, is denoted by p_E or p_C, the risk difference is given by $\Delta = p_E - p_C$, the risk ratio by $RR = \frac{p_E}{p_C}$, and the odds ratio by $OR = \frac{p_E \cdot (1-p_C)}{p_C \cdot (1-p_E)}$. The ranges of these three effect measures are given by $\Delta \in [-1, 1]$, $RR \in [0, \infty[$, and $OR \in [0, \infty[$. While the risk difference is an absolute measurement of the effect, the risk ratio and the odds ratio are relative measures. The risk ratio expresses the probability that the event of interest occurs in E relative to the associated probability in C. The odds ratio is the ratio between the odds of an event in E, i.e. $p_E/(1 - p_E)$, and that in C, i.e. $p_C/(1 - p_C)$. There is a direct relationship of the odds ratio to the regression coefficient in logistic regression (see Sect. 5.2.4). If the outcome of interest is rare, i.e., if p_E and p_C are small, the odds ratio is close to the risk ratio. However, for $p_E < p_C$, risk ratio and odds ratio are both smaller than 1 and the odds ratio is smaller than the risk ratio; for $p_E > p_C$, both measures are larger than 1 with the odds ratio being larger than the risk ratio. Therefore, for $p_E \neq p_C$ the odds ratio is always farther away from the value 1 than the risk ratio. Further characteristics of these measures and considerations on their usage can, for example, be found in Sinclair and Bracken (1994) or Schechtman (2002).

In the following, test problems, statistical tests, and related methods for sample size calculation will be presented for two-group comparisons with assessment of difference or superiority based on the risk difference, the risk ratio, and the odds ratio. Two types of tests are considered: asymptotic tests for which the distribution

© Springer Nature Switzerland AG 2020
M. Kieser, *Methods and Applications of Sample Size Calculation and Recalculation in Clinical Trials*, Springer Series in Pharmaceutical Statistics,
https://doi.org/10.1007/978-3-030-49528-2_5

Table 5.1 Sample sizes (probabilities) of a 2×2 table displaying binary data

	Event		Total
	+	−	
Group C	$x_C(p_C)$	$n_C - x_C(1 - p_C)$	n_C
Group E	$x_E(p_E)$	$n_E - x_E(1 - p_E)$	n_E
Total	$x_C + x_E$	$n_C + n_E - x_C - x_E$	$n = n_C + n_E$

of the test statistic is known as the sample size approaches infinity and for which the performance of the test is thus only approximately valid for finite sample sizes; and exact tests for which the distribution of the test statistic is known for finite sample sizes allowing, amongst others, strict control of the type I error rate.

If two groups are compared with respect to a binary outcome, the observed data can be described in a 2×2 table as shown in Table 5.1.

The (random) sum of events observed in each group follows the binomial distribution $\text{Bin}(m, p)$. It describes the probability that m independent repetitions of an experiment with binary outcome, for which the event probability is p, result in l events, $l = 0, 1, \ldots, m$. This probability is given by $\binom{m}{l} \cdot p^l \cdot (1 - p)^{m-l}$, where $p \in [0, 1]$ and $\binom{m}{l} = \frac{m!}{l! \cdot (m-l)!}$. Thus, the two-group comparison of rates is based on the following statistical model for the sum of events:

$$X_C \sim \text{Bin}(n_C, p_C) \tag{5.1a}$$

$$X_E \sim \text{Bin}(n_E, p_E), \tag{5.1b}$$

where X_C and X_E are independent.

5.2 Asymptotic Tests

5.2.1 Difference of Rates as Effect Measure

If the difference between two groups is expressed by the rate difference $p_E - p_C$, the related two-sided test problem is given by

$$
\begin{aligned}
H_0 &: p_E - p_C = 0 \\
&\text{versus} \\
H_1 &: p_E - p_C \neq 0.
\end{aligned}
\tag{5.2}
$$

For one-sided questions and the situation that superiority of group E over group C is to be demonstrated with larger rates referring to more favourable results, the test problem is

$$H_0: p_E - p_C \leq 0$$
$$\text{versus} \tag{5.3}$$
$$H_1: p_E - p_C > 0.$$

The most popular statistical test for the test problems (5.2) and (5.3) is the chi-square test. The test statistic is given by

$$\chi^2 = n \cdot \frac{(X_C \cdot (n_E - X_E) - (n_C - X_C) \cdot X_E)^2}{n_C \cdot n_E \cdot (X_C + X_E)(n_C + n_E - X_C - X_E)}. \tag{5.4}$$

Under the null hypothesis $p_E - p_C = 0$, χ^2 follows asymptotically the central chi-square distribution with one degree of freedom. An asymptotic two-sided test at level α for test problem (5.2) is therefore given by the following decision rule: "Reject H_0 if $\chi^2 > \chi^2_{1-\alpha,1}$, accept H_0 otherwise", where $\chi^2_{1-\alpha,1}$ denotes the $(1 - \alpha)$-quantile of the central chi-square distribution with one degree of freedom.

An equivalent approach to the chi-square test is to apply the so-called normal approximation test which is based on the asymptotic normality of the estimated rates; the latter is a consequence of the central limit theorem mentioned in Sect. 3.3. The test statistic of the normal approximation test is given by

$$U = \sqrt{\frac{n_C \cdot n_E}{n_C + n_E}} \cdot \frac{\hat{p}_E - \hat{p}_C}{\sqrt{\hat{p}_0 \cdot (1 - \hat{p}_0)}}, \tag{5.5}$$

where $\hat{p}_C = \frac{X_C}{n_C}$, $\hat{p}_E = \frac{X_E}{n_E}$, and $\hat{p}_0 = \frac{X_C + X_E}{n_C + n_E}$. For $p_E - p_C = 0$, U is asymptotically standard normally distributed. An asymptotic test at two-sided level α for the test problem (5.2) (or one-sided test at level $\alpha/2$ for the test problem (5.3), respectively) is therefore given by the following decision rule: "Reject H_0 if $|u| > z_{1-\alpha/2}$" ("reject H_0 if $u > z_{1-\alpha/2}$"), accept H_0 otherwise". By simple algebraic transformations, it can be seen that $U^2 = \chi^2$ holds true. Furthermore, the quantiles of the standard normal distribution and those of the central chi-square distribution with one degree of freedom are related by the equality $\left(z_{1-\alpha/2}\right)^2 = \chi^2_{1-\alpha,1}$. Therefore, the chi-square test and the normal approximation test lead to the same test decisions and to the same results in sample size calculation.

We will base the derivation of the sample size formula on the test statistic U as we are already familiar with (asymptotically) normally distributed test statistics. Then, sample size formulas are obtained which are structurally identical to those presented in Chap. 3. The only essential difference is the fact that the variance of the nominator of the U test statistic depends on the underlying rates and is therefore different under the null- and the alternative hypothesis. Hence, the derivation has to be slightly modified. We use the following notations: $r = n_E/n_C$ again denotes the ratio of the sample sizes in both groups, $p_{C,A}$ and $p_{E,A}$ are the values for p_C and p_E specified under the alternative, and $\Delta_A = p_{E,A} - p_{C,A}$. Under the null-hypothesis H_0: $p_E - p_C = 0$ it holds true that $\hat{p}_E - \hat{p}_C \sim N\left(0, \sigma_0^2\right)$, where $\sigma_0^2 = \frac{(1+r)^2}{r} \cdot \frac{1}{n} \cdot p_0 \cdot (1 - p_0)$ and with p_0 denoting the equal value of p_C and p_E. For sample size calculation, p_0 is assumed to be equal to $p_0 = \frac{p_{C,A} + r \cdot p_{E,A}}{1+r}$. Therefore, $U \approx \frac{\hat{p}_E - \hat{p}_C}{\sigma_0} \sim N(0, 1)$ holds true under H_0. Under the alternative hypothesis H_A specified by $p_{E,A} - p_{C,A} = \Delta_A$, we have $\hat{p}_E - \hat{p}_C \sim N\left(\Delta_A, \sigma_A^2\right)$ and therefore $U \approx \frac{\hat{p}_E - \hat{p}_C}{\sigma_0} \sim N\left(\frac{\Delta_A}{\sigma_0}, \frac{\sigma_A^2}{\sigma_0^2}\right)$, where $\sigma_A^2 = \frac{1+r}{n} \cdot \left(p_{C,A} \cdot \left(1 - p_{C,A}\right) + \frac{p_{E,A} \cdot (1 - p_{E,A})}{r}\right)$. For the random variable $\frac{\sigma_0}{\sigma_A} \cdot U$ it therefore holds approximately true that $\frac{\sigma_0}{\sigma_A} \cdot U \sim N\left(0, \frac{\sigma_0^2}{\sigma_A^2}\right)$ under the null-hypothesis and $\frac{\sigma_0}{\sigma_A} \cdot U \sim N\left(\frac{\Delta_A}{\sigma_A}, 1\right)$ under the specified alternative. Due to the equivalence of $U > z_{1-\alpha/2}$ and $\frac{\sigma_0}{\sigma_A} \cdot U > \frac{\sigma_0}{\sigma_A} \cdot z_{1-\alpha/2}$, we obtain from the basic equation (2.5) formulated for $\frac{\sigma_0}{\sigma_A} \cdot U$ the following equation for determining the sample size for two-sided level α or one-sided level $\alpha/2$, respectively:

$$\frac{\sigma_0}{\sigma_A} \cdot z_{1-\alpha/2} = \frac{\Delta_A}{\sigma_A} - z_{1-\beta}, \qquad (5.6)$$

or equivalently

$$\Delta_A = \sigma_0 \cdot z_{1-\alpha/2} + \sigma_A \cdot z_{1-\beta}. \qquad (5.7)$$

Note that the above equations can be seen as generalizations of Eqs. (2.9) and (2.10), where normal distribution with variance equal to 1 was assumed for the test statistic both under the null- and the alternative hypothesis. Resolving Eq. (5.7) leads to the following formula for the required total sample size:

$$n = \frac{1+r}{r}$$
$$\cdot \frac{\left(z_{1-\alpha/2} \cdot \sqrt{(1+r) \cdot p_0 \cdot (1 - p_0)} + z_{1-\beta} \cdot \sqrt{r \cdot p_{C,A} \cdot \left(1 - p_{C,A}\right) + p_{E,A} \cdot \left(1 - p_{E,A}\right)}\right)^2}{\Delta_A^2}$$

$$(5.8)$$

with $\Delta_A = p_{E,A} - p_{C,A}$ and $r = n_E/n_C$. The group-wise sample sizes are therefore given by $n_C = \left(\frac{1}{1+r}\right) \cdot n$ and $n_E = \left(\frac{r}{1+r}\right) \cdot n$.

An important difference to the analysis methods we presented for normally distributed data lies in the fact that the distribution of the test statistics on which the chi-square and the normal approximation test are based hold only asymptotically true, i.e. for "large" sample sizes. The exact distribution of the χ^2- and U test statistics can be computed by considering the probability for observing any two-by-two table which is possible for a certain sample size constellation and by summing up the probabilities of those tables for which the test statistic falls into a defined range. For example, the exact probability for rejecting the null-hypothesis when performing a one-sided normal approximation test at nominal level $\alpha/2$ can be calculated for given sample sizes n_C and n_E and probabilities p_C and p_E by the following formula:

$$\Pr\left(U \in R_{\alpha/2}\right) = \sum_{i=0}^{n_C} \sum_{j=0}^{n_E} \binom{n_C}{i} \cdot \binom{n_E}{j} p_C^i \cdot (1 - p_C)^{n_C - i}$$
$$\cdot p_E^j \cdot (1 - p_E)^{n_E - j} \cdot I_{\{u > z_{1-\alpha/2}\}}, \tag{5.9}$$

where $I_{\{.\}}$ is the indicator function and $R_{\alpha/2}$ denotes the rejection region for the normal approximation test at nominal level $\alpha/2$. For a specified alternative hypothesis, the exact power of the one-sided asymptotic normal approximation test for the sample sizes n_C and n_E can be computed by inserting the values $p_{C,A}$ and $p_{E,A}$ in (5.9). The other way round, the exact sample size for this test can be calculated by identifying for defined values of $r = n_E/n_C$, $p_{C,A}$, and $p_{E,A}$ the smallest sample sizes n_C and n_E for which the rejection probability is at least $1 - \beta$. In light of the general principle of sample size calculation presented in Sect. 2.2, this approach consists in using the asymptotic distribution of the test statistic under the null-hypothesis for defining the rejection region (as done when performing the test) but the actual (i.e. the exact) distribution under the alternative. For sample size calculation, the basic inequality (2.7) is then resolved.

Table 5.2 shows for some exemplary scenarios for $\Delta_A = 0.2$, $\alpha/2 = 0.025$, one-sided, $1 - \beta = 0.80$, and $r = 1$ or 2, respectively, the total sample sizes resulting from approximation formula (5.8) as well as the exact ones together with the related exact power values. It can be seen that the approximation formula is quite accurate for the considered cases.

Example 5.1 In the **Parkinson trial** (see Sect. 1.2.2), the patients were allocated to the active or placebo group in a ratio of 2:1 (i.e. $r = 2$). For the assumed responder rates $p_{C,A} = 0.30$ for placebo and $p_{E,A} = 0.50$ for the active treatment, respectively, Eq. (5.8) results in a total sample size of $n = 213$ $(= 142 + 71)$ for the normal approximation test at one-sided level $\alpha/2 = 0.025$ and desired power $1 - \beta = 0.80$; the exact power for this sample size amounts to 80.4%. The exact total sample size is 207 $(= 138 + 69)$ with related power 80.1%. If the total sample size 206 is allocated to the two groups in a balanced way $(= 103 + 103)$, a power of $1 - \beta = 84.7\%$ is achieved. The other way round, the exact total sample size to achieve a power of 80% for balanced allocation amounts to 188 patients, the approximation formula provides a total sample size of 186. For allocation ratios between E and C of 3:1 and 5:1, respectively, the exact and approximate sample sizes are equal and amount to 252 and 342. Allocating the threefold or fivefold sample size, respectively, to the experimental group thus leads to an increase of the total sample size by 34.0% or 81.9% as compared to balanced allocation.

Remark Brittain and Schlesselmann (1982) demonstrated that for the chi-square test or the normal approximation test, respectively, and sample size formula (5.8), a

Table 5.2 Approximate and exact total sample size (related exact power) for the normal approximation test ($\alpha/2 = 0.025$, one-sided, $1 - \beta = 0.80$)

$p_{C,A}$	$p_{E,A}$	Approximate (exact power)	Exact (exact power)
$r = n_E/n_C = 1$[a]			
0.1	0.3	124 (81.0%)	118 (80.5%)
0.2	0.4	164 (80.7%)	160 (80.1%)
0.3	0.5	186 (79.9%)	188 (80.0%)
0.4	0.6	194 (80.7%)	194 (80.7%)
$r = n_E/n_C = 2$			
0.1	0.3	147 (82.4%)	141 (80.3%)
0.2	0.4	189 (81.1%)	186 (80.3%)
0.3	0.5	213 (80.4%)	207 (80.1%)
0.4	0.6	219 (80.3%)	219 (80.3%)
0.5	0.7	207 (80.3%)	207 (80.3%)
0.6	0.8	180 (81.3%)	177 (80.5%)
0.7	0.9	132 (81.5%)	129 (80.6%)

[a]For $r = n_E/n_C = 1$, the sample sizes and related power values for $p_{C,A} = 0.5$ and $p_{E,A} = 0.7$ are equal to those for $p_{C,A} = 0.3$ and $p_{E,A} = 0.5$, those for $p_{C,A} = 0.6$ and $p_{E,A} = 0.8$ are equal to those for $p_{C,A} = 0.2$ and $p_{E,A} = 0.4$, and those for $p_{C,A} = 0.7$ and $p_{E,A} = 0.9$ are equal to those for $p_{C,A} = 0.1$ and $p_{E,A} = 0.3$

balanced allocation of the total sample size to the groups may not result in the highest power and may therefore not be optimal. However, for a given constellation of $p_{C,A}$, $p_{E,A}$, and α, the curve obtained from Eq. (5.8) expressing the approximate power in dependence of the allocation ratio is comparatively flat for a wide range of values for r. Therefore, non-optimal allocation ratios are often not essentially less efficient as compared to the optimal allocation. The same holds true for the exact sample size which is illustrated below by the example of the Parkinson trial.

Example 5.2 In the **Parkinson trial**, the specifications $p_{C,A} = 0.30$, $p_{E,A} = 0.50$, one-sided normal approximation test at level $\alpha/2 = 0.025$ and balanced allocation results in an exact total sample size of 188 patients (i.e. $n_C = n_E = 94$) with a related exact power of 80.0%. For this total sample size, the maximum power amounts to 81.6% which is achieved for the allocation $n_E = 91$, $n_C = 97$, ($r = 0.94$). For all allocation ratios $0.73 < r < 1.32$ (i.e. for $79 < n_E < 107$) the power is larger than 80% (with the exception of the constellation $n_C = 99$, $n_E = 89$, for which the power amounts to 79.8%).

Example 5.3 We learnt in Sect. 4.3.2 that, besides the original ordinal mRS, dichotomized versions of this scale are used as endpoint in stroke trials. For the **ESCAPE trial** (see Example 4.3), based on the assumptions on the probabilities displayed in Table 4.3, the dichotomization 0–2/3–6 (or 0–3/4–6, respectively) results in rates for a favourable outcome of $p_{C,A} = 0.40$ for the standard care group and $p_{E,A} = 0.546$ for the group standard care plus rapid endovascular treatment

($p_{C,A} = 0.55$ and $p_{E,A} = 0.688$). For a two-sided level $\alpha = 0.05$, desired power $1 - \beta = 0.90$ and balanced randomization, the total required sample size for analysis with the normal approximation test amounts to $572 = 2 \times 286$ ($608 = 2 \times 304$) [exact sample sizes calculated according to (5.9)]. The considerably higher sample size as compared to using the original ordinal scale, for which 378 patients are required, reflects the inefficiency of dichotomization which is due to the inherent loss of information.

5.2.2 Risk Ratio as Effect Measure

5.2.2.1 Analysis with Chi-Square Test

If the difference between groups is measured by the risk ratio, i.e. by $RR = p_E/p_C$, the related two- and one-sided test problem, respectively, are given by

$$H_0: \frac{p_E}{p_C} = 1$$
$$\text{versus} \tag{5.10}$$
$$H_1: \frac{p_E}{p_C} \neq 1$$

and

$$H_0: \frac{p_E}{p_C} \leq 1$$
$$\text{versus} \tag{5.11}$$
$$H_1: \frac{p_E}{p_C} > 1.$$

These test problems are equivalent to (5.2) and (5.3) and therefore the same statistical tests can be applied.

The approximate formula (5.8) can also be used to calculate the required sample size in case that the alternative is specified by a risk ratio $RR_A = (p_E/p_C)_A$. As for the case that the treatment effect is measured by the difference of rates, additionally to the assumed alternative the value for the rate in one group has to be defined for sample size calculation. Then, the value of the rate in the other group can be calculated by the relationship $p_{E,A} = RR_A \cdot p_{C,A}$ or $p_{C,A} = p_{E,A}/RR_A$, respectively, and can be inserted in (5.8) together with the other ingredients. An explicit approximate formula expressed in terms of the alternative RR_A can accordingly be obtained by just replacing $p_{E,A}$ by $RR_A \cdot p_{C,A}$ (or alternatively by replacing $p_{C,A}$ by $p_{E,A}/RR_A$) in (5.8):

$$n = \frac{1+r}{r}$$
$$\cdot \frac{\left(z_{1-\alpha/2} \cdot \sqrt{(1+r) \cdot p_0 \cdot (1-p_0)} + z_{1-\beta} \cdot \sqrt{r \cdot p_{C,A} \cdot (1-p_{C,A}) + RR_A \cdot p_{C,A} \cdot (1 - RR_A \cdot p_{C,A})} \right)}{p_{C,A}^2 \cdot (1 - RR_A)^2}$$

$$\tag{5.12}$$

with $RR_A = (p_E/p_C)_A$, $n_E = (r/(1+r)) \cdot n$, and $n_C = n/(1+r)$.

In a comprehensive simulation study, Pobiruchin and Kieser (2013) demonstrated that the approximation formula (5.12) and with it formula (5.8) show good accuracy over a wide range of parameter scenarios.

5.2.2.2 Analysis with Test Based on Estimated Risk Ratio

Alternatively to the chi-square test, a test statistic which is based on the estimated relative risk can be applied. By this, test decision and computation of the confidence interval for the treatment effect can be put in the same framework. If we denote the estimated risk ratio by $\widehat{RR} = \hat{p}_E/\hat{p}_C$, then the asymptotic variance of $\ln(\widehat{RR})$ is given by $\text{Var}(\ln(\widehat{RR})) = \frac{1-p_C}{n_C \cdot p_C} + \frac{1-p_E}{n_E \cdot p_E}$ (Agresti 2002, p. 73). From this, the test statistic $U = \frac{\ln(\widehat{RR})}{\hat{\sigma}_0}$ can be constructed. Here, $\hat{\sigma}_0$ is the ML-estimate of $\sigma_0 = \sqrt{\frac{(1+r)^2}{r \cdot n} \cdot \frac{1-p_0}{p_0}}$, where p_0 denotes the equal value of p_E and p_C under the null-hypothesis which is for sample size calculation set equal to $p_0 = \frac{p_{C,A} + r \cdot p_{E,A}}{1+r}$ and where $p_{C,A}$ and $p_{E,A}$ are the values for p_C and p_E specified under the alternative. Under the null-hypothesis, it holds asymptotically true that $U \approx \frac{\ln(\widehat{RR})}{\sigma_0} \sim N(0, 1)$. The following decision rule thus constitutes an asymptotic test with level α or $\alpha/2$ for the two- or one-sided null-hypothesis in (5.10) and (5.11), respectively: "Reject H_0 if $|u| > z_{1-\alpha/2}$ or $u > z_{1-\alpha/2}$, respectively, accept H_0 otherwise". Analogously as in Sect. 5.2.1, it can be shown that under the specified alternative it holds approximately true that $U \sim N\left(\frac{\ln(RR_A)}{\sigma_0}, \frac{\sigma_A^2}{\sigma_0^2}\right)$, where $RR_A = p_{E,A}/p_{C,A}$ and $\sigma_A^2 = \frac{1+r}{n} \cdot \left(\frac{1-p_{C,A}}{p_{C,A}} + \frac{1-p_{E,A}}{r \cdot p_{E,A}}\right)$. Following the same lines as in Sect. 5.2.1, an approximate sample size formula can be derived by resolving the equation

$$\ln(RR_A) = \sigma_0 \cdot z_{1-\alpha/2} + \sigma_A \cdot z_{1-\beta}.$$

The approximate required total sample size to achieve a power of $1-\beta$ for a two-sided test at level α (or a one-sided test at level $\alpha/2$) is thus given by

$$n = \frac{1+r}{r}$$
$$\cdot \frac{\left(z_{1-\alpha/2} \cdot \sqrt{(1+r) \cdot (1-p_0)/p_0} + z_{1-\beta} \cdot \sqrt{r \cdot (1-p_{C,A})/p_{C,A} + (1-p_{E,A})/p_{E,A}}\right)^2}{(\ln(RR_A))^2},$$

$$(5.13)$$

where $RR_A = p_{E,A}/p_{C,A}$ and $r = n_E/n_C$. The group-wise sample sizes can then be determined by $n_C = \left(\frac{1}{1+r}\right) \cdot n$ and $n_E = \left(\frac{r}{1+r}\right) \cdot n$.

Example 5.4 The **Parkinson trial** (see Sect. 1.2.2) assumed responder rates $p_{C,A} = 0.30$ for the placebo-group and $p_{E,A} = 0.50$ for the active treatment group. This corresponds to a risk ratio of $RR_A = p_{E,A}/p_{C,A} = 1.67$. For balanced allocation or the originally applied allocation ratio 2:1 between E and C (i.e. $r = 2$), respectively, the total required sample size resulting from the approximate formula (5.13) for a one-sided test at level $\alpha/2 = 0.025$ and a desired power $1 - \beta = 0.80$ amounts to $188 = 2 \times 94$ or $201 = 134 + 67$. If the analysis is performed with the chi-square test, the corresponding approximate sample sizes are $186 = 2 \times 93$ or $213 = 142 + 71$, respectively (see Table 5.2).

5.2.3 Odds Ratio as Effect Measure

We now consider the situation that the treatment group difference is expressed in terms of the odds ratio $OR = \frac{p_E \cdot (1 - p_C)}{p_C \cdot (1 - p_E)}$. Then, the two- and one-sided test problems for assessing a difference between groups or superiority of E versus C, respectively, are given by

$$H_0: OR = 1$$
$$\text{versus}$$
$$H_1: OR \neq 1$$

and

$$H_0: OR \leq 1$$
$$\text{versus}$$
$$H_1: OR > 1,$$

which are equivalent to

$$H_0: \ln(OR) = 0$$
$$\text{versus} \tag{5.14}$$
$$H_1: \ln(OR) \neq 0$$

and

$$H_0: \ln(OR) \leq 0$$
$$\text{versus} \tag{5.15}$$
$$H_1: \ln(OR) > 0.$$

5.2.3.1 Analysis with Chi-Square Test

As for the risk ratio, the test problems formulated in terms of the odds ratio are equivalent to those for the difference of rates and therefore the same statistical tests can be applied. If the chi-square test is used, the approximate sample size formula (5.8) can also be applied if the alternative is specified by an odds ratio $OR_A = \left(\frac{p_E \cdot (1-p_C)}{p_C \cdot (1-p_E)}\right)_A$. Again, for sample size calculation, in addition to the alternative, the value for the rate in one of the groups has to be defined. The rate in the other group is then determined by $p_{E,A} = OR_A \cdot p_{C,A} / (1 + p_{C,A} \cdot (OR_A - 1))$ or $p_{C,A} = p_{E,A} / (OR_A - p_{E,A} \cdot (OR_A - 1))$. Imputing OR_A and $p_{E,A}$ or $p_{C,A}$, respectively, in Eq. (5.8) provides approximate formulas for the required sample size when the alternative is expressed in terms of the odds ratio and when the event probability in one of the groups is specified.

5.2.3.2 Analysis with Test Based on Estimated Odds Ratio

Wang et al. (2002) proposed to use the test statistic

$$U = \frac{\ln\left(\widehat{OR}\right)}{\sqrt{\frac{1}{n_E \cdot \hat{p}_E \cdot \left(1-\hat{p}_E\right)} + \frac{1}{n_C \cdot \hat{p}_C \cdot \left(1-\hat{p}_C\right)}}} \tag{5.16}$$

with $\widehat{OR} = \frac{\hat{p}_E \cdot (1-\hat{p}_C)}{\hat{p}_C \cdot (1-\hat{p}_E)}$ for the assessment of (5.14) and (5.15). The normal approximation test statistic for the case that the treatment effect is measured by the difference or ratio of rates uses the ML-estimators of p_E and p_C under the null-hypothesis constraint $\ln(OR) = 0$ to estimate the variance of U under H_0. In contrast, the unrestricted ML-estimators are inserted in (5.16). For $\ln(OR) = 0$, it holds asymptotically true that $U \sim N(0, 1)$. Therefore, an asymptotic test at level α or $\alpha/2$ for the two- or one-sided null-hypothesis in (5.14) or (5.15), respectively, is given by the following decision rule "Reject H_0 if $|u| > z_{1-\alpha/2}$ or $u > z_{1-\alpha/2}$, respectively, accept H_0 otherwise". Under an alternative specified by values for $\ln(OR_A) = \ln\left(\frac{p_{E,A} \cdot (1-p_{C,A})}{p_{C,A} \cdot (1-p_{E,A})}\right) > 0$ and $p_{E,A}$ or $p_{C,A}$, respectively, U follows asymptotically the distribution $N(nc, 1)$ with

$$nc = \frac{\ln(OR_A)}{\sqrt{\frac{1+r}{r} \cdot \frac{1}{n} \cdot \left(\frac{1}{p_{E,A} \cdot (1-p_{E,A})} + \frac{r}{p_{C,A} \cdot (1-p_{C,A})}\right)}}.$$

Therefore, the general method presented in Sect. 2.2 for test statistics that are both under the null- and the alternative hypothesis normally distributed with variance 1 can be applied to derive an approximate sample size formula. Resolving Eq. (2.10), i.e. $nc = z_{1-\alpha/2} + z_{1-\beta}$, for n results in

$$n = \frac{1+r}{r} \cdot \left(z_{1-\alpha/2} + z_{1-\beta}\right)^2 \cdot \frac{\left(\frac{1}{p_{E,A} \cdot (1-p_{E,A})} + \frac{r}{p_{C,A} \cdot (1-p_{C,A})}\right)}{(\ln(OR_A))^2} \tag{5.17}$$

with $OR_A = \frac{p_{E,A} \cdot (1-p_{C,A})}{p_{C,A} \cdot (1-p_{E,A})}$, $n_E = (r/(1+r)) \cdot n$, and $n_C = n/(1+r)$.

Wang et al. (2002) performed simulations for $r = 2$ and various scenarios for the rates under the alternative and showed satisfactory accuracy of (5.17) in these situations.

Example 5.5 We again consider the **Parkinson trial** (see Sect. 1.2.2), where $p_{C,A} = 0.30$ and $p_{E,A} = 0.50$ were assumed which results in an odds ratio of $OR_A = 2.33$. If the test statistic (5.14) is applied at one-sided level $\alpha/2 = 0.025$ and if a power of $1 - \beta = 0.80$ is aspired, a total of $192 = 2 \times 96$ patients are required for balanced allocation, and $222 = 148 + 74$ patients are necessary for the $E{:}C = 2{:}1$ allocation originally applied in the study. These numbers are slightly higher than those required in case that the chi-square test is applied in the analysis ($n = 186$ and $n = 213$, respectively).

5.2.4 Logistic Regression

In logistic regression, the relationship between a binary response variable X and predictors of the outcome is modelled. This is done by setting up a model relating $E(X) = p$ and the predictors. As p is restricted to values $0 \le p \le 1$, a linear relationship between the outcome and the covariates (as applied in Sect. 3.4 to normally distributed outcomes) is not feasible. A common approach is to link the so-called logit of p, i.e. $\ln(p/(1-p))$, to a linear combination of the predictors. The logit maps the range $[0, 1]$ of p to $]-\infty, \infty[$ thus enabling application of a linear model. Here, the case of two binary covariates W_1 and W_2 is considered, for which the statistical model is given by

$$\ln(p/(1-p)) = \beta_0 + \beta_1 \cdot w_1 + \beta_2 \cdot w_2. \tag{5.18}$$

Equivalently, the logistic regression model (5.18) can be written as

$$p = \frac{\exp(\beta_0 + \beta_1 \cdot w_1 + \beta_2 \cdot w_2)}{1 + \exp(\beta_0 + \beta_1 \cdot w_1 + \beta_2 \cdot w_2)}. \tag{5.19}$$

If W_1 defines the treatment group allocation and if $w_1 = 1$ or 0, respectively, when the patient belongs to group E or C, then the following relationship holds true for the event rates $p_E|w_2$ and $p_C|w_2$ conditional on a fixed value w_2:

$$\ln\left(\frac{p_E|w_2 \cdot (1 - p_C|w_2)}{p_C|w_2 \cdot (1 - p_E|w_2)}\right) = \beta_1$$

or equivalently

$$OR = \exp(\beta_1).$$

It should be noted that OR is the so-called conditional odds ratio which is generally not equal to the marginal odds ratio based on the 2×2 table merged over the confounder. Equality of conditional and marginal effect measure is also called collapsibility, a property which does in general not hold true for the odds ratio. The odds ratio may even not be collapsible if the covariates are independent of treatment and do therefore not confound the effect measure (Whittemore 1978), which holds true for randomized controlled trials. In the following, we will focus on the conditional odds ratio, which is assumed to be equal for both strata formed by the confounder.

In terms of the logistic regression coefficient β_1, the two- or one-sided test problem for assessing a difference between groups or superiority of E versus C, respectively, are given by

$$\begin{aligned} &H_0\colon \beta_1 = 0 \\ &\text{versus} \\ &H_1\colon \beta_1 \neq 0 \end{aligned} \tag{5.20}$$

and

$$\begin{aligned} &H_0\colon \beta_1 \leq 0 \\ &\text{versus} \\ &H_1\colon \beta_1 > 0. \end{aligned} \tag{5.21}$$

In common software packages, the test problems (5.20) and (5.21) are addressed by applying the Wald test. For the situation described above, i.e. a binary variable W_1 indicating the treatment group allocation and a binary covariate W_2, Demidenko (2007) provided the following approximate formula for the total sample size that matches to applying a one-sided Wald test at level $\alpha/2$ in the analysis of a two-group comparison with allocation ratio $r = n_E/n_C$ and that leads to a power $1 - \beta$ for the specified alternative $\beta_{1,A} \neq 1$:

$$n = \left(z_{1-\alpha/2} + z_{1-\beta}\right)^2 \cdot \left(\frac{\sigma}{\beta_{1,A} - 1}\right)^2. \tag{5.22}$$

The variance term σ^2 is determined as follows. Let denote $\Pr(W_1 = 1) = p_{w_1}$ (i.e. $p_{w_1} = \frac{r}{1+r}$) and $\Pr(W_2 = 1) = p_{w_2}$. Then,

$$\sigma^2 = \frac{(1/a + 1/b) \cdot (1/c + 1/d)}{(1/a + 1/b + 1/c + 1/d)}$$

with

$$a = \frac{e \cdot \left(1 - p_{w_2}\right)}{(1 + e)^2 \cdot (1 + f)}$$

$$b = \frac{e \cdot f \cdot g \cdot \left(1 - p_{w_2}\right)}{(1 + e \cdot g)^2 \cdot (1 + f)}$$

$$c = \frac{e \cdot f \cdot g \cdot h \cdot p_{w_2}}{(1 + e \cdot g \cdot h)^2 \cdot (1 + f)}$$

$$d = \frac{e \cdot h \cdot p_{w_2}}{(1 + e \cdot h)^2 \cdot (1 + f)},$$

where $e = \exp(\beta_0), f = r, g = \exp(\beta_{1,A})$ and $h = \exp(\beta_2)$.

It should be noted that the power and sample size, respectively, in logistic regression analysis by use of the Wald test may show an unusual behavior: The sample size required for the Wald test in logistic regression may increase when the alternative $\beta_{1,A} = \ln(OR_A)$ moves away from the value 0 constituting the null-hypothesis (Hauck and Donner 1977; Liu 2013).

Example 5.6 We consider a variation of the **Parkinson trial** example (see Sect. 1.2.2), in which we assume that there exists a binary confounder W_2 with $\Pr(W_2 = 1) = p_{w_2} = 0.75$. Patients with confounder status $w_2 = 0$ are supposed to have response rates $(p_{C,A}|w_2 = 0) = 0.200$ and $(p_{E,A}|w_2 = 0) = 0.368$. For patients with confounder status $w_2 = 1$, response rates $(p_{C,A}|w_2 = 1) = 0.333$ and $(p_{E,A}|w_2 = 1) = 0.538$ are assumed, which results in a conditional odds ratio $OR_A = 2.33$ for both strata of the confounding variable. It should be noted that the marginal response rates in this example amount to $p_{E,A} = 0.493$ and $p_{C,A} = 0.297$ resulting in a marginal odds ratio of 2.294, which is slightly smaller than the conditional odds ratio. The quantity $e = \exp(\beta_0)$ is the odds of treatment response for patients in the control group with confounder status $w_2 = 0$, i.e. $e = \frac{0.2}{0.8} = 0.25$, while $g = \exp(\beta_{1,A}) = \frac{0.333 \cdot 0.8}{0.667 \cdot 0.2} = 1.97$. Using formula (5.22) with a two-sided level $\alpha = 0.05$ and a desired power $1 - \beta = 0.80$, a total of $198 = 2 \times 99$ patients are required for balanced allocation, and $228 = 152 + 76$ patients are necessary for $r = 2$. If the confounder is ignored, sample size calculation is based on the marginal event rates, and if formula (5.17) is applied, then the required sample size for balanced treatment allocation amounts to $200 = 2 \times 100$; in case of a 2:1 allocation between E and C, $234 = 154 + 78$ patients are required.

Remark Another approach to sample size calculation in logistic regression was pursued by Hsieh et al. (1998) which, however, does not refer to the Wald test. They derived sample size formulas for (one and multiple) normally distributed as well as binary covariates.

5.3   Exact Unconditional Tests

5.3.1   Basic Concepts

It has already been mentioned that the test statistic of the normal approximation test is not exactly but only asymptotically normally distributed. Therefore, it is not assured that the specified type I error rate is controlled when applying this test. Inserting a value $p_C = p_E = p$ in Eq. (5.9), the actual level of the normal approximation test at nominal level $\alpha/2$ can be computed. Figure 5.1 shows for $p = 0.1$, 0.3, 0.5 and $r = n_E/n_C = 2$ the actual level of the normal approximation test at nominal level 0.025 depending on the total sample size. It can be seen that the actual level approaches the nominal one for increasing sample size. However, for different values of p the course of the curve is qualitatively and quantitatively different. While the normal approximation test controls the level and is extremely conservative for small sample sizes in case of $p = 0.1$, the nominal type I error rate is exceeded for $p = 0.3$ and $p = 0.5$, and the extent of inflation is higher for $p = 0.5$ than for $p = 0.3$.

In a specific application, the value of p is unknown. Therefore, the maximum type I error inflation can only be quantified by determining for the sample sizes n_C and n_E at hand the maximum actual level over all $p \in [0, 1]$. The other way round, by correcting the nominal level, a test can be obtained which assures strict control of the specified significance level. The test decision for this level-adjusted test is based on a nominal level $\alpha^* < \alpha$ for which the maximum actual level does not exceed α for all $p \in [0, 1]$. This is the principle of constructing unconditional exact tests based on test statistics whose unmodified application may lead to an inflation or undershoot of the nominal level, respectively.

An alternative to the normal approximation test for two-group rate comparisons which strictly controls the specified level is the exact Fisher test (Irwin 1935). Here,

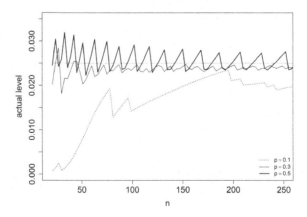

Fig. 5.1 Actual level of the normal approximation test at one-sided nominal level $\alpha/2 = 0.025$ for total sample size n, sample size allocation ratio $r = n_E/n_C = 2$, and rates $p_C = p_E = p$

the p-value applied for the test decision is calculated conditional on the observed marginal totals of the 2×2 table; the null-hypothesis is rejected if this p-value does not exceed α. Fisher's exact test strictly controls the specified type I error rate. However, it is commonly quite conservative, i.e., the actual level falls considerably below α. As a consequence the power is (quite often considerably) reduced; the other way round, the sample size required to achieve a defined power is increased. For example, for the **Parkinson trial** (see Sect. 1.2.2), Fisher's exact test was specified for the primary analysis. If responder rates of $p_{C,A} = 0.30$ and $p_{E,A} = 0.50$ are assumed, a total sample size of $n = 234$ is required for Fisher's exact test at one-sided level $\alpha/2 = 0.025$ to achieve a power of $1 - \beta = 0.80$ when an allocation ratio of $r = 2$ is applied; the total sample size amounts to $n = 213$ for the normal approximation test. Due to the pronounced conservativeness of Fisher's exact test, Lydersen et al. recommended in their tutorial that "*The traditional Fisher's exact test should practically never be used*" (Lydersen et al. 2009, p. 1174). In the following section, the Fisher-Boschloo test will be presented together with the related method for sample size calculation. This test assures control of the type I error rate while reducing the conservativeness of Fisher's test. Thus, its performance characteristics make it attractive for application.

5.3.2 Fisher-Boschloo Test

The Fisher-Boschloo test (Boschloo 1970) is based on the p-value of the conditional exact Fisher test. To reduce the conservativeness of Fisher's test, an analogous approach can be pursued as done when eliminating the level inflation of the normal approximation test by correcting the level. However, now the nominal level is increased: The Fisher-Boschloo test is based on a nominal level which is defined as the maximum level $\alpha^* > \alpha$ for which the actual level of Fisher's exact test does not exceed α for all $p \in [0, 1]$. The power of the one-sided Fisher-Boschloo test is uniformly larger as compared to the one of the one-sided exact Fisher test (Boschloo 1970). Furthermore, for the majority of scenarios investigated by Wellek (2015) the Fisher-Boschloo test shows a higher power as compared to the level-adjusted normal approximation test described in the preceding section.

As the test problems formulated in terms of the rate difference, the risk ratio, and the odds ratio are equivalent for the one- or two-sided assessment of difference or superiority, respectively, the Fisher-Boschloo test can be applied for all these effect measures. For sample size calculation, either the two rates $p_{E,A}$ and $p_{C,A}$ under the alternative have to be specified or one of the two rates $p_{E,A}$ or $p_{C,A}$, respectively, and the alternative Δ_A, RR_A, or OR_A (from which the other rate can be obtained). Determination of the exact sample size required for the Fisher-Boschloo test is performed iteratively, and it was argued that this may be compute-intensive, especially for larger sample sizes. Wellek (2015) therefore proposed an alternative method which is also based on an iterative approach but requires less computation time and provides approximate sample sizes close to the exact ones. For both methods, the approximate sample size calculated for the asymptotic normal approximation test is applied as starting value. For the Fisher-Boschloo test, Table 5.3 shows the approximate and exact total sample sizes together with the related exact power for the same scenarios as considered for the normal approximation test (see Table 5.2). The approximate sample sizes were calculated with the R program MIDN (Wellek and Ziegler 2016) and the exact ones with the R program referred to in the Appendix. For the majority of settings, the approximate sample sizes do not substantially deviate from the exact ones. Furthermore, the related power is at least as high as the desired one. However, with an ordinary PC, computation of the exact sample size using the R program referred to in the Appendix required only between two and thirteen seconds for the constellations considered in Table 5.3. Therefore, computation of the exact sample size is feasible at least for sample sizes of this magnitude. As the approximate sample size is usually higher than the actually required one, exact calculation of the sample size is the method of choice whenever the computation time allows this approach.

Example 5.7 In the **Parkinson trial** (see Sect. 1.2.2), the allocation ratio was $r = 2$ and response rates $p_{C,A} = 0.30$ for the placebo group and $p_{E,A} = 0.50$ for the active treatment group were assumed. For the one-sided Fisher-Boschloo test at level 0.025 and for desired power $1 - \beta = 0.80$, the approximate total sample size is $225 = 150 + 75$ with related power 82.7%; the exact total sample size amounts to $213 = 142 + 71$ with power 80.4%. Figure 5.2 shows for this parameter

Table 5.3 Approximate and exact total sample size (related exact power) for the Fisher-Boschloo test ($\alpha/2 = 0.025$, one-sided, $1 - \beta = 0.80$)

$p_{C,A}$	$p_{E,A}$	Approximate (exact power)	Exact (exact power)
$r = n_E/n_C = 1$[a]			
0.1	0.3	130 (81.7%)	126 (80.1%)
0.2	0.4	172 (81.4%)	168 (80.2%)
0.3	0.5	196 (81.8%)	190 (80.1%)
0.4	0.6	206 (80.8%)	204 (80.1%)
$r = n_E/n_C = 2$			
0.1	0.3	150 (82.4%)	141 (80.3%)
0.2	0.4	195 (82.2%)	186 (80.3%)
0.3	0.5	225 (82.7%)	213 (80.4%)
0.4	0.6	237 (80.8%)	219 (80.2%)
0.5	0.7	219 (81.8%)	210 (80.3%)
0.6	0.8	192 (82.2%)	186 (80.7%)
0.7	0.9	144 (81.5%)	138 (80.7%)

[a]For $r = n_E/n_C = 1$, the sample sizes and related power values for $p_{C,A} = 0.5$ and $p_{E,A} = 0.7$ are equal to those for $p_{C,A} = 0.3$ and $p_{E,A} = 0.5$, those for $p_{C,A} = 0.6$ and $p_{E,A} = 0.8$ are equal to those for $p_{C,A} = 0.2$ and $p_{E,A} = 0.4$, and those for $p_{C,A} = 0.7$ and $p_{E,A} = 0.9$ are equal to those for $p_{C,A} = 0.1$ and $p_{E,A} = 0.3$

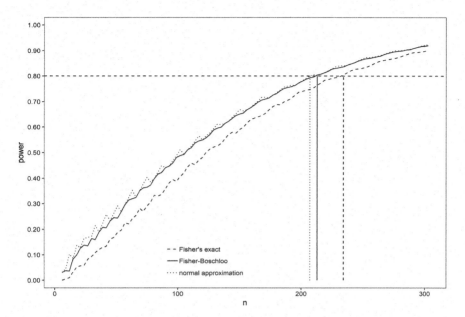

Fig. 5.2 Power of the normal approximation test, Fisher's exact test, and the Fisher-Boschloo test at one-sided level $\alpha/2 = 0.025$ for alternative $p_{C,A} = 0.30, p_{E,A} = 0.50$ and sample size allocation ratio $r = n_E/n_C = 2$ depending on the total sample size n

constellation the power of the normal approximation test, Fisher's exact test, and the Fisher-Boschloo test in dependence of the total sample size n. It can be seen that the power of the Fisher-Boschloo test is very similar to that of the normal approximation test. However, while the former assures control of the type I error rate, the latter does not (see Fig. 5.1). This constitutes a substantial advantage of the Fisher-Boschloo test. Furthermore, it should be noted that the fact that the type I error rate may be inflated by the normal approximation test makes the power comparison unfair to the disadvantage of the Fisher-Boschloo test. The conservativeness of Fisher's exact test is reflected by a substantially lower power as compared to the two competitors.

If the total sample size of 212 patients is allocated to the two groups in a balanced way (i.e. $r = 1$), the achieved power is 84.2%. The other way round, for balanced allocation an exact total sample size of $n = 190$ is required for $1 - \beta = 0.80$. For the allocation ratios $r = 3$ or $r = 5$, respectively, the exact total sample sizes increase to 252 or 342 and thus by 32.6% or 80.0% as compared to the sample size required for balanced allocation. The amount of increase is therefore similar to that observed for the normal approximation test (see Example 5.1) and for normally distributed outcomes (see Sect. 3.2).

Chapter 6
Comparison of Two Groups for Time-to-Event Outcomes and Test for Difference or Superiority

6.1 Background and Notation

6.1.1 Time-to-Event Data

Time-to-event outcomes measure the time from a defined start (for example randomization or begin of therapy) until occurrence of an event of interest (for example death, relapse, or improvement of condition by a pre-specified extent). In the literature, time-to-event analysis is often synonymously denoted as survival analysis. We will in the following also use this term and related notations bearing in mind that the statements hold true for any type of event. It is a specific characteristic of time-to-event outcomes that usually the event of interest—and with it the time to occurrence—cannot be observed for all participants of a clinical trial. Reasons are, for example, limited follow-up period or drop-out before the end of the study. The resulting data are denoted as censored observations, and the information they provide has to be adequately taken into account in the analysis. For example, in case that the interesting event has not occurred for a participant until the end of the follow-up time, it is known that the time to event is some value larger than the censoring time and that the participant was at risk for the event until censoring. This is the reason why analysis of time-to-event data and with it sample size calculation for this type of data require specific methods. Hereafter, we will use the following notations:

T: continuous random variable representing the time from start until the event of interest or censoring, respectively

$F(t) = \Pr(T \leq t)$: distribution function of T

$S(t) = \Pr(T > t) = 1 - F(t)$: survival function of T

$f(t) = \frac{d}{dt}F(t) = -\frac{d}{dt}S(t)$: density function of T

$\lambda(t)$: hazard function (or hazard rate), i.e. instantaneous risk of dying right after time t given to be alive at time t.

© Springer Nature Switzerland AG 2020

M. Kieser, *Methods and Applications of Sample Size Calculation and Recalculation in Clinical Trials*, Springer Series in Pharmaceutical Statistics, https://doi.org/10.1007/978-3-030-49528-2_6

The following relationship holds true:

$$
\begin{aligned}
\lambda(t) &= \lim_{\Delta t \to 0} \frac{\Pr(t \le T \le t + \Delta t | T \ge t)}{\Delta t} \\
&= \frac{1}{P(T \ge t)} \cdot \lim_{\Delta t \to 0} \frac{\Pr(t \le T \le t + \Delta t)}{\Delta t} \\
&= \frac{f(t)}{S(t)} \\
&= -\frac{d}{dt} \ln(S(t)).
\end{aligned}
\tag{6.1}
$$

Therefore, $S(t)$ determines $\lambda(t)$ and *vice versa*. Note that while $S(t)$ is always decreasing in t and attains values between 0 and 1, $\lambda(t)$ can principally attain any positive value and can have various shapes. For two survival functions $S_C(t)$ and $S_E(t)$ with related hazard rates $\lambda_C(t)$ and $\lambda_E(t)$, the hazard ratio is defined as $\theta(t) = \lambda_E(t)/\lambda_C(t)$. The hazard ratio attains positive values.

Example 6.1 Assuming Constant Hazard Rate—The Exponential Distribution
The simplest distribution for time-to-event data is defined by assuming that the hazard rate is constant over time, i.e., $\lambda(t) = \lambda$. From (6.1), the related survival function is $S(t) = \exp(-\lambda t)$ which is denoted as exponential distribution with parameter λ. The mean and median survival time for the exponential distribution are given by $\mu = 1/\lambda$ and median $= (\ln 2)/\lambda$, respectively. Denoting for two exponential survival functions hazard rates with λ_C and λ_E, respectively, and using the corresponding notation for the means and medians, the following equalities hold true for the hazard ratio:

$$
\theta = \lambda_E/\lambda_C
\tag{6.2a}
$$

$$
= \mu_C/\mu_E
\tag{6.2b}
$$

$$
= \mathrm{median}_C/\mathrm{median}_E
\tag{6.2c}
$$

$$
= \ln(S_E(t))/\ln(S_C(t)) \text{for all } t.
\tag{6.2d}
$$

Therefore, in case of the exponential distribution, the hazard ratio equals to the inverse ratio of the mean and median survival times. Moreover, it is equal to the ratio of the logarithm of the survival functions at any fixed time t.

Example 6.2 Survival Functions with Proportional Hazards Assuming the hazard rate to be constant over time as for the exponential distribution is quite restrictive. A more general class of survival models allows the hazard rate to be time-dependent but assumes that the hazard ratio of two survival functions is constant over time, i.e. $\theta = \frac{(\lambda_E(t))}{(\lambda_C(t))} = $ const. From (6.1) and $S(0) = 1$ as well as $S(\infty) = 0$ it follows that Eq. (6.2d) does not only hold true for exponentially distributed survival data but also for the more general case of proportional hazards. Furthermore, (6.2d) is equivalent to the following relationship between two survival functions under the proportional hazards assumption:

$$
S_E(t) = (S_C(t))^{\theta} \quad \text{for all } t.
\tag{6.3}
$$

6.1.2 Sample Size Calculation for Time-to-Event Data

The distribution characteristics of the test statistics applied for comparing survival functions between groups depend on the number of events that have been observed. Due to censoring, this number is usually smaller than the number of patients included in the trial. Therefore, sample size calculation for time-to-event outcomes consists of two steps:

Step 1 Calculation of the required number of events.
Step 2 Calculation of the required number of patients to be included in order to achieve the required number of events.

The first step is conducted by applying the general methodology presented in Chap. 2 tailored to the specific statistical test applied. This results in a required total number of events d. For step 2, the probability has to be calculated that for a randomly selected study patient an event will occur during the course of the trial. Denoting this probability by $\Pr(D)$, the required total number of patients is given by

$$n = \frac{d}{\Pr(D)}. \tag{6.4}$$

Aside from the survival functions, the probability $\Pr(D)$ depends on the duration and pattern of accrual as well as on the follow-up period. This is an important difference to other types of outcomes: In addition to assumptions on treatment effect and variability of the endpoint, further suppositions on the mode of patient recruitment have to be made for sample size calculation.

We assume in the following that patients enter the trial over an accrual period a and that the last patient entering the trial is observed for a follow-up period f. Hence, the follow-up period of the patients included in the trial lies between f (last patient entered) and the total study duration $a + f$ (first patient entered). For a patient with survival function S, the probability of observing an event during the study is then given by (Collett 2003, p. 307ff)

$$\Pr(D) = \int_0^a \Pr(\text{event and entering at } t)dt$$

$$= 1 - \int_0^a S(a + f - t) \cdot \Pr(\text{entry at } t)dt. \tag{6.5}$$

If the survival function $S(t)$ and the accrual pattern $\Pr(\text{entry at } t)$ are known, (6.5) can be evaluated and thus $\Pr(D)$ can be determined. Else, and this is the situation usually faced in practice, realistic assumptions on these quantities have to be made. If not stated otherwise, it is in the following assumed that the patients enter the trial with constant rate during the accrual period, i.e., the distribution of entry times is supposed to be uniform with $\Pr(\text{entry at } t) = 1/a$ for all $t \in [0, a]$. Then (6.5) results in

$$\Pr(D) = 1 - \frac{1}{a} \cdot \int_0^a S(a + f - t)dt = 1 - \frac{1}{a} \cdot \int_f^{a+f} S(t)dt. \qquad (6.6)$$

In two-group comparisons with survival functions $S_C(t)$ and $S_E(t)$ and sample size allocation ratio $r = n_E/n_C$, the overall probability of observing an event during the trial for a randomly selected study patient is given by

$$\begin{aligned} \Pr(D) &= \frac{1}{1+r} \cdot (\Pr_C(D) + r \cdot \Pr_E(D)) \\ &= 1 - \frac{1}{a \cdot (1+r)} \cdot \int_f^{a+f} (S_C(t) + r \cdot S_E(t))dt. \end{aligned} \qquad (6.7)$$

Note that $\Pr(D)$ accounts for censoring due to the limited follow-up period but not for censoring that occurs for other reasons such as drop-out or occurrence of an event that makes observation of the target event impossible. To compensate for such loss of information, the same considerations apply as for other types of outcome which are discussed in Sect. 1.3.

In the following two sections, methods for determining d and $\Pr(D)$ and with it n will be presented for two situations that play a central role in the analysis of time-to-event, namely exponentially distributed time-to-event data and survival functions with proportional hazards.

6.2 Exponentially Distributed Time-to-Event Data

We assume two groups C and E with d_C and d_E events and exponentially distributed survival times with hazard rates λ_C and λ_E, respectively:

$$T_{Cj} \sim \exp(\lambda_C), \quad j = 1, \ldots, d_C$$

$$T_{Ej} \sim \exp(\lambda_E), \quad j = 1, \ldots, d_E$$

where $\exp(\lambda)$ denotes the exponential distribution with hazard rate λ. If the study aim is to demonstrate a difference between the survival functions $S_C(t)$ and $S_E(t)$, the related two-sided test problem is given by

$$\begin{aligned} H_0&: \lambda_C - \lambda_E = 0 \\ &\text{versus} \\ H_1&: \lambda_C - \lambda_E \neq 0, \end{aligned} \qquad (6.8)$$

which can equivalently be formulated in terms of the hazard ratio $\theta = \lambda_E/\lambda_C$ as

$$H_0: \theta = 1$$
$$\text{versus}$$
$$H_1: \theta \neq 1.$$

Calculation of the required number of events

We denote by $\overline{T}_C$ and $\overline{T}_E$ the estimators $\overline{T}_C = \frac{1}{d_C} \cdot \sum_{j=1}^{d_C} T_{Cj}$ and $\overline{T}_E = \frac{1}{d_E} \cdot \sum_{j=1}^{d_E} T_{Ej}$ of the expected survival times $1/\lambda_C$ and $1/\lambda_E$, respectively. Taking into account that $E(T_{ij}) = 1/\lambda_i$ and $\text{Var}(T_{ij}) = 1/\lambda_i^2$, $i = C, E$, $j = 1, \ldots, d_i$, and applying the central limit theorem as well as the delta method (see, for example, Bishop et al. 1975, p. 486ff), it can be concluded that $\ln(\overline{T}_i) \sim N(-\ln \lambda_i, \sigma_i^2)$ holds approximately true, where $\sigma_i^2 = \frac{1}{d_i \cdot \lambda_i^2} \cdot \left(\frac{d}{dx}(\ln(x))\right)_{x=1/\lambda_i^2} = \frac{1}{d_i}$, $i = C, E$. Consequently, the test statistic

$$Z = \sqrt{\frac{d_C \cdot d_E}{d_C + d_E}} \cdot \left(\ln(\overline{T}_C) - \ln(\overline{T}_E)\right)$$

is approximately normally distributed as $Z \sim N(\sqrt{\frac{d_C \cdot d_E}{d_C + d_E}} \cdot \ln \theta, 1)$. Thus, we can apply Eq. (2.10) for deriving the approximation formula for the required number of events proposed by George and Desu (1974). If we denote by $r_d = d_E/d_C$ the ratio of the number of events in both groups and by $d = d_C + d_E$ the total number of events, the non-centrality parameter for a specified alternative $\theta_A \neq 1$ is given by $nc = \sqrt{\frac{r_d}{(1+r_d)^2} \cdot d} \cdot \ln \theta_A$. Thus, Eq. (2.10) results in

$$\sqrt{\frac{r_d}{(1+r_d)^2}} \cdot d \cdot \ln \theta_A = z_{1-\alpha/2} + z_{1-\beta}.$$

Therefore, the approximate required total number of events to achieve a power of $1 - \beta$ for a two-sided test at level α (or one-sided test at level $\alpha/2$) is given by the following formula:

$$d = \frac{(1+r_d)^2}{r_d} \cdot \frac{(z_{1-\alpha/2} + z_{1-\beta})^2}{(\ln(\theta_A))^2} \tag{6.9}$$

with $\theta_A = (\lambda_E/\lambda_C)_A$ and $r_d = d_E/d_C$, i.e. $d_C = \left(\frac{1}{1+r_d}\right) \cdot d$ and $d_E = \left(\frac{r_d}{1+r_d}\right) \cdot d$. Note that George and Desu (1974) derived the above approximation formulas for the special case of balanced event numbers, i.e. for $r_d = 1$. For this setting, they also provided a method for calculating the exact sample size which does not make use of distribution approximations. They showed that for common values of α, $1 - \beta$, and θ_A ($\alpha = 0.01$ and 0.05, $1 - \beta = 0.95, 0.90$, and 0.80, and θ_A in the range between 1.5 and 3.0), the approximate number of events per group provided by the above Eq. (6.9) does not differ from the exact ones by more than one event.

Calculation of the required number of patients

Under the assumption of constant recruitment rate over the accrual interval $[0, a]$ and with subsequent follow-up period f, Eq. (6.7) can be evaluated by employing $S_C(t) = \exp(-\lambda_C \cdot t)$ and $S_E(t) = \exp(-\lambda_E \cdot t)$. This results in

$$\mathrm{Pr}_i(D) = 1 - \frac{1}{a \cdot \lambda_i} \cdot (\exp(-\lambda_i \cdot f) - \exp(-\lambda_i \cdot (a + f))), i = C, E \quad (6.10)$$

and thus

$$\mathrm{Pr}(D) = 1 - \frac{1}{a \cdot (1 + r)} \cdot \left[\frac{1}{\lambda_C} \left[\exp(-\lambda_C \cdot f) - \exp(-\lambda_C \cdot (a + f)) \right] \right.$$
$$\left. + \frac{r}{\lambda_E} \cdot \left[\exp(-\lambda_E \cdot f) - \exp(-\lambda_E \cdot (a + f)) \right] \right]. \quad (6.11)$$

It is important to note that in order to be able to perform these calculations, it is not sufficient to specify the alternative θ_A. Additionally, a value for one of the parameters $\lambda_{C,A}$ or $\lambda_{E,A}$ has to be assumed to calculate the other one via the relationship $\theta_A = \lambda_{E,A}/\lambda_{C,A}$. The required total sample size is then given by $n = d/\mathrm{Pr}(D)$, where d and $\mathrm{Pr}(D)$ are calculated according to (6.9) and (6.11), respectively.

6.3 Time-to-Event Data with Proportional Hazards

In this section, we assume two groups C and E with underlying survival functions $S_C(t)$ and $S_E(t)$, respectively, for which the proportional hazards assumption holds true, i.e. $\theta = \lambda_E(t)/\lambda_C(t)$ is constant over time. Under this assumption, the two survival functions are related by $S_E(t) = (S_C(t))^\theta$ for all t, from which it follows that $S_C(t)$ and $S_E(t)$ do not cross. The two-sided test problem is then the same as in Sect. 6.2 for exponentially distributed time-to-event data, namely $H_0: \theta = 1$ versus $H_1: \theta \neq 1$. The log-rank test is commonly applied for comparing time-to-event data between two groups. It is a non-parametric test that does not require specific assumptions about the underlying distribution of the time-to-event data. In the proportional hazards setting, it is asymptotically the most powerful rank-invariant test (Peto and Peto 1972). The basic idea behind the log-rank test is to compare the number of events observed in one group to the related expected number of events under the assumption that the null-hypothesis is true. If the difference is "large enough", H_0 can be rejected. The following notations are used:

m_C: number of event time points in group C
m_E: number of event time points in group E
$t_{1,C} < t_{2,C} < \ldots < t_{m_C,C}$: ordered event time points in group C
$t_{1,E} < t_{2,E} < \ldots < t_{m_E,E}$: ordered event time points in group E
$n_{i,t_{j,k}}$: number of patients at risk in group i at time point $t_{j,k}$ with $i, k = C, E, j = 1, \ldots, m_C$ or $1, \ldots, m_E$, respectively.

Under the proportional hazards assumption, the test statistic of the log-rank test is then given by

$$L = \frac{W}{\hat{\sigma}} \tag{6.12}$$

with W being the partial score function and $\hat{\sigma}^2$ the information function:

$$W = \sum_{j=1}^{m_E} \frac{n_{C,t_j,E}}{n_{E,t_j,E} + n_{C,t_j,E}} - \sum_{j=1}^{m_C} \frac{n_{E,t_j,C}}{n_{E,t_j,C} + n_{C,t_j,C}} \tag{6.13}$$

$$\hat{\sigma}^2 = \sum_{j=1}^{m_E} \frac{n_{E,t_j,E} \cdot n_{C,t_j,E}}{\left(n_{E,t_j,E} + n_{C,t_j,E}\right)^2} + \sum_{j=1}^{m_C} \frac{n_{E,t_j,C} \cdot n_{C,t_j,C}}{\left(n_{E,t_j,C} + n_{C,t_j,C}\right)^2}. \tag{6.14}$$

If the number of event times is not too small, L is approximately standard normally distributed under the null-hypothesis. Therefore, an asymptotic level-α test for the two-sided test problem is given by the decision rule "Reject H_0, if $|l| > z_{1-\alpha/2}$, accept H_0 otherwise"; in the one-sided approach, $H_0: \theta \geq 1$ is rejected at level $\alpha/2$ if $l > z_{1-\alpha/2}$.

The sample size formulas proposed by Schoenfeld (1981, 1983) and Freedman (1982) which are presented in the following sections relate to the log-rank test for alternatives defined by a constant hazard ratio $\theta_A \neq 1$ for the groups to be compared.

6.3.1 Approach of Schoenfeld

Calculation of the required number of events
To derive a formula for the required number of events, the (approximate) distribution of L under a specified alternative $\theta_A = \lambda_E(t)/\lambda_C(t)$ is required. Schoenfeld (1981) showed that under a specified alternative H_A it holds approximately true that $L \sim N(\sqrt{\frac{r_d}{(1+r_d)^2}} \cdot d \cdot \ln(\theta_A), 1)$, where $r_d = d_E/d_C$ and $d = d_C + d_E$ with $d_i = \sum_{j=1}^{m_i} d_{ij}$ and d_{ij} denoting the number of events in group i at time point j, $i = C, E, j = 1, \ldots, m_i$. Thus, the situation underlying Eq. (2.10) applies with $nc = \sqrt{\frac{r_d}{(1+r_d)^2}} \cdot d \cdot \ln(\theta_A)$. Solving the equation $nc = z_{1-\alpha/2} + z_{1-\beta}$ for d results in the approximation formula for the required total number of events proposed by Schoenfeld (1981):

$$d = \frac{(1 + r_d)^2}{r_d} \cdot \frac{\left(z_{1-\alpha/2} + z_{1-\beta}\right)^2}{(\ln(\theta_A))^2} \tag{6.15}$$

with $\theta_A = \lambda_E(t)/\lambda_C(t)$ and $r_d = d_E/d_C$. Remarkably, this expression is identical to Eq. (6.9) obtained for exponentially distributed survival times although we relaxed the strong assumption of exponential time-to-event data and now only assume proportionality of hazards. However, the probability for an event during the trial and with it the required sample size depends explicitly on the shape of the survival distributions and therefore Eq. (6.10) does only cover the special case of exponential time-to-event data. In the next section, it is described how to deal with this issue by following the proposal of Schoenfeld (1983).

Calculation of the required number of patients
Under the proportional hazards assumption, the survival functions S_C and S_E are not specified and thus the integral (6.6) cannot be evaluated directly. Schoenfeld (1983) proposed to approximate the integral by applying Simpson's rule resulting in

$$\Pr(D) = 1 - \frac{1}{6} \cdot (S(f) + 4 \cdot S(a/2 + f) + S(a + f)). \qquad (6.16)$$

Specifying the required three values for one of the treatment groups, say $S_C(f)$, $S_C(a/2 + f)$, and $S_C(a + f)$, the related values for the other groups can be determined by using the relationship $S_E(t) = (S_C(t))^{\theta_A}$ for all t. Then, the overall probability to observe an event for a randomly selected patient can be approximated by

$$\begin{aligned}
\Pr(D) = 1 &- \frac{1}{6 \cdot (1 + r)} \cdot \Big[S_C(f) + r \cdot (S_C(f))^{\theta_A} \\
&+ 4 \cdot (S_C(a/2 + f) + r \cdot (S_C(a/2 + f))^{\theta_A}) \\
&+ S_C(a + f) + r \cdot (S_C(a + f))^{\theta_A} \Big].
\end{aligned} \qquad (6.17)$$

The total number of required patients according to Schoenfeld's approach is then given by $n = d / \Pr(D)$, where d and $\Pr(D)$ are calculated according to (6.15) and (6.17), respectively.

6.3.2 Approach of Freedman

Freedman (1982) used an alternative approach to derive an approximate distribution of L under the alternative hypothesis. He showed that under a specified alternative θ_A approximately $L \sim N\left(\sqrt{r_d \cdot d} \cdot \frac{(\theta_A - 1)}{(r_d \cdot \theta_A + 1)}, \sigma_A^2\right)$ holds true with $\sigma_A^2 = \frac{(1 + r_d)^2}{(r_d \cdot \theta_A + 1)^2} \cdot \theta_A$. Therefore, the test statistic is normally distributed under the null- and the alternative hypothesis with variance 1 and σ_A^2, respectively. As a consequence, the same approach as the one used in Sect. 5.2.1 can be applied for deriving the sample size formula. Inserting the respective quantities in Eq. (5.7) yields

$$\sqrt{r_d \cdot d} \cdot \frac{(\theta_A - 1)}{(r_d \cdot \theta_A + 1)} = z_{1-\alpha/2} + \frac{(1 + r_d) \cdot \sqrt{\theta_A}}{(r_d \cdot \theta_A + 1)} \cdot z_{1-\beta}.$$

Applying the approximation $(1 + r_d) \cdot \sqrt{\theta_A}/(r_d \cdot \theta_A + 1) \approx 1$ and solving the equation for d leads to the approximation formula for the total number of events to achieve a power of $1 - \beta$ for a two-sided log-rank test at level α (or one-sided test at level $\alpha/2$) for a specified alternative $\theta_A \neq 1$ which was proposed by Freedman (1982):

$$d = \frac{1}{r_d} \cdot (z_{1-\alpha/2} + z_{1-\beta})^2 \cdot \frac{(r_d \cdot \theta_A + 1)^2}{(\theta_A - 1)^2} \tag{6.18}$$

with $\theta_A = \lambda_E(t)/\lambda_C(t)$ and $r_d = d_E/d_C$.

Calculation of the required number of patients
Freedman (1982) proposed to estimate $\Pr(D)$ by the probability of an event at the mean follow-up time, which is equal to $a/2 + f$ under the assumption of constant accrual rate, i.e.

$$\Pr(D) = 1 - S(a/2 + f). \tag{6.19}$$

When assuming only proportionality of hazards, the survival functions $S_C(t)$ and $S_E(t)$ are not explicitly known. However, if a value for one of the treatment groups is assumed, say $S_C(a/2 + f)$, the relationship $S_E(t) = (S_C(t))^{\theta_A}$ can be used to calculate the respective value for the other. For a sample size allocation ratio $r = n_E/n_C$ this results in

$$\Pr(D) = 1 - \frac{1}{1+r} \cdot \left[S_C(a/2 + f) + r \cdot (S_C(a/2 + f))^{\theta_A} \right]. \tag{6.20}$$

The required total number of patients according to Freedman's approach is then given by $n = d/\Pr(D)$, where d and $\Pr(D)$ are calculated according to (6.18) and (6.20), respectively.

Example 6.3 Colon cancer is one of the most common malignancies. When this disease is discovered early, it is highly treatable. However, a large portion of colon cancer patients show synchronous metastases at diagnosis. At the time when the **SYNCHRONOUS trial** was planned it was unclear which of the two principal treatment strategies for such colon cancer patients without tumor-related symptoms requiring urgent surgery was most efficient: chemotherapy without previous surgical resection of the primary colon tumor or resection of the primary tumor followed by chemotherapy. This uncertainty was reflected in conflicting recommendations made in guidelines for practitioners. The SYNCHRONOUS trial compared these two treatments in a multicenter, randomized, controlled parallel-group design (Rahbari et al. 2012). The primary endpoint was overall survival. The accrual period and the follow-up period amount to 24 months or 36 months, respectively. Based on results of a literature search, a median overall survival time of 20 months was expected for chemotherapy without previous resection of the primary tumor, and an improvement to 26 months by surgery was considered as clinically relevant and realistic. Exponentially distributed survival times were assumed for which the above assumptions correspond to a hazard ratio of $\theta_A = \frac{\lambda_E}{\lambda_C} = \frac{\text{median}_C}{\text{median}_E} = \frac{20}{26} = 0.769$. The two-sided log-rank test at level $\alpha = 0.05$ was specified for the analysis, a power of $1-\beta = 0.85$ was desired, and balanced randomization was implemented ($r = 1$). According to the approach of Schoenfeld, a total of 522 deaths are required, and the probability that a randomly selected patient dies during the study is calculated as 0.761 according to (6.17); this results in a total number of patients of $686 = 2 \times 343$. For the approach of Freedman, a total of 528 deaths are necessary and $\Pr(D) = 0.766$ according to (6.20) resulting in a total sample size of $690 = 2 \times 345$. If the test for exponentially distributed survival times is used in the analysis, the required number of events is 522. Calculation of $\Pr(D)$ by (6.10) results in a value of 0.761 leading to a total sample size of $686 = 2 \times 343$ (as for Schoenfeld's approach). Note that, in contrast to the approach for exponentially distributed survival times, for the determination of $\Pr(D)$ according to Schoenfeld and Freedman, respectively, only the values at the required three or one time point(s) are determined via the exponential distribution.

Remarks to the methods presented in this chapter

1. Both the sample sizes provided by the Schoenfeld and the Freedman formula are approximate as various approximations have been made when deriving the expressions for d and $\Pr(D)$. Note that three values of the survival function of one group have to be specified in Schoenfeld's approach (time points f, $a/2+f$, and $a+f$) while only one value (at the mean follow-up time $a/2+f$) is required for Freedman's formula. If the assumptions about the survival function hold (approximately) true, Schoenfeld's formula for $\Pr(D)$ can therefore be expected to be more accurate than the one proposed by Freedman.
2. Hsieh (1992) compared the accuracy of the sample size formulas proposed by Schoenfeld and Freedman by simulations. Under an exponential model, Schoenfeld's approach provided more accurate results in case of balanced allocation while Freedman's formula performs better when the allocation ratio is towards the reciprocal of the hazard ratio.
3. Abel et al. (2015) investigated the consequences of deviations from the various assumptions made when deriving the sample size formulas of Schoenfeld and Freedman. They showed that even small departures may lead to heavily over- or underpowered trials if sample size is calculated with these formulas. As a consequence, they recommend to use formula-based approaches only for a first orientation and to complement them by simulations that consider various scenarios. For this reason, we do not present further analytical methods that get rid of some of the restrictive assumptions made above with respect to accrual and follow-up and that allow dealing with more complex scenarios. As a more flexible alternative, we describe the simulation-based approach in Chap. 16.
4. According to formulas (6.15) and (6.18) for the required total number of events according to the approach of Schoenfeld and Freedman, respectively, a balanced number of events (i.e. $r_d = 1$; Schoenfeld) or an event ratio of $r_d = 1/\theta$ (Freedman) are optimal.
5. When calculating the required total number of events according to the approach of Schoenfeld or Freedman, respectively, the ratio of the number of events in the

two groups $r_d = d_E/d_C$ is in formulas (6.15) or (6.18) commonly replaced by the sample size allocation ratio $r = n_E/n_C$. However, these quantities usually differ under alternatives $\theta_A \neq 1$ and are equal only if there is no censoring, i.e., if all patients are observed until the event occurs.

6. Hsieh (1992) investigated the power of the log-rank test depending on the sample size ratio by simulation studies. It turned out that the power curve of the log-rank test is quite flat for sample size ratios $r = n_E/n_C$ between one and the reciprocal of the hazard ratio $\theta^{-1} = \lambda_C(t)/\lambda_E(t)$ and that equal sample size allocation may not be optimal. Furthermore, for unbalanced group allocation, the allocation of patients should tend towards balanced numbers of events, i.e., more patients should be allocated to the group with the smaller event rate.

7. During the course of clinical trials, patients may switch from the assigned treatment to the other study treatment or to a non-trial treatment, or they may discontinue any further treatment. For analysis with the log-rank test, Luo et al. (2019) provided sample size calculation methods that enable to take into account the dilution of the treatment effect caused by such behavior. The derived formula for the required number of events can be seen as a generalization of formula (6.15) proposed by Schoenfeld (1981).

Remarks to extensions to the methods presented in this chapter

1. As already mentioned above, the log-rank test is asymptotically the most powerful rank-invariant test under proportional hazards. However, its power may drop markedly if the proportional hazards assumption is violated. As a generalization to the common log-rank test, weighted log-rank tests have been proposed to increase the power in case that non-proportional hazards are expected. Here, pre-specified weights are assigned to time points thus emphasizing certain parts of the survival curves. Examples are the weighting class proposed by Fleming and Harrington (1981) and the optimal piecewise weights given by Xu et al. (2016). This approach has recently attracted afresh attention due to the development of immunotherapies in oncology. As the immune system requires some time to be activated, delayed treatment effects are typically observed in this area leading to lagged separation of the time-to-event curves. Hasegawa (2014) and Wu and Wei (2020) proposed methods for sample size calculation when analyzing clinical trials in this field by a weighted or piecewise weighted log-rank test, respectively.

2. In clinical trials with time-to-event endpoints, the treatment effect is mostly captured by the hazard ratio. However, this measure is only interpretable in case of proportional hazards. Another measure of the treatment effect that does not rely on this assumption is the restricted mean survival time (RMST). It is defined as the mean survival time within a defined time segment, which is equal to the area under the survival curve within this time frame. Besides its easy interpretability, the RMST has further appealing features (Royston and Parmar 2013). Tests assessing the difference in RMST between two groups and related sample size calculation methods were proposed by Royston and Parmar (2013) and by Luo et al. (2019). For these tests, estimation and testing lead to coherent results even in case of non-proportional hazards.

3. The proportional hazards model proposed by Cox (1972) is one of the most commonly applied methods for the analysis of time-to-event data. Here, the hazard function for an individual i is modelled as a function of k explanatory covariates with values c_{ij}, $j = 1, \ldots, k$, in the following way:

$$\lambda_i(t) = \lambda_0(t) \cdot \exp(\beta_1 \cdot c_{i1} + \beta_2 \cdot c_{i2} + \cdots + \beta_k \cdot c_{ik}),$$

where $\lambda_0(t)$ denotes the so-called baseline hazard function and β_j, $j = 1, \ldots, k$, are the coefficients of the covariates. Modelling time-to-event data as a function of explanatory variables allows to investigate the influence of the latter on the hazard function. Furthermore, estimates for the hazard function and with it the survival function can be obtained from the modelling approach. For the most simple situation that only the treatment group affiliation is included in the model which is defined by $c_{i1} = 0$ or $c_{i1} = 1$ if the patient is in group C or E, respectively, the hazard ratio between individuals in group E and group C is given by $\theta = \lambda_E(t)/\lambda_C(t) = \exp(\beta_1)$. Therefore, the test problem $H_0: \theta = 1$ versus $H_1: \theta \neq 1$ can equivalently be assessed by considering the test problem $H_0: \beta_1 = 0$ versus $H_1: \beta_1 \neq 0$. Standard asymptotic tests for hypotheses about the β-coefficients are the likelihood-ratio test, the Wald test, and the score test (Therneau and Grambsch 2000, p. 53ff). These tests are asymptotically equivalent, and it can be shown that in case of a single covariate (i.e. $k = 1$) which is categorical the score test is identical to the log-rank test (Collett 2003, p. 106ff). Therefore, the sample size formulas derived for the log-rank test hold also true if the analysis is performed by applying a Cox proportional hazard model with treatment group allocation as single covariate. The advantage of the latter is that, in addition to the test decision, it provides point and interval estimates for the hazard ratio. For the general Cox regression model with a binary treatment group variable and optionally further covariates, Scosyrev and Glimm (2019) presented various approximate approaches for sample size calculation and investigated their accuracy.

4. In this chapter, constant or proportional hazards were assumed. An alternative way of modelling time-to-event data is the additive hazards model (Aalen 1989; Lin and Ying 1994). This model may be easier to interpret for clinicians as the alternative is specified in terms of the rate of events per time. Therefore, in cases where this model fits better, it may be an attractive alternative. McDaniel et al. (2016) and Su (2017) presented statistical tests and related formulas for calculating the required sample size for this approach.

5. When analyzing time-to-event data, a competing risk situation may be present. Here, a patient is at risk of several mutually exclusive events and competing risks potentially prevent the occurrence of the event of primary interest. For example, in a cardiology trial with the primary endpoint "death due to myocardial infarction", the event "death due to causes other than myocardial infarction" constitutes a competing risk. Latouche and Porcher (2007) provide an overview and comparison of methods for sample size calculation for time-to-event endpoints in the

presence of competing risks and give recommendations for application. The situation of proportional hazards as well as the case of anticipated deviations from this assumption are considered, and for both scenarios the approaches are based on Schoenfeld's formula (1983).

6. The log-rank test and Cox' proportional hazards model analyze the time to occurrence of the first event of interest. However, in some settings the same type of event may occur repeatedly. Ignoring this information may be inappropriate as important clinical information would be discarded. The Andersen-Gill model (Andersen and Gill 1982) extends the original Cox model for the analysis of recurrent events. Ingel and Jahn-Eimermacher (2014) and Tang and Fitzpatrick (2019) derived sample size formulas for this approach.

7. If it is expected that for a considerable portion of the patients the event of interest will not occur even after a very long period of follow-up, for example due to cure, this can be taken into account in planning and analysis by applying so-called cure models (see, for example, Othus et al. (2012) and references given there). Here, the population is considered as a mixture of the "susceptible" and the "cured" patients. Xiong and Wu (2017) provided a sample size formula for the weighted log-rank test under the assumption of a proportional hazards cure model. Liu et al. (2018) proposed a cure model, an associated weighted log-rank test, and a related method for sample size calculation for the situation of a (randomly) delayed treatment effect (see Remark 1 above).

Chapter 7
Comparison of More Than Two Groups and Test for Difference

7.1 Background and Notation

In many clinical trials, more than two interventions are investigated simultaneously thus enabling a head-to-head comparison of different therapeutic options. For example, in a trial that aims at demonstrating efficacy of a new treatment to placebo, an active comparator, that is already approved and established in the disease area, may be additionally included to allow a direct comparison (cf. Chap. 9). Other examples are dose-finding studies where different dosages of a drug are compared against each other and usually also to placebo. In the following sections, methods for sample size calculation are presented when comparing more than two groups and when assessing the global null-hypothesis that states that there is no difference between the groups. Throughout this chapter, we denote the total sample size by n and the allocation ratios by $r_i = n_i/n_1, i = 1, \ldots, k$, where n_i denotes the sample size in group i and $k > 2$ the number of groups to be compared.

7.2 Normally Distributed Outcomes

We assume random samples $X_{ij}, i = 1, \ldots, k, j = 1, \ldots, n_i$, of the independent normally distributed random variables $X_i \sim N(\mu_i, \sigma^2), i = 1, \ldots, k$. The test problem for the assessment of a difference between the k groups is given by

$$H_0: \mu_1 = \mu_2 = \cdots = \mu_k$$
$$\text{versus} \tag{7.1}$$
$$H_1: \mu_u \neq \mu_v \quad \text{for at least one pair } u, v \in \{1, \ldots, k\}, u \neq v.$$

The null-hypothesis can be tested by using the F test statistic

© Springer Nature Switzerland AG 2020
M. Kieser, *Methods and Applications of Sample Size Calculation and Recalculation in Clinical Trials*, Springer Series in Pharmaceutical Statistics,
https://doi.org/10.1007/978-3-030-49528-2_7

$$F = \frac{\left(\sum_{i=1}^{k} n_i \cdot \left(\bar{X}_{i.} - \bar{X}_{..}\right)^2\right)/(k-1)}{\left(\sum_{i=1}^{k} \sum_{j=1}^{n_i} \left(X_{ij} - \bar{X}_{i.}\right)\right)^2/(n-k)}, \tag{7.2}$$

where $\bar{X}_{i.}$ denotes the mean in group i and $\bar{X}_{..}$ the overall mean of the sample pooled over all patients and groups. Under the null-hypothesis, F follows the central F-distribution with $(k-1, n-k)$ degrees of freedom. Under the alternative hypothesis H_A specified by $\mu_{i,A}, i = 1, \ldots, k$, F follows the non-central F-distribution with the same number of degrees of freedom and non-centrality parameter $nc = \frac{\sum_{i=1}^{k} n_i \cdot (\mu_{i,A} - \bar{\mu}_A)^2}{\sigma^2} = n \cdot \frac{\left(\sum_{i=1}^{k} r_i \cdot (\mu_{i,A} - \bar{\mu}_A)^2\right)}{\left(\sum_{i=1}^{k} r_i\right) \cdot \sigma^2}$, where $r_i = n_i/n_1$ and $\bar{\mu}_A = \frac{\sum_{i=1}^{k} r_i \cdot \mu_{i,A}}{\left(\sum_{i=1}^{k} r_i\right)}$. For balanced allocation, i.e. $r_i = 1$ for all $i = 1, \ldots, k$, the expression for the non-centrality parameter is given by $nc = \left(\frac{n}{k}\right) \cdot \frac{\left(\sum_{i=1}^{k} (\mu_{i,A} - \bar{\mu}_A)^2\right)}{\sigma^2}$ with $\bar{\mu}_A = \left(\sum_{i=1}^{k} \mu_{i,A}\right)/k$. If the distribution functions of the central and non-central F-distribution are denoted by $F_{F;k-1,n-k,0}$ and $F_{F;k-1,n-k,nc}$, respectively, the inequality for determining the sample size follows directly from the basic inequality (2.7):

$$F^{-1}_{F;k-1,n-k,0}(1-\alpha) \leq F^{-1}_{F;k-1,n-k,nc}(\beta). \tag{7.3}$$

The required sample size is given by the smallest integer n fulfilling (7.3).

Example 7.1 The **ChroPac trial** (see Sect. 1.2.1) compared two surgical techniques in patients with chronic pancreatitis. When planning the study, a clinically relevant difference of $\Delta_A = 10$ and a standard deviation of $\sigma = 20$ were assumed. For balanced allocation, the required total sample size for the two-sided t-test at level $\alpha = 0.05$ and desired power $1 - \beta = 0.90$ amounts to $n = 172 \, (= 2 \times 86)$. For illustrative purposes, we assume that three surgical interventions were compared. Under the assumption that the population mean of the third intervention is equal to the average of values assumed for the two other groups in the ChroPac trial (or equal to the population mean of one of these groups, respectively), solution of (7.1) results for the same values for α and $1 - \beta$ in total sample sizes of $n = 309 \, (= 3 \times 103)$ (or $n = 231 \, (= 3 \times 77)$) required for the F-test. For the first scenario with $n = 309$, $nc = 12.875$ and $F^{-1}_{F;2,306,0}(1 - 0.05) = F^{-1}_{F;2,306,12.875}(0.098) = 3.025$. The actual power for this sample size therefore amounts to 0.902. A graphical display of this setting is given in Fig. 7.1.

7.3　Continuous Outcomes

We generalize the situation addressed in Sect. 4.2 to $k > 2$ groups by now considering random samples $X_{ij}, i = 1, \ldots, k, j = 1, \ldots, n_i$, of independent continuous random variables X_i with distribution functions $F_i, i = 1, \ldots, k$. The relative effect

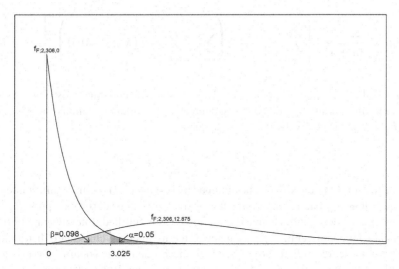

Fig. 7.1 Density function of the test statistic of the F-test under the null-hypothesis $\left(f_{F;2,306,0}\right)$ and under the specified alternative hypothesis $\left(f_{F;2,306,12.875}\right)$ for Example 7.1

between group l and m is defined by $\pi_{lm} = \Pr(X_l > X_m) + 0.5 \cdot \Pr(X_l = X_m)$ which equals to $\Pr(X_l > X_m)$ for continuous data. We consider the general two-sided test problem

$$H_0: F_1(t) = F_2(t) = \cdots = F_k(t) \text{ for all } t$$
$$\text{versus} \tag{7.4}$$
$$H_1: F_l(t) \neq F_m(t) \text{ for at least one } l, m = 1, \ldots, k, l \neq m \text{ and one } t.$$

Let denote $\hat{\pi}_{lm} = \frac{1}{n_l \cdot n_m} \cdot \sum_{u=1}^{n_l} \sum_{v=1}^{n_m} \delta(X_{lu} - X_{mv})$ with $\delta(z) = \begin{cases} 1 \text{ if } z > 0 \\ 0 \text{ if } z \leq 0 \end{cases}$. Then the test statistic of the asymptotic Kruskal-Wallis test is given by

$$U^2 = \frac{12 \cdot n^2}{(n+1)} \cdot \left[\sum_{l=1}^{k} \frac{r_l}{\sum_{u=1}^{k} r_u} \cdot \left(\sum_{\substack{m=1 \\ m \neq l}}^{k} \frac{r_m}{\sum_{u=1}^{k} r_u} \cdot \left(\hat{\pi}_{lm} - 0.5 \right) \right)^2 \right], \tag{7.5}$$

where $r_i = n_i/n_1$, $i = 1, \ldots, k$. Under the null-hypothesis, U^2 follows asymptotically the central chi-square distribution with $k - 1$ degrees of freedom. As shown by Fan et al. (2011), under an alternative specified by values $\pi_{lm,A}$, $l, m = 1, \ldots, k, l \neq m$, of which at least one is different from 0.5, U^2 follows approximately the non-central chi-square distribution with $k - 1$ degrees of freedom and non-centrality parameter

$$nc = \frac{12 \cdot n^2}{(n+1)} \cdot \left[\sum_{l=1}^{k} \frac{r_l}{\sum_{u=1}^{k} r_u} \cdot \left(\sum_{\substack{m=1 \\ m \neq l}}^{k} \frac{r_m}{\sum_{u=1}^{k} r_u} \cdot \left(\pi_{lm,A} - 0.5 \right) \right)^2 \right]. \qquad (7.6)$$

If the distribution functions of the central and non-central chi-square distribution with non-centrality parameter nc and $k - 1$ degrees of freedom are denoted by $F_{\chi^2;k-1,0}$ and $F_{\chi^2;k-1,nc}$, respectively, the basic inequality (2.7) reads

$$F_{\chi^2;k-1,0}^{-1}(1 - \alpha) \leq F_{\chi^2;k-1,nc}^{-1}(\beta) \qquad (7.7)$$

and the required sample size can be obtained by solving (7.7) for the smallest integer. An equivalent approach to determine the sample size can be derived by pursuing the following considerations. The basic equation (2.5) is fulfilled if nc given by (7.6) is set equal to the non-centrality parameter of the non-central chi-square distribution with $(k - 1)$ degrees of freedom for which the β-quantile is equal to the $(1 - \alpha)$-quantile of the respective central chi-square distribution. If the latter non-centrality parameter is denoted by $\tau(\alpha, 1 - \beta, k - 1)$ and the approximation $(n + 1) \approx n$ is inserted in Eq. (7.6), the total sample size required to achieve a power of $1 - \beta$ at a specified alternative with the two-sided Kruskal-Wallis test is approximately given by (Fan et al. 2011)

$$n = \frac{\tau(\alpha, 1 - \beta, k - 1)}{12 \cdot \left[\sum_{l=1}^{k} \frac{r_l}{\sum_{u=1}^{k} r_u} \cdot \left(\sum_{\substack{m=1 \\ m \neq l}}^{k} \frac{r_m}{\sum_{u=1}^{k} r_u} \cdot \left(\pi_{lm,A} - 0.5 \right) \right)^2 \right]}. \qquad (7.8)$$

For balanced allocation, i.e. $r_i = 1$ for all $i = 1, \ldots, k$, Eq. (7.8) reads

$$n = \frac{k^3 \cdot \tau(\alpha, 1 - \beta, k - 1)}{12 \cdot \left[\sum_{l=1}^{k} \left(\sum_{\substack{m=1 \\ m \neq l}}^{k} \left(\pi_{lm,A} - 0.5 \right) \right)^2 \right]}. \qquad (7.9)$$

Remark For $k = 2$, inserting $r_1 = 1$, $r_2 = r$, and $\pi_{12,A} = 1 - \pi_{21,A} = \pi_A$, Eq. (7.8) results in $n = \frac{(1+r)^2}{r} \cdot \frac{\tau(\alpha,1-\beta)}{12 \cdot (\pi_A - 0.5)^2}$. As $\tau(\alpha, 1 - \beta, 1) \approx \left(z_{1-\alpha/2} + z_{1-\beta} \right)^2$, the sample sizes resulting from this formula are approximately equal to those obtained from the formula of Noether (1987) presented in Sect. 4.2.

Example 7.2 For illustrative purposes, we assume as in Example 7.1 that the **ChroPac trial** was performed with three treatment groups and consider the same situation as above, namely treatment group differences Δ_{lm} between group l and m of $\Delta_{12} = 5$ (or $\Delta_{12} = 10$, respectively) and $\Delta_{13} = 10$ leading to $\Delta_{23} = 5$ (or $\Delta_{23} = 0$). For normally distributed outcomes, $\pi_{lm} = \Phi\left(\frac{\Delta_{lm}}{\sqrt{2} \cdot \sigma}\right)$ holds true (compare Example 4.1). This results for $\sigma = 20$ in $\pi_{12} = 0.570$ (or $\pi_{12} = 0.638$), $\pi_{13} = 0.638$, and $\pi_{23} = 0.570$ (or $\pi_{23} = 0.5$). Furthermore, for $1 - \beta = 0.90$, $\tau(0.05, 0.90, 2) =$

12.654 as $F^{-1}_{\chi^2;2,0}(1 - 0.05) = F^{-1}_{\chi^2;2,12.654}(0.10)$. Therefore, for balanced allocation and a desired power of 0.90 when applying the two-sided Kruskal-Wallis test at level $\alpha = 0.05$, the total required sample size determined according to formula (7.8) amounts to 330 $(= 3 \times 110)$ $(252 = 3 \times 84)$.

7.4 Binary Outcomes

In this section, we assume that the random variables X_i representing the sum of events in group i are independent and binomially distributed as $X_i \sim Bin(n_i, p_i)$. Firstly, a global test for demonstrating that there is a difference between any of the k proportions $p_i, i = 1, \ldots, k$, and related methods for sample size calculation are presented. After that, the situation is considered that a trend in proportions over the groups shall be detected.

7.4.1 Chi-Square Test

The test problem for assessing whether there is any difference in the rates between the groups can be formulated as follows:

$$H_0: p_1 = p_2 = \cdots = p_k$$
$$\text{versus} \tag{7.10}$$
$$H_1: p_u \neq p_v \text{ for at least one pair } u, v \in \{1, \ldots, k\}, u \neq v,$$

where $p_i, i = 1, \ldots, k$, denotes the probability for the event of interest in group i. We denote by $\hat{p}_i = \frac{X_i}{n_i}$ the estimated rate in group i and by $\hat{p} = \frac{\sum_{i=1}^k X_i}{n}$ the estimated average rate over all groups. Then, H_0 can be tested by application of the chi-square test with test statistic

$$\chi^2 = \frac{\sum_{i=1}^k n_i \cdot (\hat{p}_i - \hat{p})^2}{\hat{p} \cdot \left(1 - \hat{p}\right)}. \tag{7.11}$$

Under the null-hypothesis, χ^2 follows asymptotically the central chi-square distribution with $k - 1$ degrees of freedom. Under the alternative hypothesis H_A specified by the rates $p_{i,A}, i = 1, \ldots, k$, χ^2 follows approximately the non-central chi-square distribution with the same number of degrees of freedom and non-centrality parameter $nc = n \cdot \frac{\sum_{i=1}^k r_i \cdot (p_{i,A} - \bar{p}_A)^2}{\left(\sum_{i=1}^k r_i\right) \cdot \bar{p}_A \cdot (1 - \bar{p}_A)}$, where $r_i = \frac{n_i}{n_1}, i = 1, \ldots, k$, and $\bar{p}_A = \left(\sum_{i=1}^k r_i \cdot p_{i,A}\right) / \left(\sum_{i=1}^k r_i\right)$ (Lachin 1977). For balanced sample sizes, i.e. $r_i = 1, i = 1, \ldots, k$, the expression for the non-centrality parameter simplifies to $nc = \left(\frac{n}{k}\right) \cdot \frac{\sum_{i=1}^{k}(p_{i,A}-\bar{p}_A)^2}{\bar{p}_A \cdot (1-\bar{p}_A)}$. Note the structural analogy concerning the test statistics and non-centrality parameters of the F-test presented in Sect. 7.2 and the chi-square test considered here. If the distribution function of the central and non-central chi-square distribution with nc degrees of freedom is denoted by $F_{\chi^2;k-1,0}$ and $F_{\chi^2;k-1,nc}$, respectively, the inequality for determining the sample size for the rate comparison between k groups can be directly obtained from the basic inequality (2.7):

$$F_{\chi^2;k-1,0}^{-1}(1-\alpha) \le F_{\chi^2;k-1,nc}^{-1}(\beta). \tag{7.12}$$

The required sample size is given by the smallest integer fulfilling (7.12). Equivalently, as in the preceding section, the sample size can be determined by equating nc to the non-centrality parameter $\tau(\alpha, 1-\beta, k-1)$ of the non-central chi-square distribution with $(k-1)$ degrees of freedom for which the β-quantile equals to the $(1-\alpha)$-quantile of the related central chi-square distribution. This leads to the equation

$$n = \tau(\alpha, 1-\beta, k-1) \cdot \frac{\left(\sum_{i=1}^{k} r_i\right) \cdot \bar{p}_A \cdot (1-\bar{p}_A)}{\sum_{i=1}^{k} r_i \cdot (p_{i,A}-\bar{p}_A)^2}, \tag{7.13}$$

which simplifies for the case of balanced allocation to

$$n = \tau(\alpha, 1-\beta, k-1) \cdot \frac{k \cdot \bar{p}_A \cdot (1-\bar{p}_A)}{\sum_{i=1}^{k}(p_{i,A}-\bar{p}_A)^2}. \tag{7.14}$$

Example 7.3 In the **Parkinson trial** (see Sect. 1.2.2), responder rates of $p_{C,A} = 0.30$ and $p_{E,A} = 0.50$ were assumed for the placebo and the active treatment group, respectively. For balanced sample size allocation, two-sided level $\alpha = 0.05$, and desired power $1 - \beta = 0.80$, the exact total sample size required for the chi-square test amounts to 188 ($= 2 \times 94$) patients (see Table 5.2). If another active treatment had been included in the trial and a responder rate of 0.40 (or 0.50, respectively) had been assumed for this group, a required sample size of $n = 348$ ($= 3 \times 116$) (or $n = 267$ ($= 3 \times 89$)) results from (7.12) or (7.14) (with $\tau(0.05, 0.80, 2) = 9.6347$), respectively.

7.4.2 Cochran-Armitage Test

A common model for the probabilities $p_i, i = 1, \ldots, k$, is that of a trend of the response probabilities. Here, it is assumed that the p_i are increasing or decreasing, respectively, over the ordered groups. For modelling, scores $s_i, i = 1, \ldots, k$, which are monotonically increasing, are assigned expressing the ordering of the k groups. For example, in a dose-finding study, the scores may by chosen as the dose of the investigated drug administered in the corresponding group. For the derivation of the Cochran-Armitage test, which is commonly applied in this situation, it is assumed that the probabilities are linear functions of the scores, i.e. $p_i = \alpha + \beta \cdot s_i$, $i = 1, \ldots, k$. For $\beta = 0$, the null-hypothesis H_0: $p_1 = p_2 = \cdots = p_k$ holds. For $\beta > 0$, $p_1 < p_2 < \cdots < p_k$, and for $\beta < 0$ the relationship $p_1 > p_2 > \cdots > p_k$ holds true. Therefore, monotonically increasing or decreasing, respectively, response probabilities over the groups can be demonstrated by assessing the following two-sided test problem

$$H_0: \beta = 0$$
$$\text{versus} \tag{7.15}$$
$$H_1: \beta \neq 0.$$

Applying weighted least squares estimation results in (Lachin 2011)

$$\hat{\beta} = \frac{\sum_{i=1}^{k} s_i \cdot n_i \cdot \left(\hat{p}_i - \hat{\bar{p}} \right)}{\sum_{i=1}^{k} (s_i - \bar{s})^2 \cdot n_i}$$

$$\widehat{\text{Var}}_{H_0}\left(\hat{\beta} \right) = \frac{\hat{\bar{p}} \cdot \left(1 - \hat{\bar{p}} \right)}{\sum_{i=1}^{k} (s_i - \bar{s})^2 \cdot n_i},$$

where $\hat{p}_i = \frac{X_i}{n_i}$, $\hat{\bar{p}} = \frac{\sum_{i=1}^{k} X_i}{n}$, and $\bar{s} = \frac{\sum_{i=1}^{k} s_i \cdot n_i}{n}$. This leads to the Cochran-Armitage test statistic $\chi_{CA}^2 = \hat{\beta}^2 / \widehat{\text{Var}}_{H_0}\left(\hat{\beta} \right)$ (Cochran 1954; Armitage 1955) which can be written as (Lachin 2011)

$$\chi_{CA}^2 = \frac{\left(\sum_{i=1}^{k} n_i \hat{p}_i \cdot (s_i - \bar{s}) \right)^2}{\hat{\bar{p}} \cdot \left(1 - \hat{\bar{p}} \right) \cdot \sum_{i=1}^{k} (s_i - \bar{s})^2 \cdot n_i}. \tag{7.16}$$

Under H_0, χ_{CA}^2 follows asymptotically the central chi-square distribution with one degree of freedom. Under an alternative specified by rates $p_{i,A}$, $i = 1, \ldots, k$, fulfilling $p_{i,A} = \alpha + \beta_A \cdot s_i$, $i = 1, \ldots, k$, for some $\beta_A \neq 0$, χ_{CA}^2 follows approximately the non-central chi-square distribution with one degree of freedom and non-centrality parameter $nc = n \cdot \frac{(\sum_{i=1}^{k} r_i \cdot p_{i,A} \cdot (s_i - \bar{s}))^2}{(\sum_{i=1}^{k} r_i) \cdot \bar{p}_A (1 - \bar{p}_A) \cdot (\sum_{i=1}^{k} r_i \cdot (s_i - \bar{s})^2)}$, with $r_i = n_i / n_1$, $\bar{s} = \left(\sum_{i=1}^{k} r_i \cdot s_i \right) / \left(\sum_{i=1}^{k} r_i \right)$, and $\bar{p}_A = \left(\sum_{i=1}^{k} r_i \cdot p_{i,A} \right) / \left(\sum_{i=1}^{k} r_i \right)$ (Lachin 2011). For balanced sample sizes, i.e. $r_i = 1$ for $i = 1, \ldots, k$, the expression for the non-centrality parameter simplifies to $nc = \left(\frac{n}{k} \right) \cdot \frac{\left(\sum_{i=1}^{k} p_{i,A} \cdot (s_i - \bar{s}) \right)^2}{\bar{p}_A (1 - \bar{p}_A) \cdot \left(\sum_{i=1}^{k} (s_i - \bar{s})^2 \right)}$. According to the basic inequality (2.7), the required sample size can be obtained by determining the smallest integer n such that the following inequality is fulfilled:

$$F^{-1}_{\chi^2;1,0}(1 - \alpha) \leq F^{-1}_{\chi^2;1,nc}(\beta), \tag{7.17}$$

where $F_{\chi^2;1,0}$ and $F_{\chi^2;1,nc}$, respectively, denote the central and non-central chi-square distribution with one degree of freedom.

Analogously to the preceding sections, the sample size can alternatively be calculated by resolving the equation $nc = \tau(\alpha, 1 - \beta, 1)$ for n, where $\tau(\alpha, 1 - \beta, 1)$ is the non-centrality parameter of the non-central chi-square distribution with one degree of freedom for which the β-quantile is equal to the $(1 - \alpha)$-quantile of the central chi-square distribution with one degree of freedom. This leads to the equation

$$n = \tau(\alpha, 1 - \beta, 1) \cdot \frac{\left(\sum_{i=1}^{k} r_i\right) \cdot \bar{p}_A \cdot (1 - \bar{p}_A) \cdot \left(\sum_{i=1}^{k} r_i \cdot (s_i - \bar{s})^2\right)}{\left(\sum_{i=1}^{k} r_i \cdot p_{i,A} \cdot (s_i - \bar{s})\right)^2}, \tag{7.18}$$

which simplifies for balanced samples to

$$n = \tau(\alpha, 1 - \beta, 1) \cdot \frac{k \cdot \bar{p}_A \cdot (1 - \bar{p}_A) \cdot \left(\sum_{i=1}^{k}(s_i - \bar{s})^2\right)}{\left(\sum_{i=1}^{k} p_{i,A} \cdot (s_i - \bar{s})\right)^2}. \tag{7.19}$$

In case that a one-sided approach is pursued, formulas (7.17) and (7.18) provide the required sample size if a one-sided level $\alpha/2$ is applied.

Example 7.4 In a randomized, double-blind, placebo-controlled trial, the efficacy and safety of magnesium sulfate-rich natural mineral water in patients with **functional constipation** was investigated (Dupont et al. 2014). In the planning stage of the trial, recommendations promoted sufficient water intake for these patients. Furthermore, there existed some limited evidence that magnesium and sulfate may alleviate the discomfort. The patients included in this study had to drink 1.5 L water per day during four weeks. Depending on the group allocation they received 1.5 L low-mineral water (group 1), 0.5 L magnesium-sulfate-rich natural mineral water plus 1 L low-mineral water (group 2), or 1 L magnesium-sulfate-rich natural mineral water plus 0.5 L low-mineral water (group 3). Primary outcome was treatment response after one week which was operationalized based on the occurrence of two objective criteria. Sample size calculation was performed for balanced groups and for a one-sided Cochran-Armitage test at level $\alpha/2 = 0.05$ with a desired power of $1 - \beta = 0.90$ and assuming a linear trend for the response probabilities with $p_{1,A} = 0.10$, $p_{2,A} = 0.20$, and $p_{3,A} = 0.30$. Assigning the scores $s_i = i, i = 1, \ldots, 3$, application of (7.17) (or equivalently (7.19) with $\tau(0.10, 0.90, 1) = 8.564$) results in a total sample size of $207 = 3 \times 69$. For the more convenient one-sided level $\alpha/2 = 0.025$ ($\tau(0.05, 0.90, 1) = 10.507$), $255 = 3 \times 85$ patients would be required. If an allocation ratio of 1:2:2 would have been applied, $265 = 53 + 106 + 106$ or $325 = 65 + 130 + 130$ patients were necessary for a one-sided level of $\alpha/2 = 0.05$ or 0.025, respectively. We assume for illustration that instead of the Cochran-Armitage test the three-group chi-square test presented in Sect. 7.4.1 is applied in the analysis. The required total sample size for balanced groups and a one-sided level of $\alpha/2 = 0.05$ (0.025) would than amount to $252 = 3 \times 84$ with $\tau(0.10, 0.90, 2) = 10.458$ ($306 = 3 \times 102$; $\tau(0.05, 0.90, 2) = 12.654$), and the corresponding number for 1:2:2 allocation is $325 = 65 + 130 + 130$ ($390 = 78 + 156 + 156$). This demonstrates that ignoring the trend in response probabilities over the treatment groups and applying the "omnibus" chi-square test leads to considerably higher sample sizes as compared to the Cochran-Armitage test which is tailored to detect these specific alternatives.

Remarks

1. Nam (1987) presented a method for sample size calculation for a continuity-corrected version of the Cochran-Armitage test. If the continuity correction is omitted, this test statistic is just the square root of χ^2_{CA}. Continuity correction aims at achieving a better approximation of the continuous test statistic to the exact discrete one. However, this leads to a more (and in some scenarios extremely) conservative test with related higher sample sizes.
2. As noted by Lachin (2011), the probabilities assumed under the alternative need not be exactly a linear function of the scores. Instead, the method for sample size calculation allows for some departure from exact linearity. Nam (1987) argued that in case of unknown relationship between the response probabilities and the groups, assuming a linear model when calculating the sample size may be a good choice as it approximates many monotone curves.

7.5   Time-to-Event Outcomes

Ahnn and Anderson (1995) for balanced sample size and Halabi and Singh (2004) for unequal allocation generalized the approach of Schoenfeld to comparisons of $k > 2$ groups by means of the log-rank test. We assume the proportional hazards assumption to hold true and use the notation $\theta_i = \frac{\lambda_i}{\lambda_1}, i = 1, \ldots, k$. Then the test problem is given by

$$H_0: \theta_1 = \theta_2 = \cdots = \theta_k = 1$$

versus   (7.20)

$$H_1: \theta_i \neq 1 \text{ for at least one } i = 2, \ldots, k.$$

As in Sect. 6.3, we denote by m_i the number of event time points in group i and by $t_{1,i} < t_{2,i} < \cdots < t_{m_i,i}$ the ordered time points at which events occur in group $i, i = 1, \ldots, k$. The number of individuals at risk in group i at time $t_{j,i}$ is $n_{i,t_{j,i}}, i = 1, \ldots, k, j = 1 \ldots, m_i$, and the total number of events in group i is $d_i, i = 1, \ldots, k$. The log-rank test statistic can then be written as $L = U^t \cdot I^{-1} \cdot U$ with $(k-1)$-dimensional vector $U = (U_2, \ldots, U_k)^t$ and $(k-1) \times (k-1)$ matrix $I = [I_{il}]$ with components

$$U_i = d_i - \sum_{v=1}^{k} \sum_{j=1}^{m_v} \frac{n_{i,t_{j,v}}}{\sum_{g=1}^{k} n_{g,t_{j,v}}}, \quad i = 2, \ldots, k, \tag{7.21}$$

and

$$I_{il} = \delta_{il} \cdot \left(\sum_{v=1}^{k} \sum_{j=1}^{m_v} \frac{n_{i,t_{j,v}}}{\sum_{g=1}^{k} n_{g,t_{j,v}}} \right) - \sum_{v=1}^{k} \sum_{j=1}^{m_v} \frac{n_{i,t_{j,v}} \cdot n_{l,t_{j,v}}}{\left(\sum_{g=1}^{k} n_{g,t_{j,v}} \right)^2}, \quad i, l = 2, \ldots, k, \tag{7.22}$$

where $\delta_{il} = 1$ for $i = l$ and $\delta_{il} = 0$ for $i \neq l$. Under the null-hypothesis, L follows asymptotically the central chi-square distribution with $k - 1$ degrees of freedom. Under an alternative specified by values $\theta_{i,A}, i = 2, \ldots, k$, L follows approximately the non-central chi-square distribution with $k - 1$ degrees of freedom and non-centrality parameter $nc = \frac{d}{(\sum_{u=1}^{k} r_u)^2} \cdot [\sum_{i=2}^{k} ((\sum_{u=1}^{k} r_u) - r_i^2) \cdot (\ln \theta_{i,A})^2 - 2 \cdot \sum_{i=2}^{k} \sum_{\substack{v=2 \\ i<v}}^{k} r_i \cdot r_v \cdot (\ln \theta_{i,A}) \cdot (\ln \theta_{v,A})]$. Using the same notation as in the preceding sections, the basic inequality (2.7) thus reads

$$F_{\chi^2;k-1,0}^{-1}(1 - \alpha) \leq F_{\chi^2;k-1,nc}^{-1}(\beta) \tag{7.23}$$

with nc as given above.

Again, an equivalent approach to sample size calculation is by setting nc equal to $\tau(\alpha, 1 - \beta, k - 1)$, which is the non-centrality parameter of the non-central chi-square distribution with $k - 1$ degrees of freedom for which the β-quantile is equal to the $(1 - \alpha)$-quantile of the related central chi-square distribution.

The required total number of events is then given by

$$d = \left(\sum_{u=1}^{k} r_u \right)^2$$

$$\cdot \frac{\tau(\alpha, 1 - \beta, k - 1)}{\sum_{i=2}^{k} \left(\left(\sum_{u=1}^{k} r_u \right) - r_i^2 \right) \cdot (\ln \theta_{i,A})^2 - 2 \cdot \sum_{i=2}^{k-1} \sum_{\substack{v=2 \\ i<v}}^{k} r_i \cdot r_v \cdot (\ln \theta_{i,A}) \cdot (\ln \theta_{v,A})} \tag{7.24}$$

with $r_i = n_i/n, i = 1, \ldots, k$. For balanced allocation, i.e. $r_i = 1$ for all $i = 1, \ldots, k$, formula (7.24) reads

$$d = \frac{k^2 \cdot \tau(\alpha, 1 - \beta, k - 1)}{(k - 1) \cdot \sum_{i=2}^{k} (\ln \theta_{i,A})^2 - 2 \cdot \sum_{i=2}^{k-1} \sum_{\substack{v=2 \\ i<v}}^{k} (\ln \theta_{i,A}) \cdot (\ln \theta_{v,A})}. \tag{7.25}$$

The required sample size can then be calculated from $n = d/\Pr(D)$. Here, d is calculated from (7.23) or (7.24), respectively, and $\Pr(D) = \frac{\sum_{i=1}^{k} r_i \cdot \Pr_i(D)}{\sum_{i=1}^{k} r_i}$, where $\Pr_i(D)$ is the probability that an event occurs for a randomly selected patient in group $i, i = 1, \ldots, k$. Under the same assumptions as made in Chap. 6 (constant accrual rate over accrual period a and last entered patient observed for a follow-up period f), $\Pr_i(D)$ can be calculated by applying the methods presented there. This means that Eq. (6.10) is applied when assuming exponentially distributed survival times and Eq. (6.16) (Schoenfeld's approach) or (6.20) (Freedman's approach), respectively, are used for survival functions with proportional hazards. For the latter situation, the value of the survival function is specified at three or one time point(s) for one group and the proportional hazards assumption is exploited to calculate the related values for the other $k - 1$ groups.

Example 7.5 The **SYNCHRONOUS trial** compared two treatment strategies for colon cancer patients with unresectable metastases (see Example 6.3). The patients were assumed to be accrued with constant rate during a period of 24 months and the follow-up for the last patient entering the trial was 36 months. Exponentially distributed survival times with medians of 20 months and 26 months, respectively, were assumed. The desired power for a two-sided log-rank test at $\alpha = 0.05$ and balanced allocation of the sample size was $1 - \beta = 0.85$. According to the approach by Schoenfeld, 261 events per group and 343 patients per group ($\Pr(D) = 0.761$) were required. We assume for illustration purposes that a third group would have been included with an assumed median survival time of 23 months (or 26 months, respectively) and that again balanced allocation is applied, i.e. $r_i = 1$ for all $i = 1, \ldots, k$. Then $\theta_{2,A} = \lambda_2/\lambda_1 = 0.769$, $\theta_{3,A} = \lambda_3/\lambda_1 = 0.870$ (or $\theta_{2,A} = \theta_{3,A} = 0.769$). Inserting these quantities in (7.23) or (7.25), respectively, results in a required number of events per group of 317 (or 239). If the complete shape of the exponential distribution with parameters defined by the above median survival times is used to calculate the probability that an event occurs during the study for a randomly selected patient, this results in $\Pr(D) = 0.761 (0.746)$. Following the proposals of Schoenfeld and Freedman, respectively, and employing only three or one value(s) for one group and calculating the related values for the other by applying the proportional hazards assumption leads to the values $0.761 (0.746)$ and $0.766 (0.751)$. The related required sample sizes per group are 417, 417, and 414 (319, 319, and 317).

Remark For $k = 2$, it holds true that $r_1 = 1$ and $r_2 = r$, and Eq. (7.24) thus reads $d = \frac{(1+r)^2}{r} \cdot \frac{\tau(\alpha, 1-\beta, 1)}{(\ln \theta_{2,A})^2}$. As $\tau(\alpha, 1 - \beta, 1) \approx (z_{1-\alpha/2} + z_{1-\beta})^2$, the sample size formula for this special case is approximately equal to formula (6.15) proposed by Schoenfeld.

Chapter 8
Comparison of Two Groups and Test for Non-Inferiority

8.1 Background and Notation

The gold standard for establishing efficacy of a new treatment is by demonstrating superiority to placebo. However, due to the increasing number of therapies with proven efficacy, using placebo as control treatment may no longer be justifiable. The ICH E10 Guideline *Choice of control group in clinical trials* clearly states whether or not inclusion of a placebo group in a clinical study is defensible: "*In cases where an available treatment is known to prevent serious harm, such as death or irreversible morbidity in the study population, it is generally inappropriate to use a placebo control. … In other situations, when there is no serious harm, it is generally considered ethical to ask patients to participate in a placebo-controlled trial, even if they may experience discomfort as a result, provided the setting is noncoercive and patients are fully informed about available therapies and the consequences of delaying treatment*" (ICH 2001, p. 16). In situations where inclusion of a placebo arm is not appropriate, the experimental treatment is to be compared with a therapy for which efficacy has already been demonstrated in the past. Frequently, the new treatment is not expected to be more efficacious than this active control but that there is an advantage with respect to another aspect, for example, a better safety profile or a more comfortable way of administration. Then, it is accepted that, due to this advantage, efficacy of the new therapy is allowed to be "slightly" smaller; however, the extent of inferiority has to be clinically irrelevant. If θ denotes the parameter expressing the difference between the experimental (E) and the control (C) group and larger values of θ refer to more advantageous efficacy of E, a general formulation of the test problem for the assessment of non-inferiority is given by $H_0: \theta \leq \delta$ versus $H_1: \theta > \delta$, where δ denotes the non-inferiority margin defining the maximum amount of acceptable decrease in efficacy of E compared to C. Besides the necessity of specifying a non-inferiority margin, clinical trials aiming to demonstrate non-inferiority show a number of differences to superiority studies concerning design, conduct, analysis, interpretation, and reporting. For details on these issues, it is referred to the

© Springer Nature Switzerland AG 2020
M. Kieser, *Methods and Applications of Sample Size Calculation and Recalculation in Clinical Trials*, Springer Series in Pharmaceutical Statistics,
https://doi.org/10.1007/978-3-030-49528-2_8

comprehensive literature on this topic [see, for example, D'Agostino et al. (2003); Piaggio et al. (2012); Mauri and D'Agostino (2017)]. Note that the assessment of non-inferiority relates intrinsically to a one-sided test. It is recommended in regulatory guidelines to set the type I error for one-sided testing at half of the conventional level used for two-sided tests, i.e. to use $\alpha/2 = 0.025$ for non-inferiority tests if the conventional α-level of 0.05 is applied [see, for example, ICH (1998), p. 25, and CPMP (2000), p. 4].

In this chapter, methods for sample size calculation for two-arm non-inferiority trials are presented for the same scale levels and measures of treatment effect as considered in Chaps. 3, 4, 5 and 6 for trials aiming to demonstrate difference or superiority.

8.2 Normally Distributed Outcomes

For normally distributed outcomes, both the difference of means $\theta = \mu_E - \mu_C$ and the ratio of means $\theta = \mu_E/\mu_C$ are applied as measures to compare groups. A number of factors should be taken into account when deciding which measure to chose [see, for example, Sun et al. (2017), who give a summary of recommendations]. When testing for difference or superiority, respectively, the null-hypothesis of equal means is the same when expressed in terms of the difference and ratio of means and can therefore be assessed by the same test, for example by the two-sample t-test. Sample size calculation for alternatives formulated in terms of the ratio of means can also be performed using the methods from Chap. 3 just by specifying an alternative $\theta_A = (\mu_E/\mu_C)_A \neq 1$ and a value for $\mu_{E,A}$ or $\mu_{C,A}$, respectively (which then defines $\mu_{C,A}$ or $\mu_{E,A}$ and with it $(\mu_E - \mu_C)_A$). This is different in the non-inferiority setting, and for this reason we present in the following sample size calculation methods for both measures in separate sections.

In this chapter, we throughout assume the same statistical model as for the assessment of difference or superiority for normally distributed outcomes, namely (3.1a) and (3.1b), and we restrict to considering the practically relevant situation of unknown variance σ^2.

8.2.1 Difference of Means as Effect Measure

In this section, we consider the situation that the treatment effect is expressed in terms of the difference of means, i.e. $\theta = \mu_E - \mu_C$. The test problem for demonstrating non-inferiority of E versus C is then given by

$$H_0: \mu_E - \mu_C \leq -\delta$$
$$\text{versus} \tag{8.1}$$
$$H_1: \mu_E - \mu_C > -\delta$$

with pre-specified non-inferiority margin $\delta > 0$. The null-hypothesis can be tested by applying the shifted t-test with test statistic

$$T = \sqrt{\frac{n_C \cdot n_E}{n_C + n_E}} \cdot \frac{\bar{X}_E - \bar{X}_C + \delta}{S} \tag{8.2}$$

with the pooled standard deviation S defined in Sect. 3.3 by (3.8a) and (3.8b). Note that the test statistic is analogous to that of the common t-test with just in the nominator "$\bar{X}_E - \bar{X}_C$" replaced by "$\bar{X}_E - \bar{X}_C + \delta$". Therefore, sample size calculation is also very similar to that for the test for difference or superiority, respectively, and reduces to the latter if δ is set equal to zero. For $\mu_E - \mu_C = -\delta$, T follows the central t-distribution with $n_C + n_E - 2$ degrees of freedom. Hence, H_0 can be rejected at level $\alpha/2$ if $t > t_{1-\alpha/2, n_C + n_E - 2}$. Under an alternative hypothesis H_A defined by a value $\Delta_A = (\mu_E - \mu_C)_A > -\delta$, T follows the non-central t-distribution with the same number of degrees of freedom and non-centrality parameter

$$nc = \sqrt{\frac{r}{(1+r)^2}} \cdot n \cdot \left(\frac{\Delta_A + \delta}{\sigma}\right) \tag{8.3}$$

with $r = n_E/n_C$ and $n = n_C + n_E$. Therefore, the required sample size for the shifted t-test can be calculated by just replacing "Δ_A" in the procedure for the t-test for superiority by "$\Delta_A + \delta$". The exact sample size can be calculated by iteratively solving inequality (3.10) while inserting the non-centrality parameter (8.3) given above, i.e., by determining the smallest integer n fulfilling $F_{T;n-2,0}^{-1}(1 - \alpha/2) \leq F_{T;n-2,nc}^{-1}(\beta)$. Analogously to the considerations made for the (unshifted) t-test, a good approximation of the total sample size can be obtained by the Guenther-Schouten formula adapted to the non-inferiority test problem:

$$n = \frac{(1+r)^2}{r} \cdot \left(z_{1-\alpha/2} + z_{1-\beta}\right)^2 \cdot \left(\frac{\sigma}{\Delta_A + \delta}\right)^2 + \frac{\left(z_{1-\alpha/2}\right)^2}{2} \tag{8.4}$$

with $\Delta_A = (\mu_E - \mu_C)_A$ and $r = n_E/n_C$, i.e. $n_C = \left(\frac{1}{1+r}\right) \cdot n$ and $n_E = \left(\frac{r}{1+r}\right) \cdot n$.

Example 8.1 **Chronic obstructive pulmonary disease** (COPD) is a chronic inflammatory lung disease characterized by airflow limitation. Bronchodilators such as the long-acting beta-agonist formoterol aim at controlling symptoms. As COPD symptoms occur frequently in the morning, bronchodilators should be inhaled early in the day. Furthermore, once-daily inhalation improves compliance as compared to more complicated treatment regimens. In a randomized, double-blind, parallel-group, multicenter trial in patients with moderate to severe COPD, it was investigated whether inhalation of a single morning dose of 24 µg of the bronchodilator formoterol (E) is not relevantly inferior to two divided doses of 12 µg formoterol inhaled in the morning and in the evening (C) (Welte et al. 2008). The primary efficacy endpoint was the change of pre-dose forced expiratory volume in one second (FEV$_1$) from baseline to the examination after 12 weeks of treatment for which a non-inferiority margin of 100 ml was defined. Analysis was planned to be performed with the shifted t-test at one-sided level $\alpha/2 = 0.025$, and sample size calculation was based on the assumption of equal efficacy of E and C (i.e. $\Delta_A = 0$) and a common standard deviation of $\sigma = 275$ ml. The treatment groups should be balanced (i.e. $r = 1$), and a power of $1 - \beta = 0.90$ was aspired. Both exact and approximate calculations result in a sample size of 160 per group.

Remarks

1. Frequently, sample size calculations for non-inferiority trials are performed under the assumption that experimental and control treatment are in fact equally effi-cacious with respect to the primary endpoint, i.e. for $\Delta_A = 0$. Then, the formula for determining the sample size for testing superiority and non-inferiority differ just by employing Δ_A or δ, respectively. As in superiority trials $|\Delta_A|$ defines a clinically relevant amount but $|\delta|$ constitutes by definition a clinically irrelevant one in the non-inferiority setting, $|\Delta_A| > |\delta|$. Thus, leaving all other quantities unchanged, the sample size required for non-inferiority assessment is larger under the assumption of equal efficacy of E and C. More precisely: If the non-inferiority margin is given by ε times the clinically relevant effect assumed for superiority testing against placebo, $0 < \varepsilon < 1$, according to (8.4) the total sample size for a non-inferiority trial comparing E and C is approximately $(1/\varepsilon)^2$ times the one required when investigating E in a placebo-controlled superiority trial (for example, quadrupled if $\varepsilon = 0.5$). This is the reason why active-controlled non-inferiority trials generally need higher resources than corresponding comparisons to placebo.

2. If we assume the same statistical model as for the assessment of difference or superiority with an ANCOVA, namely (3.12a) and (3.12b), a test for non-inferiority and the related sample size calculation method can be derived anal-ogously to the t-test. The test problem for demonstrating non-inferiority of the experimental treatment E to the control C is given by

$$H_0: \tau_E - \tau_C \leq -\delta$$
$$\text{versus} \tag{8.5}$$
$$H_1: \tau_E - \tau_C > -\delta$$

with pre-specified non-inferiority margin $\delta > 0$. In analogy to the preceding section, the shifted ANCOVA test statistic

$$T_{ANCOVA} = \frac{\hat{\tau}_E - \hat{\tau}_C + \delta}{\sqrt{\widehat{\text{Var}}(\hat{\tau}_E - \hat{\tau}_C)}},$$

can be used for testing H_0, where $\widehat{\text{Var}}(\hat{\tau}_E - \hat{\tau}_C)$ is defined by (3.14). For $\tau_E - \tau_C = -\delta$, T_{ANCOVA} follows the central t-distribution with $n_C + n_E - 3$ degrees of freedom, and under an alternative defined by $\Delta_A = (\tau_E - \tau_C)_A > -\delta$ it follows the non-central t-distribution with the same number of degrees of freedom and non-centrality parameter

$$nc = \sqrt{\frac{r}{(1+r)^2} \cdot n} \cdot \left(\frac{\Delta_A + \delta}{\sqrt{1 - \rho^2} \cdot \sigma} \right). \tag{8.6}$$

The exact total sample size can therefore be determined by iteratively resolving inequality (3.15) for n, whereby now the non-centrality parameter given by (8.6) is to be inserted, i.e., by determining the smallest integer n fulfilling $F_{T;n-3,0}^{-1}(1 - \alpha/2) \leq F_{T;n-3,nc}^{-1}(\beta)$.

An approximate sample size formula can be obtained by adapting Eq. (3.18) to the non-inferiority test problem:

$$n = \frac{(1+r)^2}{r} \cdot \left(z_{1-\alpha/2} + z_{1-\beta} \right)^2 \cdot \frac{(1 - \rho^2) \cdot \sigma^2}{(\Delta_A + \delta)^2} + \frac{\left(z_{1-\alpha/2} \right)^2}{2} \tag{8.7}$$

with $\Delta_A = (\tau_E - \tau_C)_A$ and $r = n_E/n_C$. The group-wise sample sizes are then given by $n_C = \left(\frac{1}{1+r} \right) \cdot n$ and $n_E = \left(\frac{r}{1+r} \right) \cdot n$.

8.2.2 Ratio of Means as Effect Measure

In clinical trials, the non-inferiority margin is frequently expressed as a portion of the (unknown) population mean of the control treatment. Then, the margin $\gamma > 0$ can be expressed as $\gamma = f \cdot \mu_C$ for a specified value f, $0 < f < 1$. The test problem for the assessment of non-inferiority can now be written as

$$H_0: \mu_E - \mu_C \leq -f \cdot \mu_C$$
$$\text{versus}$$
$$H_1: \mu_E - \mu_C > -f \cdot \mu_C$$

which is equivalent to

$$H_0: \frac{\mu_E}{\mu_C} \leq \delta$$
$$\text{versus} \tag{8.8}$$
$$H_1: \frac{\mu_E}{\mu_C} > \delta$$

with $\delta = 1 - f$, $0 < \delta < 1$. In the following, methods for sample size calculation for test problem (8.8) are presented, i.e. for the effect measure $\theta = \mu_E/\mu_C$.

The null-hypothesis in (8.8) can be tested by applying the test statistic

$$T = \sqrt{\frac{n_C \cdot n_E}{n_C + \delta^2 \cdot n_E}} \cdot \frac{\bar{X}_E - \delta \cdot \bar{X}_C}{S} \tag{8.9}$$

with S denoting the pooled estimator of σ [see (3.8a) and (3.8b) in Sect. 3.3]. For $\mu_E/\mu_C = \delta$, T follows the central t-distribution with $n_C + n_E - 2$ degrees of freedom. Therefore, H_0 can be rejected at level $\alpha/2$ if $t > t_{1-\alpha/2, n_c+n_E-2}$. Under an alternative hypothesis specified by $\theta_A = (\mu_E/\mu_C)_A > \delta$, T follows the non-central t-distribution with the same number of degrees of freedom and non-centrality parameter

$$nc = \sqrt{\frac{r}{(1+r)\cdot\left(1+\delta^2\cdot r\right)}} \cdot n \cdot \left(\frac{\mu_{E,A} - \delta\cdot\mu_{C,A}}{\sigma}\right) \tag{8.10a}$$

$$= \sqrt{\frac{r}{(1+r)\cdot\left(1+\delta^2\cdot r\right)}} \cdot n \cdot \left(\frac{\theta_A - \delta}{CV}\right), \tag{8.10b}$$

where $r = n_E/n_C$ and $CV = \sigma/\mu_C$ denotes the coefficient of variation (Hauschke et al. 1999). Hence, the exact total sample size can again be calculated by iteratively resolving the inequality (3.10) applied for the non-shifted t-test but now inserting the non-centrality parameter (8.10a) or (8.10b), respectively. Note that when version (8.10a) is used for expressing the non-centrality parameter, it is not sufficient to define an alternative $\theta_A = (\mu_E/\mu_C)_A$ but additionally one of the expectations has to be specified (which then defines the other one via θ_A).

The statistically optimal allocation ratio can be determined by maximizing for fixed n the value of the non-centrality parameter. Derivating the function $nc(r)$ for r shows that a unique maximum is achieved for $r = n_E/n_C = 1/\delta > 1$. Therefore, from a statistical viewpoint, allocating more patients to the experimental treatment is advantageous. As a by-product, more information, especially on safety, can then be obtained for the new treatment. This is sensible as in non-inferiority trials the comparator is an established control with already well-known characteristics.

The Guenther-Schouten method modified to the current setting leads to the following approximation formulas for the total sample size:

$$n = \frac{(1+r)\cdot\left(1+\delta^2\cdot r\right)}{r} \cdot \left(z_{1-\alpha/2} + z_{1-\beta}\right)^2 \cdot \left(\frac{\sigma}{\mu_{E,A} - \delta\cdot\mu_{C,A}}\right)^2 + \frac{\left(z_{1-\alpha/2}\right)^2}{2} \tag{8.11a}$$

$$= \frac{(1+r)\cdot\left(1+\delta^2\cdot r\right)}{r} \cdot \left(z_{1-\alpha/2} + z_{1-\beta}\right)^2 \cdot \left(\frac{CV}{\theta_A - \delta}\right)^2 + \frac{\left(z_{1-\alpha/2}\right)^2}{2} \tag{8.11b}$$

with $\theta_A = (\mu_E/\mu_C)_A$ and $r = n_E/n_C$, i.e. $n_C = \left(\frac{1}{1+r}\right)\cdot n$ and $n_E = \left(\frac{r}{1+r}\right)\cdot n$.

Example 8.2 A considerable proportion of long bone fractures fail to heal properly. The **ORTHOUNION trial** was initiated to evaluate the efficacy of two doses of autologous human bone marrow-derived expanded mesenchymal stromal cell (hBM-MSC) treatments and iliac crest autograft (ICA) to enhance bone healing in patients with long bone nonunion in a multicenter, open, three-arm, randomized setting (Gómez-Barrena et al. 2018). The primary objective was to demonstrate that the combined hBM-MSC group including both doses is superior to ICA. If the related null-hypothesis can be rejected, non-inferiority of the lower dose versus the higher dose of hBM-MSC was to be assessed. The non-inferiority assessment was to be based on the score calculated from the REBORNE radiological scale at 12 months after surgery. This validated scale evaluates the presence or absence of radiological consolidations. The outcome was assumed to be normally distributed, the ratio of means was used as effect measure, and the non-inferiority margin was set to $\delta = 0.90$. For sample size calculation, equal means were assumed, i.e. $(\mu_E/\mu_C)_A = 1$, and the coefficient of variation was set to $CV = 0.135$. If a power of $1 - \beta = 0.90$ was desired for the non-inferiority test at one-sided level $\alpha/2 = 0.025$ with balanced groups $(r = 1)$, exact and approximate sample size calculation results in a total sample size of $72 = 2 \times 36$. The optimal allocation ratio is $r = 1/\delta = 1.11$ for which the same total sample size of $n = 72$ is required as for balanced groups. The allocation $n_E = 38$ and $n_C = 34$ $(r = 1.12)$ results in a marginally higher power as compared to balanced groups $(0.9038$ vs. $0.9031)$. In practice, such a minor advantage would hardly lead to abandoning balanced allocation.

8.3 Continuous and Ordered Categorical Outcomes

We consider the same situation as in Sect. 4.2, i.e. mutually independent random variables $X_{C1}, \ldots, X_{Cn_C}$ and $X_{E1}, \ldots, X_{En_E}$ with distribution functions F_C and F_E, respectively. The parameter of interest is again the relative effect $\pi = \Pr(X_E > X_C) + 0.5 \cdot \Pr(X_E = X_C)$. For $F_E = F_C$, i.e., if there is no difference between E and C, $\pi = 0.5$ holds true. Therefore, differences between the treatments can be expressed by deviations of π from the value 0.5. If higher values of the outcome indicate favorable results, the test problem for the assessment of non-inferiority of E versus C can be formulated as follows:

$$H_0: \pi \leq 0.5 - \delta$$
$$\text{versus} \tag{8.12}$$
$$H_1: \pi > 0.5 - \delta,$$

where $\delta \in {]}0, 0.5{[}$ denotes the non-inferiority margin defining a clinically irrelevant amount of inferiority of E versus C. The approach proposed by Munzel and Hauschke (2003), which is presented in the following, includes both continuous and ordered categorical data.

As in Sect. 4.3.1, we define the function $\delta(z) = \begin{cases} 1 & \text{if } z > 0 \\ 0.5 & \text{if } z = 0 \\ 0 & \text{if } z < 0 \end{cases}$. An unbiased estimator of π is then given by $\hat{\pi} = \frac{1}{n_C \cdot n_E} \cdot \sum_{i=1}^{n_E} \sum_{j=1}^{n_C} \delta(X_{Ei} - X_{Cj})$. Furthermore, asymptotically (i.e. for $n = n_C + n_E \to \infty$, where n_E/n_C converges to a constant different from zero) $\sqrt{n} \cdot (\hat{\pi} - E(\hat{\pi})) \sim N(0, \sigma^2)$, where $\sigma^2 = n \cdot \left(\frac{\sigma_C^2}{n_C} + \frac{\sigma_E^2}{n_E}\right)$ with

$\sigma_C^2 = \text{Var}(F_E(X_{C1}))$ and $\sigma_E^2 = \text{Var}(F_C(X_{E1}))$. Furthermore, consistent estimators of σ_i^2, $i = C, E$, (i.e. estimators that asymptotically converge in probability to the true values) are given by

$$\hat{\sigma}_i^2 = \frac{1}{(n - n_i) \cdot (n_i - 1)} \cdot \sum_{j=1}^{n_i} \left(R_{ij} - R_{ij}^{(i)} - \bar{R}_i + \frac{n_i + 1}{2} \right)^2,$$

where R_{ij} is the rank of X_{ij} among all random variables $X_{C1}, \ldots, X_{Cn_C}, X_{E1}, \ldots, X_{En_E}$, $i = C, E, j = 1, \ldots, n_i$, $R_{ij}^{(i)}$ is the rank of X_{ij} in the sample $X_{i1}, \ldots, X_{in_i}$, and $\bar{R}_i = \frac{1}{n_i} \cdot \sum_{j=1}^{n_i} R_{ij}$ (Brunner and Munzel 2000; Munzel and Hauschke 2003). Therefore, an asymptotic test of H_0 can be based on the test statistic $U = \frac{\sqrt{n} \cdot (\hat{\pi} - (0.5 - \delta))}{\hat{\sigma}}$ which is asymptotically standard normally distributed for $\pi = 0.5 - \delta$. Therefore, H_0 can be rejected at one-sided level $\alpha/2$ if $u > z_{1-\alpha/2}$; otherwise, H_0 is accepted. Under an alternative specified by a value $\pi_A > 0.5 - \delta$, $\sqrt{n} \cdot (\hat{\pi} - (0.5 - \delta)) \sim N(\sqrt{n} \cdot (\pi_A - (0.5 - \delta)), \sigma^2)$ holds approximately true. Applying Eq. (2.10) and inserting the allocation ratio $r = n_E/n_C$ leads to the following approximate formula for the total sample size required to achieve a power of $1 - \beta$ for a one-sided test at level $\alpha/2$:

$$n = \frac{(1 + r)}{r} \cdot (z_{1-\alpha/2} + z_{1-\beta})^2 \cdot \frac{(\sigma_E^2 + r \cdot \sigma_C^2)}{(\pi_A - (0.5 - \delta))^2}. \tag{8.13}$$

For sample size calculation, σ_C^2 and σ_E^2 have to be calculated for the alternative specified by a value for π_A. For this, the distribution functions $F_{C,A}$ und $F_{E,A}$ assumed under the alternative have to be defined.

For ordered categorical data, the following expressions were given by Munzel and Hauschke (2003). As in Sect. 4.3, the probability that category k is observed in treatment group i, is denoted by p_{ik}, $i = C, E, k = 1, \ldots, M$. Then

$$\pi = \sum_{k=1}^{M} p_{Ck} \cdot \left(\frac{\sum_{l=1}^{k} p_{El} + \sum_{l=1}^{k-1} p_{El}}{2} \right)$$

$$\sigma_C^2 = \sum_{k=1}^{M} p_{Ck} \cdot \left(\frac{\sum_{l=1}^{k} p_{El} + \sum_{l=1}^{k-1} p_{El}}{2} \right)^2 - \pi^2$$

$$\sigma_E^2 = \sum_{k=1}^{M} p_{Ek} \cdot \left(\frac{\sum_{l=1}^{k} p_{Cl} + \sum_{l=1}^{k-1} p_{Cl}}{2} \right)^2 - (1 - \pi)^2.$$

For $k = 1$ and upper boundary of the inner summation equal to $k - 1$, the inner sum is equal to zero.

Remarks

1. Brunner and Munzel (2000) showed by simulations that the normal approximation for the test statistic U works well for $n_i \geq 50, i = C, E$. To avoid an inflation of the type I error rate for smaller sample sizes, the test decision should be based on the $(1 - \alpha/2)$-quantile of the central t-distribution with the number of degrees of freedom being equal to $\left(\sum_{i=C,E} S_i^2/(n - n_i) \right)^2 / \sum_{i=C,E} \left(S_i^2/(n - n_i) \right)^2 /(n_i - 1)$ (so-called Satterthwaite-Smith-Welch approximation) which provides satisfactorily accurate results for $n_i \geq 15, i = C, E$.

2. The variance estimator proposed by Munzel and Hauschke (2003) presented above is derived under the alternative hypothesis. Zhang et al. (2015) presented a test statistic and a corresponding method for sample size calculation where the variance is estimated under the null-hypothesis (see related approach for binary data in Sect. 8.4). However, their simulations showed that this test exceeds the nominal level in some situations. This may explain the observed gain in power as compared to the approach of Munzel and Hauschke (2003). Therefore, application of this method is only recommended in scenarios for which it can be shown that there is no or at most an acceptably slight inflation of the type I error rate.

3. Schmidtmann et al. (2018) provided methods for power and sample size calculations for nonparametric non-inferiority assessment for the situation that values for the primary endpoint are missing due to death of patients.

4. Specification of an appropriate non-inferiority margin in case that the treatment group difference is measured by the relative effect $\Pr(X_E > X_C) + 0.5 \cdot \Pr(X_E = X_C)$ is even more challenging than when it is captured on the original scale of the outcome. Considerations on the choice of δ in the nonparametric setting are presented by Munzel and Hauschke (2003) and Zhang et al. (2015).

8.4 Binary Outcomes

Throughout this section, we assume the statistical model described in (5.1a) and (5.1b) and the situation that higher rates correspond to a more favorable outcome. As for the assessment of difference or superiority, respectively, we consider tests and related sample size calculation methods for the difference of rates, the risk ratio, and the odds ratio as effect measures. References for recommendations concerning the choice of appropriate effect measures in non-inferiority trials with binary outcomes can be found in Siqueira et al. (2015).

8.4.1 Asymptotic Tests

8.4.1.1 Difference of Rates as Effect Measure

For binary outcomes and the treatment effect expressed in terms of the difference of rates, the test problem for the assessment of non-inferiority of E versus C is given by

$$
\begin{aligned}
H_0&: p_E - p_C \leq -\delta \\
&\text{versus} \\
H_1&: p_E - p_C > -\delta
\end{aligned}
\tag{8.14}
$$

with pre-defined non-inferiority margin $\delta > 0$. The null-hypothesis can be tested by applying the shifted version of the normal approximation test

$$
U = \frac{\hat{p}_E - \hat{p}_C + \delta}{\hat{\sigma}_0},
\tag{8.15}
$$

where $\hat{\sigma}_0^2$ denotes an appropriate estimate of the variance of $(\hat{p}_E - \hat{p}_C)$ under the null hypothesis restriction $p_E - p_C = -\delta$. Various proposals exist concerning the variance estimator to be inserted in the test statistic (8.15). In an extensive empirical investigation, Roebruck and Kühn (1995) showed that the test and the related method for sample size calculation proposed by Farrington and Manning (1990) can be recommended for settings commonly met in clinical non-inferiority trials. Here,

$$
\hat{\sigma}_0^2 = \frac{1}{n_E} \cdot \left(r \cdot \check{p}_C \cdot (1 - \check{p}_C) + \check{p}_E \cdot (1 - \check{p}_E) \right),
\tag{8.16}
$$

where $\check{p}_C$ and $\check{p}_E$ are the maximum likelihood estimates of p_C and p_E under the null hypothesis restriction $\check{p}_E - \check{p}_C = -\delta$. These restricted maximum likelihood estimators are given by

$$
\check{p}_E = 2 \cdot u \cdot \cos(w) - b/(3 \cdot a)
\tag{8.17a}
$$

$$
\check{p}_C = \check{p}_E + \delta,
\tag{8.17b}
$$

where

$$
\begin{aligned}
u &= \text{sgn}(v) \cdot \sqrt{b^2/(3 \cdot a)^2 - c/(3 \cdot a)} \\
v &= b^3/(3 \cdot a)^3 - (b \cdot c)/(6 \cdot a^2) + d/(2 \cdot a) \\
w &= (1/3) \cdot \left(\pi + \cos^{-1}(v/u^3) \right)
\end{aligned}
$$

with

$$a = 1 + (1/r)$$
$$b = -\left(1 + (1/r) + \hat{p}_E + (1/r) \cdot \hat{p}_C - \delta \cdot ((1/r) + 2)\right)$$
$$c = \delta^2 - \delta \cdot \left(2 \cdot \hat{p}_E + (1/r) + 1\right) + \hat{p}_E + (1/r) \cdot \hat{p}_C$$
$$d = \hat{p}_E \cdot \delta \cdot (1 - \delta)$$

and $\mathrm{sgn}(x)$ denoting the sign of x.

For $p_E - p_C = -\delta$, U is asymptotically standard normally distributed, and hence H_0 can be rejected at level $\alpha/2$ if $u > z_{1-\alpha/2}$. For the derivation of an (approximate) sample size formula, the same considerations can be applied as in Sect. 5.2.1. Under an alternative hypothesis specified by $p_{E,A} - p_{C,A} = \Delta_A > -\delta$, it holds true asymptotically that $\hat{p}_E - \hat{p}_C + \delta \sim N(\Delta_A + \delta, \sigma_A^2)$ with $\sigma_A^2 = \frac{1}{n_E} \cdot \left(r \cdot p_{C,A} \cdot (1 - p_{C,A}) + p_{E,A} \cdot (1 - p_{E,A})\right)$. The approximate sample size formula can then be obtained by solving Eq. (5.7) adapted to the non-inferiority setting, i.e.

$$\Delta_A + \delta = \sigma_0 \cdot z_{1-\alpha/2} + \sigma_A \cdot z_{1-\beta}.$$

Here, σ_0 is obtained by inserting in (8.16) the values $\tilde{p}_C$ and $\tilde{p}_E$ that result when replacing in the computations defined by (8.17a) and (8.17b) $\hat{p}_E$ and $\hat{p}_C$ by $p_{E,A}$ and $p_{C,A}$, respectively (i.e. by employing the large sample approximations of the restricted maximum likelihood estimators):

$$n = \frac{1+r}{r}$$
$$\cdot \frac{\left(z_{1-\alpha/2} \cdot \sqrt{r \cdot \tilde{p}_C \cdot (1 - \tilde{p}_C) + \tilde{p}_E \cdot (1 - \tilde{p}_E)} + z_{1-\beta} \cdot \sqrt{r \cdot p_{C,A} \cdot (1 - p_{C,A}) + p_{E,A} \cdot (1 - p_{E,A})}\right)^2}{(\Delta_A + \delta)^2}$$

$$(8.18)$$

with $\Delta_A = p_{E,A} - p_{C,A}$ and $r = n_E/n_C$, i.e. $n_C = \left(\frac{1}{1+r}\right) \cdot n$ and $n_E = \left(\frac{r}{1+r}\right) \cdot n$.

Example 8.3 Transcatheter aortic valve replacement is applied widespread for selected patients with severe aortic stenosis. Notwithstanding the advances in this field, established transcatheter aortic valve replacement devices have limitations and risks. The **REPRISE III trial** is a randomized multicenter study which was conducted to demonstrate non-inferiority of a mechanically expanded valve (E) to an approved self-expanding valve (C) in high-risk patients with aortic stenosis undergoing transcatheter aortic valve replacement (Feldman et al. 2018). Two co-primary endpoints were defined for which the related null-hypotheses had both to be rejected to achieve the trial aim ("all or none" success criterion; see Sect. 11.3.2a). The primary safety endpoint (endpoint 1) was the 30 days composite rate of all-cause mortality, stroke, life-threatening and major bleeding events, stage 2/3 acute kidney injury, and major vascular complications. The primary effectiveness endpoint (endpoint 2) was the 1-year composite rate of all-cause mortality, disabling stroke, and moderate or greater paravalvular leak based on care laboratory assessment. The term "composite" means that an event for this endpoint happens if an event occurs for at least one of the components. For a comprehensive overview on the planning and analysis of clinical trials with composite endpoints, the interested reader is referred to the monograph by Rauch et al. (2018). For sample size calculation, the expected rates were derived from previous trials that enrolled patients similar to the ones to be recruited in the current trial. The assumed rates were $p_{E,1,A} = p_{C,1,A} = 0.40$ (endpoint 1) and $p_{E,2,A} = p_{C,2,A} = 0.32$ (endpoint 2), respectively, and the non-inferiority margins were defined as $\delta_1 = 0.105$ and $\delta_2 = 0.095$. Sample size calculations were performed for a one-sided type I error rate of $\alpha/2 = 0.025$ for each endpoint and a desired power of $(1 - \beta)_1 = 0.85$ for endpoint 1 and $(1 - \beta)_2 = 0.80$ for endpoint 2. Note that while the related sample size assures rejection of each of the single hypotheses with the desired power, the power to reject both, which is the aim of the study, is not addressed by this approach: Depending on the correlation between the test statistics, the "conjunctive power" may be considerably smaller than the lower one of these values (see Sect. 11.3). The patients were to be randomized with an allocation ratio $E{:}C$ of 2:1 ($r = 2$). It should be mentioned that the presented test problem, statistical test, and approximate sample size formula are tailored to the situation that higher rates are advantageous. Therefore, obvious modifications are to be made to apply the approximate sample size formula in the current situation where the rates refer to harmful events. The restricted estimates of the rates to be inserted in Eq. (8.16) for endpoint 1 and 2, respectively, are $\tilde{p}_{E,1} = 0.332$ and $\tilde{p}_{C,1} = 0.437$ and $\tilde{p}_{E,2} = 0.26$ and $\tilde{p}_{C,2} = 0.355$. This results in total sample sizes of 856 and 817 for endpoint 1 and 2, respectively, which have to be rounded up to $858 = 572 + 286$ and $819 = 546 + 273$ to comply with the allocation ratio $r = 2$. Figure 8.1a shows the required sample size depending on the allocation ratio for the planning assumptions of the REPRISE III trial. The optimum allocation ratio lies slightly above 1 for both endpoints ($r = 1.06$ and $r = 1.12$ for endpoint 1 and 2, respectively) and requires considerably smaller total sample sizes of $777 = 400 + 377$ and $753 = 411 + 342$ compared to 2:1 allocation. For balanced groups, the required total sample size amounts to $778 = 2 \times 389$ and $756 = 2 \times 378$ for endpoint 1 and 2, respectively. Figure 8.1b depicts the power achieved for the total sample size calculated for the allocation ratio 2:1 depending on r. For balanced allocation, the power amounts to 0.88 and 0.83 for endpoint 1 and 2, respectively, instead of 0.85 and 0.80 achieved for $r = 2$.

8.4.1.2 Risk Ratio as Effect Measure

If the treatment group difference is expressed in terms of the ratio of rates, the test problem for the assessment of non-inferiority of E versus C is given by

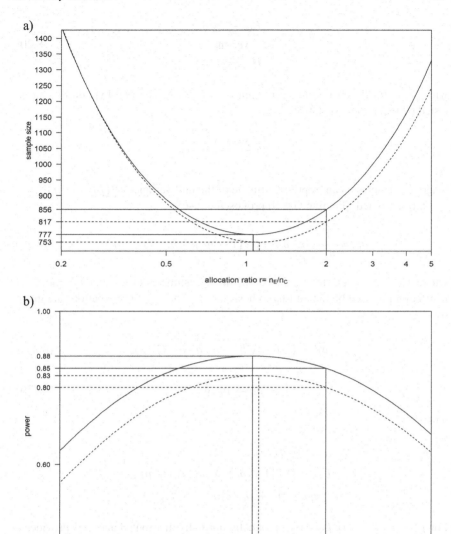

Fig. 8.1 **a** Total sample size depending on the allocation ratio $r = n_E/n_C$ (see Example 8.3; solid line: endpoint 1, dashed line: endpoint 2), **b** Power for the sample size calculated to achieve a power of $1 - \beta = 0.85$ (endpoint 1) or 0.80 (endpoint 2), respectively, for allocation ratio $r = n_E/n_C = 2$ depending on the allocation ratio r (see Example 8.3; solid line: endpoint 1, dashed line: endpoint 2)

$$H_0: \frac{p_E}{p_C} \le \delta$$
$$\text{versus} \tag{8.20}$$
$$H_1: \frac{p_E}{p_C} > \delta$$

with pre-defined non-inferiority margin $0 < \delta < 1$. The null-hypothesis can be tested by using the test statistic

$$U = \frac{\hat{p}_E - \delta \cdot \hat{p}_C}{\hat{\sigma}_0}, \tag{8.21}$$

where $\hat{\sigma}_0^2$ denotes an appropriate estimate of the null variance of $\left(\hat{p}_E - \delta \cdot \hat{p}_C\right)$.

Farrington and Manning (1990) proposed to use

$$\hat{\sigma}_0^2 = \frac{1}{n_E} \cdot \left(r \cdot \delta^2 \cdot \check{p}_C \cdot \left(1 - \check{p}_C\right) + \check{p}_E \cdot \left(1 - \check{p}_E\right)\right), \tag{8.22}$$

where $\check{p}_C$ and $\check{p}_E$ are the maximum likelihood estimates of p_C and p_E under the restriction imposed by the null-hypothesis, i.e. $\check{p}_E / \check{p}_C = \delta$. The solutions are given by

$$\check{p}_E = \left(-b - \sqrt{b^2 - 4 \cdot a \cdot c}\right)/(2 \cdot a) \tag{8.23a}$$

$$\check{p}_C = \check{p}_E / \delta, \tag{8.23b}$$

where

$$a = 1 + (1/r)$$
$$b = -\left(\delta \cdot \left(1 + \hat{p}_C/r\right) + (1/r) + \hat{p}_E\right)$$
$$c = \delta \cdot \left(\hat{p}_E + \hat{p}_C/r\right).$$

For $p_E - \delta \cdot p_C = 0$, U is asymptotically normally distributed and thus provides a test with approximate level $\alpha/2$ if H_0 is rejected for $u > z_{1-\alpha/2}$. Under an alternative hypothesis H_A specified by $p_{E,A} - \delta \cdot p_{C,A} > 0$, $\hat{p}_E - \delta \cdot \hat{p}_C \sim N\left(p_{E,A} - \delta \cdot p_{C,A}, \sigma_A^2\right)$ with $\sigma_A^2 = \frac{1}{n_E} \cdot \left(r \cdot \delta^2 \cdot p_{C,A} \cdot \left(1 - p_{C,A}\right) + p_{E,A} \cdot \left(1 - p_{E,A}\right)\right)$. The same arguments as for the situation when the treatment group difference is measured in terms of the difference of rates can be applied to derive an approximate sample size formula:

$$n = \frac{1+r}{r}$$
$$\cdot \frac{\left(z_{1-\alpha/2} \cdot \sqrt{r \cdot \delta^2 \cdot \tilde{p}_C \cdot (1 - \tilde{p}_C) + \tilde{p}_E \cdot (1 - \tilde{p}_E)} + z_{1-\beta} \cdot \sqrt{r \cdot \delta^2 \cdot p_{C,A} \cdot \left(1 - p_{C,A}\right) + p_{E,A} \cdot \left(1 - p_{E,A}\right)}\right)^2}{\left(p_{E,A} - \delta \cdot p_{C,A}\right)^2}, \tag{8.24}$$

where $\tilde{p}_C$ and $\tilde{p}_E$ are obtained by replacing in the calculations induced by (8.23a) and (8.23b) the sample rates $\hat{p}_E$ and $\hat{p}_C$ by $p_{E,A}$ and $p_{C,A}$, respectively. Again, $r = n_E/n_C$ and therefore $n_C = \left(\frac{1}{1+r}\right) \cdot n$ and $n_E = \left(\frac{r}{1+r}\right) \cdot n$.

Example 8.4 Effective thromboprophylaxis is essential for patients undergoing hip replacement. Low-molecular-weight heparins, one of the established prophylactic techniques, require subcutaneous injection which prompted to develop oral anticoagulant agents. The **ADVANCE-3 trial** was conducted as a randomized, double-blind, multicenter study to demonstrate non-inferiority of apixaban (E), an orally active specific factor Xa inhibitor, to the subcutaneously administered low-molecular-weight heparin enoxaparin (C) (Lassen et al. 2010). The double-dummy technique was applied to achieve blinding: patients randomized to the experimental group received placebo subcutaneous study medication in addition to the active oral study medication and vice versa for the patients in the control group. The primary outcome was a composite endpoint with the components adjudicated asymptomatic or symptomatic deep-vein thrombosis, nonfatal pulmonary embolism, or death from any cause during the intended treatment period. Note that, as in Example 8.3, test problem, statistical test, and sample size calculation are to be modified accordingly as compared to the description above to deal with the situation that smaller event rates are advantageous here. The non-inferiority test problem was formulated in terms of the risk ratio with a non-inferiority margin $\delta = 1.25$ for the ratio p_E/p_C. The Farrington-Manning test at one-sided level $\alpha = 0.025$ was applied. True event rates of $p_{E,A} = 0.0385$ and $p_{C,A} = 0.0550$ were assumed for sample size calculation (i.e. $\theta_A = 0.70$) and balanced allocation was implemented ($r = 1$). An uncommon value of $1 - \beta = 0.92$ was specified for the power. This is due to the fact that three test problems were formulated for the confirmatory analysis which were to be tested in a hierarchical way and that the sample size should be high enough to provide sufficient power not only for the primary outcome considered here. The restricted estimates of the rates are given by $\tilde{p}_E = 0.0519$ and $\tilde{p}_C = 0.0415$, respectively, and the approximation formula (8.24) provides a required total sample size of $2826 = 2 \times 1413$. If this sample size would be allocated with a ratio $E{:}C$ of 2:1 ($r = 2; n = 1884+942$), a power of 0.88 results.

Remark Hoover and Blackwelder (2001) considered the test problem $H_0: \frac{p_E}{p_C} \geq \delta$ versus $H_1: \frac{p_E}{p_C} < \delta$ with non-inferiority margin $0 < \delta < 1$. In this setting, it is the aim of the trial to demonstrate that the experimental treatment substantially reduces an adverse outcome. They showed that the optimal sample size allocation ratio is approximately given by $r = n_E/n_C = 1/\sqrt{\delta} > 1$. Therefore, allocating more patients to the experimental treatment may increase efficiency. Furthermore, they showed that if the optimal allocation ratio is not a simple fraction, using more convenient allocation schemes leads to very similar performance. For example, for $0.10 \leq \delta \leq 0.17$ the allocation ratio $r = 3$ can be used instead of the optimal ratio, for $0.17 < \delta \leq 0.33$ the ratio $r = 2$ is a good alternative, and for $0.33 < \delta \leq 0.5$ sample size allocation with the ratio $r = 3/2$ can be applied without major loss in efficiency as compared to the optimal allocation.

8.4.1.3 Odds Ratio as Effect Measure

If the odds ratio is used as summary measure to compare the groups, the test problem for assessing non-inferiority of E versus C is given by

$$H_0: \frac{p_E \cdot (1-p_C)}{p_C \cdot (1-p_E)} \leq \gamma$$

$$\text{versus}$$

$$H_1: \frac{p_E \cdot (1-p_C)}{p_C \cdot (1-p_E)} > \gamma$$

with non-inferiority margin $0 < \gamma < 1$. Equivalently, the test problem can be written as

$$H_0: \ln\left(\frac{p_E \cdot (1-p_C)}{p_C \cdot (1-p_E)}\right) \leq \delta$$

$$\text{versus} \tag{8.25}$$

$$H_1: \ln\left(\frac{p_E \cdot (1-p_C)}{p_C \cdot (1-p_E)}\right) > \delta$$

with $\delta = \ln(\gamma) < 0$.

Wang et al. (2002) proposed to use the shifted version of the superiority test statistic (5.16) for the assessment of (8.25):

$$U = \frac{\ln\left(\widehat{OR}\right) - \delta}{\sqrt{\frac{1}{n_E \cdot \hat{p}_E \cdot \left(1-\hat{P}_E\right)} + \frac{1}{n_C \cdot \hat{p}_C \cdot (1-\hat{p}_C)}}}. \tag{8.26}$$

For $\ln(OR) - \delta = 0$, U follows asymptotically the standard normal distribution. Under an alternative specified by values for $\ln(OR_A) = \ln\frac{p_{E,A} \cdot (1-p_{C,A})}{p_{C,A} \cdot (1-p_{E,A})} > \delta$ and $p_{E,A}$ (or $p_{C,A}$, respectively), U follows asymptotically the normal distribution $N(nc, 1)$ with

$$nc = \frac{\ln(OR_A) - \delta}{\sqrt{\frac{1+r}{r} \cdot \frac{1}{n} \cdot \left(\frac{1}{p_{E,A} \cdot (1-p_{E,A})} + \frac{r}{p_{C,A} \cdot (1-p_{C,A})}\right)}}.$$

Therefore, the general methodology given in Sect. 2.2 can be applied to obtain an approximate sample size formula. Solving Eq. (2.10), i.e. $nc = z_{1-\alpha/2} + z_{1-\beta}$, for n results in

$$n = \frac{1+r}{r} \cdot \left(z_{1-\alpha/2} + z_{1-\beta}\right)^2 \cdot \frac{\left(\frac{1}{p_{E,A} \cdot (1-p_{E,A})} + \frac{r}{p_{C,A} \cdot (1-p_{C,A})}\right)}{(\ln(OR_A) - \delta)^2} \tag{8.27}$$

with $OR_A = \frac{p_{E,A} \cdot (1-p_{C,A})}{p_{C,A} \cdot (1-p_{E,A})}$ and $r = n_E/n_C$, i.e. $n_C = \left(\frac{1}{1+r}\right) \cdot n$ and $n_E = \left(\frac{r}{1+r}\right) \cdot n$.

Analogously to the case of testing for difference or superiority, by inserting in the variance term the rates assumed under the specified alternative, it is implicitly assumed that the variances of the estimated log odds ratio are equal under the null- and the alternative hypothesis.

Example 8.5 For patients infected with **hepatitis C virus (HCV)** genotype 2 or 3, treatment with peginterferon alfa-2a plus ribavirin for 24 weeks, sustained virologic response rates of about 80% were observed in clinical trials (Hadziyannis et al. 2004; Manns et al. 2001). As similar response rates were reported for shorter treatment durations, a multi-national, randomized trial was performed that aimed at demonstrating non-inferiority of 16 weeks of treatment (E) with peginterferon alfa-2a and ribavirin to 24 weeks of treatment (C) in patients infected with HCV genotype 2 or 3 (Shiffman 2007). The trial was double-blind until week 16, when the investigators were informed about the patients' treatment group assignments. For sample size calculation, equal rates of sustained virologic response (= primary endpoint) in the experimental and the control group of $p_{E,A} = p_{C,A} = 0.80$ were assumed. The non-inferiority margin for the odds ratio was fixed at $\gamma = 0.70$ (i.e. $\delta = \ln(\gamma) = -0.357$). Balanced sample size allocation was implemented, and a one-sided type I error rate of $\alpha/2 = 0.025$ was specified. If the asymptotic test based on the test statistic (8.26) is applied, $1544 = 2 \times 772$ patients are required according to the approximate sample size formula (8.27) to assure a power of $1 - \beta = 0.80$. If the sample size allocated to the experimental treatment is twice of that assigned to the control (i.e., $r = n_E/n_C = 2$), the total sample size needed is $1737 = 1158 + 579$.

Remark Siqueira et al. (2015) compared the accuracy of formula (8.27) with another formula based on the score test for test problem (8.25). Both sample size formulas work well for non-inferiority margins $\gamma \geq 0.5$ and alternatives $\theta_A \in [1, 2.5]$. For other scenarios, formula (8.27) may over- or underestimate the sample size depending on the values of γ and θ_A. In summary, it was concluded that neither approximation formula is accurate in all scenarios but that overall both lead to satisfactory sample size calculations.

8.4.2 Exact Unconditional Tests

8.4.2.1 Basic Concepts

In Sect. 5.3, the principle of constructing exact unconditional tests was sketched for the case of superiority tests and the setting that these tests are based on the test statistic of the normal approximation test or the p-value of Fisher's exact test, respectively. The test statistics are used for ranking the 2×2 tables in the sampling space such that the region of tables that are at least as extreme as the observed one can be defined. Then, the exact p-value (and with it the related unconditional exact test) can be defined as the maximum probability of tables that are more extreme than the observed one, where the maximum is taken over all parameter constellations fulfilling H_0. By taking the maximum (i.e. considering the worst case constellation), the nuisance parameter (for superiority tests: $p_C = p_E = p$) is eliminated. The same approach can be followed for constructing unconditional exact tests for the assessment of non-inferiority. In principle, any reasonable test statistic can be applied for ordering the sampling space. However, searching for the maximum over the whole null-space is computationally demanding and it is not self-evident that for arbitrary test statistics maximization over its boundary is sufficient. In this context, Röhmel and Mansmann (1999) showed for a quite general formulation of the test problem for two-group comparisons of rates that it is sufficient to search for the maximum on the boundary of the null-space if the so-called Barnard convexity condition is satisfied for the rejection region. Roughly speaking, fulfillment of this condition means that if for observed values x_C and x_E the null-hypothesis is rejected, then this is also true for realizations that even stronger support the alternative hypotheses. Barnard (1947) introduced this condition as a logical requirement for rejection regions. Röhmel and Mansmann (1999) showed that, for example, the test statistic of the normal approximation test and the p-value of Fisher's exact test lead to rejection regions fulfilling the convexity condition.

In the following sections, sample size calculation methods will be described that match to unconditional tests based on test statistics which were already introduced previously. Ordering of the sample space based on these test statistics leads to rejection regions that satisfy Barnard's convexity condition. Thus, for these tests, maximization can be restricted to the (one-dimensional) boundary of the null-space reducing the computational complexity considerably.

8.4.2.2 Difference of Rates as Effect Measure

Chan (1998) used the test statistic of Farrington and Manning (1990) based on the observed rate difference [see Sect. 8.4.1.1, (8.15), (8.16), (8.17a) and (8.17b)] to derive an exact unconditional test for the assessment of the test problem H_0: $p_E - p_C \leq -\delta$ versus H_1: $p_E - p_C > -\delta$, $\delta > 0$. Röhmel (2005) showed that this test statistic fulfills the convexity condition, and thus the rejection region can be determined by restricting maximization to the interval $[0, 1 - \delta]$. For an alternative specified by values $p_{C,A}$ and $p_{E,A}$ and for given sample size, the power of the related exact unconditional test can be obtained by evaluating the probability that the test statistic of an observed table falls in the rejection region under this parameter scenario (Chan 2002). By inversion, the sample size required to achieve a desired power $1 - \beta$ can be determined.

Example 8.6 The **REPRISE III trial** (see Example 8.3) considered two primary endpoints in confirmatory analysis. For sample size calculation, the assumed rates were $p_{E,1,A} = p_{C,1,A} = 0.40$ and $p_{E,2,A} = p_{C,2,A} = 0.32$. As in Example 8.3, test problem and sample size calculation have to be adapted accordingly to take into account that the event of interest is harmful and that thus small event rates are advantageous. The non-inferiority margins $\delta_1 = 0.105$ and $\delta_2 = 0.095$, respectively, were defined to be applied, and the sample size was to be allocated with ratio $r = n_E/n_C = 2$. Testing at one-sided type I error rate $\alpha/2 = 0.025$ should provide a power of $(1 - \beta)_1 = 0.85$ for endpoint 1 and $(1 - \beta)_2 = 0.80$ for endpoint 2. For the unconditional exact test based on the Farrington and Manning test statistic, the total required sample size amounts to $870 = 580 + 290$ and $831 = 554 + 277$ for endpoint 1 and 2, respectively (compared to $858 = 572 + 286$ and $819 = 546 + 273$ required for the related asymptotic test which, however, may not control the specified type I error rate). If sample size allocation is balanced, the total sample size needed is $n = 2 \times 391 = 782$ for endpoint 1 and $n = 2 \times 386 = 772$ for endpoint 2.

8.4.2.3 Risk Ratio as Effect Measure

The test statistic proposed by Farrington and Manning (1990) for an asymptotic test assessing the test problem $H_0: \frac{p_E}{p_C} \leq \delta$ versus $H_1: \frac{p_E}{p_C} > \delta$ [see Sect. 8.4.1.2, (8.21), (8.22), (8.23a) and (8.23b)] can be used to construct a related exact unconditional test (Chan 2003). Sample size calculation for this test can be performed along the same lines as sketched in the preceding section for the difference of rates as effect measure.

Example 8.7 When planning the **ADVANCE-3 trial** (see Example 8.4), event rates of $p_{E,A} = 0.0385$ and $p_{C,A} = 0.0550$ were assumed. Balanced sample size allocation was applied and a power of $1 - \beta = 0.92$ was aspired for a one-sided test at level $\alpha/2 = 0.025$ with non-inferiority margin $\delta = 1.25$ and specified alternative $(p_E/p_C)_A = 0.70$. For the exact unconditional test based on the Farrington-Manning test statistic, the required total sample size amounts to $2862 = 2 \times 1431$ patients. For the sample size allocation $r = n_E/n_C = 2$, a total sample size of $3486 = 1162 + 2324$ is needed.

8.4.2.4 Odds Ratio as Effect Measure

Similar to Fisher's exact test for the assessment of difference or superiority of rates, an exact conditional p-value can be calculated when assessing the test problem $H_0: \frac{p_E \cdot (1 - p_C)}{p_C \cdot (1 - p_E)} \leq \delta$ versus $H_1: \frac{p_E \cdot (1 - p_C)}{p_C \cdot (1 - p_E)} > \delta$ by using the so-called extended hypergeometric distribution (see, for example, Wellek (2010), p. 172ff). Röhmel and Mansmann (1999) stated that using this p-value as ordering criterion leads to a rejection region that satisfies the convexity condition. Therefore, again the computational effort can be noticeably reduced as only the boundary of the null-space has to be considered. Sample size calculation can then be conducted analogously as described in Sect. 8.4.2.2 for the non-inferiority test problem formulated in terms of the difference of rates.

Example 8.8 When planning the **HCV trial** (see Example 8.5), equal sustained virologic response rates of $p_{E,A} = p_{C,A} = 0.80$ were assumed for the experimental and the control treatment. The non-inferiority margin for the odds ratio was fixed at 0.70, and a one-sided test at level $\alpha/2 = 0.025$ should provide a power of $1-\beta = 0.80$ to demonstrate non-inferiority of E versus C. For balanced sample size allocation and application of the Fisher-Boschloo test for non-inferiority, the required exact total sample size is $1570 = 2 \times 785$ (as compared to 1542 for the asymptotic test). If the sample size is allocated with ratio $E{:}C = 2{:}1$ (i.e. $r = 2$), the exact total sample size amounts to $1818 = 1212 + 606$ (as compared to 1737 for the asymptotic test). Note that the sample size comparison between the asymptotic test and the exact unconditional test is not fair: In contrast to the latter, the asymptotic test may show an inflation of the type I error rate that entails an illegitimate gain in power. It should also be mentioned that calculation of the exact sample size needed for the Fisher-Boschloo non-inferiority test for $r = 1$ and 2, respectively, required about 3 and 7 min when using the R program provided in the Appendix. Hence, exact computations are feasible even for such large sample sizes.

8.5 Time-to-Event Outcomes

In this section, we assume survival functions $S_C(t)$ and $S_E(t)$ with proportional hazards, i.e. $\theta = \frac{\lambda_E(t)}{\lambda_C(t)}$ is constant over time, where occurrence of an event is unfavorable. Therefore, $\theta < 1$ ($\theta > 1$) corresponds to an advantage (disadvantage) of E as compared to C.

For proportional hazards, the test problem for demonstrating non-inferiority of E versus C can be formulated as follows:

$$H_0{:}\ \theta \geq \delta$$
$$\text{versus} \tag{8.28}$$
$$H_1{:}\ \theta < \delta$$

with pre-defined boundary of the non-inferiority margin $\delta > 1$. Jung et al. (2005) proposed a modification of the log-rank test statistic for the non-inferiority setting which is defined as the ratio of the partial score function $W(\theta)$ and the square root of the information function $\hat{\sigma}^2(\theta)$. The same notations as in Sect. 6.3 are used:

$t_{1,C} < t_{2,C} < \ldots < t_{m_C,C}$: ordered event time points for group C
$t_{1,E} < t_{2,E} < \ldots < t_{m_E,E}$: ordered event time points for group E
$n_{i,t_{j,k}}$: number of patients at risk in group i at time point $t_{j,k}$ with $i, k = C, E, j = 1, \ldots, m_C$ or $1, \ldots, m_E$, respectively.

Under the proportional hazards assumption, the test statistic of the log-rank test for the non-inferiority setting is given by

$$L(\theta) = \frac{W(\theta)}{\hat{\sigma}(\theta)} \tag{8.29}$$

with $W(\theta)$ being the partial score function and $\hat{\sigma}^2(\theta)$ the information function:

$$W(\theta) = \theta \cdot \sum_{j=1}^{m_E} \frac{n_{C,t_j,E}}{n_{E,t_j,E} + \theta \cdot n_{C,t_j,E}} - \sum_{j=1}^{m_C} \frac{n_{E,t_j,C}}{n_{E,t_j,C} + \theta \cdot n_{C,t_j,C}} \qquad (8.30)$$

$$\hat{\sigma}^2(\theta) = \theta \cdot \sum_{j=1}^{m_E} \frac{n_{E,t_j,E} \cdot n_{C,t_j,E}}{\left(n_{E,t_j,E} + \theta \cdot n_{C,t_j,E}\right)^2} + \theta \cdot \sum_{j=1}^{m_C} \frac{n_{E,t_j,C} \cdot n_{C,t_j,C}}{\left(n_{E,t_j,C} + \theta \cdot n_{C,t_j,C}\right)^2}. \qquad (8.31)$$

Note that for $\theta = 1$ the above test statistic of the log-rank test for the non-inferiority setting is equal to that for assessing difference or superiority (see Sect. 6.3).

Asymptotically, $\hat{\sigma}^2(\theta)$ approaches $\sigma^2(\theta) = \frac{r_d}{(1+r_d)} \cdot \theta \cdot d$, where d is the total number of events and $r_d = d_E/d_C$. For $\theta = \delta$, $L(\delta) = W(\delta)/\hat{\sigma}(\delta)$ follows asymptotically the standard normal distribution. Hence, H_0 is rejected at level $\alpha/2$ if $l(\delta) > z_{1-\alpha/2}$ and is accepted otherwise. Jung et al. (2005) derived an approximate sample size formula for the specific alternative $\theta = 1$, for which the expectation of $W(1)$ is approximately given by $\gamma = \frac{r_d}{(1+r_d)} \cdot \frac{(\delta-1)}{(1+\delta \cdot r_d)} \cdot d$. The approximate expressions for the variance of W under the null- and alternative hypothesis, σ_0^2 and σ_1^2, respectively, are given by $\hat{\sigma}^2(\delta) = \frac{r_d \cdot \delta}{(1+\delta \cdot r_d)^2} \cdot d$ and $\hat{\sigma}^2(1) = \frac{r_d}{(1+r_d)^2} \cdot d$. The sample size formula can be derived analogously to Sect. 5.2.1, and inserting the corresponding quantities in (5.7) leads to the following equation:

$$\frac{r_d}{(1+r_d)} \cdot \frac{(\delta-1)}{(1+\delta \cdot r_d)} \cdot d = z_{1-\alpha/2} \cdot \sqrt{\frac{r_d \cdot \delta}{(1+r_d \cdot \delta)^2} \cdot d} + z_{1-\beta} \cdot \sqrt{\frac{r_d}{(1+r_d)^2} \cdot d}. \qquad (8.32)$$

Solving this equation for d leads to an approximate formula for the required total number of events:

$$d = \frac{(1+r_d)^2}{r_d} \cdot \frac{\left(z_{1-\alpha/2} \cdot \sqrt{\delta} + z_{1-\beta} \cdot (1+\delta \cdot r_d)/(1+r_d)\right)^2}{(\delta-1)^2}. \qquad (8.33)$$

The required number of patients can then be determined according to one of the methods discussed in Sect. 6.3 for superiority testing. Note that $S_E(t) = S_C(t)$ holds true under the alternative considered here which is used when calculating $\Pr(D)$.

Example 8.9 Pulmonary vein isolation by means of catheter ablation is recommended according to guidelines as treatment for patients with drug-refractory atrial fibrillation. The most commonly applied method is radiofrequency ablation followed by cryoballoon ablation. The **FIRE AND ICE trial** (Kuck et al. 2018) was performed as a multicenter, randomized, open-label trial with blinded endpoint assessment to demonstrate non-inferiority of cryoballoon ablation (E) versus radiofrequency ablation (C). Primary endpoint was the time to first documentation of a clinical failure occurring more than 90 days after the index ablation procedure. Clinical failure was defined as recurrence of atrial fibrillation, occurrence of atrial flutter or atrial tachy-cardia, prescription of antiarrhythmic drugs, or repeat ablation. Proportional hazards were assumed for the primary endpoint, where for the hazard ratio a non-inferiority margin of $\delta = 1.43$ was specified and a value of $\theta_A = 1$ was assumed for sample size calculation. The log-rank test for non-inferiority according to Jung et al. (2005) was specified to be applied at level $\alpha/2 = 0.025$ and the aspired power at the alternative $\theta_A = 1$ was $1 - \beta = 0.80$. For balanced events in the two groups ($r_d = 1$), this leads to a required total number of events of $d = 246$. If double as much patients would have been allocated to the experimental treatment ($r_d = 2$), $288 = 192 + 96$ events for the primary endpoint are necessary. Moreover, a uniform accrual over a period of $a = 18$ months was assumed. The follow-up period was $f = 12$ months, and exponential time-to-event data with an event-free 1-year survival rate of 0.70 in both groups were expected, i.e. $\lambda = -\ln(0.70)/12 = 0.0297$ (Fürnkranz et al. 2014). Calculation of $\Pr(D)$ by Eq. (6.10) based on these assumptions results in a value of 0.458 which leads to a total sample size of 538 for $r = 1$ and $627 = 418 + 209$ for $r = 2$.

Remarks

1. Chow et al. (2003) proposed another approximate sample size formula for non-inferiority assessment based on the same test statistic (8.29) and for the same alternative $\theta = 1$. For the derivation, approximate equality of $S_E(t)$ and $S_C(t)$ under the null-hypothesis was assumed resulting in

$$d = \frac{(1 + r_d)^2}{r_d} \cdot \frac{\left(z_{1-\alpha/2} + z_{1-\beta}\right)^2}{(\ln \delta)^2}. \tag{8.34}$$

   Jung et al. (2005) showed by simulations that both formulas (8.33) and (8.34) show in general good performance. For non-inferiority margins "far" above 1 and unbalanced allocation, the formula by Chow et al. (2003) tends to result in too small ($r_d > 1$) or too large ($r_d < 1$) event numbers, respectively, where the extent of this trend is increasing with increasing δ.
2. Jung and Chow (2012) derived an approximate sample size formula that is not limited to the alternative $\theta_A = 1$ but covers arbitrary alternatives $\theta_A < \delta$. The formula and the computations get more complicated as the expressions for the quantities γ, σ_0^2, and σ_1^2 introduced above are then functions of the survival distributions under the alternative and the censoring distribution. Thus, they have to be calculated using numerical methods.
3. The additive hazards model was already mentioned in Sect. 6.3 as an alternative to the assumption of constant or proportional hazards. Sample size formulas for related non-inferiority test problems were presented by McDaniel et al. (2016).
4. The situation of competing risks which is common in time-to-event analyses was already described in Sect. 6.3. Methods for sample size calculation for non-inferiority trials in case of competing risks were proposed by Han et al. (2018).

Chapter 9
Comparison of Three Groups in the Gold Standard Non-Inferiority Design

9.1 Background and Notation

In Chap. 8, we considered two-arm trials that aim at demonstrating non-inferiority of a new experimental treatment (E) relative to a reference treatment (R). Starting point for the non-inferiority design was the availability of efficacious treatments in the indication under investigation making placebo control ethically unfeasible in certain situations. However, in case that non-inferiority of E versus R is shown in a two-arm trial, this provides in most cases only an indirect proof of efficacy of E and further assumptions have to be made to demonstrate efficacy. Basically, these premises rely on the so-called assay sensitivity defined as the capability of a clinical trial to distinguish between treatments of different efficacy. The ICH E10 Guideline on the choice of the control group (ICH 2001) mentions the three-arm trial including not only E and R but also a placebo group (P) as a useful approach allowing the assessment of assay sensitivity in active control trials. There are a number of situations where inclusion of a placebo arm in a non-inferiority trial is reasonable (Koch and Röhmel 2004). Furthermore, proposals exist to decide when using placebo in clinical trials for methodological reasons is ethically justifiable (Millum and Grady 2013). Due to the rich amount of information that can be gained, the three-arm design comparing P, E, and R is denoted as the "gold standard" non-inferiority design in the literature.

To illustrate the methods for sample size calculation for the "gold standard" non-inferiority design, we consider normally distributed endpoints; at the end of this chapter, references to related procedures for other outcome scales are given. We therefore assume in the following

$$X_{Pj} \sim N\left(\mu_P, \sigma^2\right), \quad j = 1, \ldots, n_P \tag{9.1a}$$

$$X_{Ej} \sim N\left(\mu_E, \sigma^2\right), \quad j = 1, \ldots, n_E \tag{9.1b}$$

© Springer Nature Switzerland AG 2020
M. Kieser, *Methods and Applications of Sample Size Calculation and Recalculation in Clinical Trials*, Springer Series in Pharmaceutical Statistics,
https://doi.org/10.1007/978-3-030-49528-2_9

$$X_{Rj} \sim N\left(\mu_R, \sigma^2\right), \quad j = 1, \ldots, n_R, \tag{9.1c}$$

where $X_{ij}, i = P, E, R, j = 1, \ldots, n_i$, are independent and σ^2 is known. As in the previous chapters, it is assumed that higher values correspond to beneficial outcomes.

Two different approaches can be distinguished for the confirmatory analysis of trials with "gold standard" non-inferiority design:

(i) net effect approach: assessment of multiple test problems formulated for pairwise comparisons of P, E, and R

(ii) fraction effect approach: non-inferiority assessment based on a test problem formulated in terms of an effect retention of E versus R relative to P.

9.2 Net Effect Approach

In the net effect approach, the non-inferiority margin for the comparison of the experimental treatment and the reference is expressed in terms of the difference of the effects of E and R. In this section, the associated test problems, analysis strategies, and methods for sample size calculation are given.

Test problem and test procedure

In a three-arm "gold standard" design, the following test problems are in principle of interest:

- assessment of superiority of E versus P:

$$
\begin{array}{c}
H_0^{EP}: \mu_E - \mu_P \leq 0 \\
\text{versus} \\
H_1^{EP}: \mu_E - \mu_P > 0
\end{array}
\tag{9.2a}
$$

- assessment of non-inferiority of E versus R:

$$
\begin{array}{c}
H_0^{ER}: \mu_E - \mu_R \leq -\delta \\
\text{versus} \\
H_1^{ER}: \mu_E - \mu_R > -\delta
\end{array}
\tag{9.2b}
$$

with pre-specified non-inferiority margin $\delta > 0$

- assessment of superiority of R versus P:

$$
\begin{array}{c}
H_0^{RP}: \mu_R - \mu_P \leq 0 \\
\text{versus} \\
H_1^{RP}: \mu_R - \mu_P > 0.
\end{array}
\tag{9.2c}
$$

Rejection of H_0^{EP} demonstrates efficacy of E as well as assay sensitivity of the trial. To make the trial successful, non-inferiority of E versus R has to be shown additionally by rejecting H_0^{ER}. Therefore, the following intersection-union test problem is to be assessed in confirmatory analysis:

$$H_0: H_0^{EP} \cup H_0^{ER}$$
$$\text{versus} \tag{9.3}$$
$$H_1: H_1^{EP} \cap H_1^{ER}.$$

The global null-hypothesis H_0 can be rejected at level α if both H_0^{EP} and H_0^{ER} can be rejected by associated tests at local level α.

It is controversial whether demonstration of efficacy of R in the current trial is mandatory, i.e. additional rejection of H_0^{RP}. The motivation to initiate a three-arm trial with "gold standard" design is to demonstrate that the experimental treatment is efficacious and has at least comparable efficacy to that of an established reference. One may therefore argue that the study aim is reached when H_0 in (9.3) can be rejected. However, there are indications where failure to show superiority of the reference to placebo will challenge the validity of the trial making rejection of H_0^{RP} to an inevitable condition in these situations (Hauschke and Pigeot 2005). We therefore also consider the related intersection-union test problem which is given by

$$\tilde{H}_0: H_0^{EP} \cup H_0^{ER} \cup H_0^{RP}$$
$$\text{versus} \tag{9.4}$$
$$\tilde{H}_1: H_1^{EP} \cap H_1^{ER} \cap H_1^{RP}.$$

The null-hypotheses H_0^{EP}, H_0^{ER}, and H_0^{RP} can be tested applying the z-test statistics

$$Z_{EP} = \sqrt{\frac{n_E \cdot n_P}{n_E + n_P}} \cdot \frac{\bar{X}_E - \bar{X}_P}{\sigma}, \tag{9.5a}$$

$$Z_{ER} = \sqrt{\frac{n_E \cdot n_R}{n_E + n_R}} \cdot \frac{\bar{X}_E - \bar{X}_R + \delta}{\sigma}, \tag{9.5b}$$

$$Z_{RP} = \sqrt{\frac{n_R \cdot n_P}{n_R + n_P}} \cdot \frac{\bar{X}_R - \bar{X}_P}{\sigma}, \tag{9.5c}$$

where $\bar{X}_i = \frac{1}{n_i} \cdot \sum_{j=1}^{n_i} X_{ij}$, $i = P, E, R$. The null-hypotheses H_0 or $\tilde{H}_0$, respectively, can be rejected at one-sided level $\alpha/2$ if $Z_{EP} > z_{1-\alpha/2}$ and $Z_{ER} > z_{1-\alpha/2}$, or if $Z_{EP} > z_{1-\alpha/2}$, $Z_{ER} > z_{1-\alpha/2}$, and $Z_{RP} > z_{1-\alpha/2}$.

Sample size calculation

(a) *Study aim: rejection of global null-hypothesis H_0*

For sample size calculation, an alternative $H_{1,A}$ has to be specified by fixing $\Delta_A = (\Delta_{EP,A}, \Delta_{ER,A})$ with $\Delta_{EP,A} = (\mu_E - \mu_P)_A$ and $\Delta_{ER,A} = (\mu_E - \mu_R)_A$. The power for rejecting H_0 is then given by

$$\text{Pr}_{H_{1,A}}\left(Z_{EP} > z_{1-\alpha/2} \text{ and } Z_{ER} > z_{1-\alpha/2}\right)$$

$$= \text{Pr}_{H_{1,A}}\left(Z_{EP} - \sqrt{\frac{n_E \cdot n_P}{n_E + n_P}} \cdot \left(\frac{\Delta_{EP,A}}{\sigma}\right) > z_{1-\alpha/2} - \sqrt{\frac{n_E \cdot n_P}{n_E + n_P}} \cdot \left(\frac{\Delta_{EP,A}}{\sigma}\right)\right)$$

and

$$Z_{ER} - \sqrt{\frac{n_E \cdot n_R}{n_E + n_R}} \cdot \left(\frac{\Delta_{ER,A} + \delta}{\sigma}\right) > z_{1-\alpha/2} - \sqrt{\frac{n_E \cdot n_R}{n_E + n_R}} \cdot \left(\frac{\Delta_{ER,A} + \delta}{\sigma}\right)$$

$$= \text{Pr}_{H_{1,A}}\left(Z_{EP}^* > c_{EP}^* \text{ and } Z_{ER}^* > c_{ER}^*\right).$$

Under $\mathbf{\Delta}_A$, the vector of centered test statistics $\mathbf{Z}^* = \left(Z_{EP}^*, Z_{ER}^*\right)$ with $Z_{EP}^* = Z_{EP} - \sqrt{\frac{n_E \cdot n_P}{n_E + n_P}} \cdot \left(\frac{\Delta_{EP,A}}{\sigma}\right)$ and $Z_{ER}^* = Z_{ER} - \sqrt{\frac{n_E \cdot n_R}{n_E + n_R}} \cdot \left(\frac{\Delta_{ER,A} + \delta}{\sigma}\right)$ follows the two-dimensional standardized normal distribution with expectation zero and off-diagonal elements of the covariance matrix being equal to $\rho_{12} = \text{Corr}(Z_{EP}, Z_{ER}) = \sqrt{\frac{n_R \cdot n_P}{(n_E + n_P) \cdot (n_E + n_R)}} = \sqrt{\frac{r_R}{(1+r_E) \cdot (r_E + r_R)}}$, where $r_E = n_E/n_P$ and $r_R = n_R/n_P$. The total required sample size to assure a power of $1 - \beta$ can therefore be computed by determining the smallest integer n_P fulfilling

$$\text{Pr}_{H_{1,A}}\left(Z_{EP}^* > c_{EP}^*, Z_{ER}^* > c_{ER}^*\right) \geq 1 - \beta, \tag{9.6}$$

where $c_{EP}^* = z_{1-\alpha/2} - \sqrt{\frac{r_E}{1+r_E} \cdot n_P} \cdot \left(\frac{\Delta_{EP,A}}{\sigma}\right)$ and $c_{ER}^* = z_{1-\alpha/2} - \sqrt{\frac{r_E \cdot r_R}{r_E + r_R} \cdot n_P} \cdot \left(\frac{\Delta_{ER,A} + \delta}{\sigma}\right)$ and by setting (eventually after rounding up to the next integer) $n_E = r_E \cdot n_P$, $n_R = r_R \cdot n_P$, and $n = n_P + n_E + n_R$. Note that after obvious modifications this approach corresponds to calculating the sample size for a Dunnett-like test and a conjunctive power of $1 - \beta$ according to the method presented in Sect. 11.4.2a. For details on the terms and methods it is referred to Chap. 11.

(b) *Study aim: rejection of global null-hypothesis $\tilde{H}_0$*

Power and required sample size, respectively, can be determined analogously to the previous paragraph just by including the comparison of R versus P with related specified component of the alternative $\Delta_{RP,A} = (\mu_R - \mu_P)_A$ and centered test statistic $Z_{RP}^* = Z_{RP} - \sqrt{\frac{n_R \cdot n_P}{n_R + n_P}} \cdot \left(\frac{\Delta_{RP,A}}{\sigma}\right)$. Then, under $\tilde{H}_{1,A}$ the vector of test statistics $\tilde{\mathbf{Z}}^* = \left(Z_{EP}^*, Z_{ER}^*, Z_{RP}^*\right)$ follows the three-dimensional standardized normal distribution with expectation zero and off-diagonal elements of the covariance matrix being equal to $\tilde{\rho}_{12} = \rho_{12}$, $\tilde{\rho}_{13} = \sqrt{\frac{r_E \cdot r_R}{(1+r_E) \cdot (1+r_R)}}$, and $\tilde{\rho}_{23} = -\sqrt{\frac{r_E}{(r_E + r_R) \cdot (1+r_R)}}$. The total sample size can now again be determined by numerical integration: It is given by the smallest integer fulfilling the inequality

$$\text{Pr}_{\tilde{H}_{1,A}}\left(Z_{EP} > z_{1-\alpha/2}, Z_{ER} > z_{1-\alpha/2}, Z_{RP} > z_{1-\alpha/2}\right) \geq 1 - \beta,$$

which is equivalent to resolving

$$\text{Pr}_{\tilde{H}_{1,A}}\left(Z_{EP}^* > c_{EP}^*, Z_{ER}^* > c_{ER}^*, Z_{RP}^* > c_{RP}^*\right) \geq 1 - \beta \qquad (9.7)$$

for the smallest integer n_P and setting (eventually after rounding up to the next integer) $n_E = r_E \cdot n_P$, $n_R = r_R \cdot n_P$ and $n = n_P + n_E + n_R$. Here, c_{EP}^* and c_{ER}^* are defined as in the previous section and $c_{RP}^* = z_{1-\alpha/2} - \sqrt{\frac{r_R}{1+r_R} \cdot n_P} \cdot \left(\frac{\Delta_{RP,A}}{\sigma}\right)$.

The expressions (9.6) and (9.7), respectively, can also be used to determine the optimal allocation ratios r_E and r_P minimizing the total required sample size. For ethical reasons, the sample size of the placebo group should be at most as large as those for the active treatment groups, i.e. $r_E = n_E/n_P \geq 1$ and $r_R = n_R/n_P \geq 1$. Furthermore, it is desirable to obtain as much information as possible on efficacy and safety of the new experimental therapy and therefore $n_E \geq n_R$ is aspired, i.e. $r_E \geq r_R$. By a grid search under these constraints, Stucke and Kieser (2012) showed that the total sample size can be considerably decreased when applying optimal allocation instead of using balanced groups.

Example 9.1 The **ELECT-TDCS trial** was performed in the "gold standard" non-inferiority design and compared transcranial direct-current stimulation (tDCS; E), a noninvasive brain-stimulating technique, with the selective serotonin-reuptake inhibitor escitalopram (R) and with placebo (P) in patients with unipolar depression (Brunoni et al. 2017). Sham tDCS or oral placebo, respectively, was included in the respective groups thus achieving a double-blind conduct of the study. The 17-item Hamilton Depression Rating Scale (HAM-D; Hamilton 1960), which is a well-established and widely used depression assessment scale, was used as primary measure of efficacy, and the change between baseline and week 10 (end of therapy) was the primary endpoint. The ELECT-TDCS trial was planned according to the fraction effect approach (see following Sect. 9.3). For illustrative purposes, we assume that the net effect approach was applied and base the assumptions on the explanations given in the study protocol (Brunoni et al. 2015). According to Montgomery (1994), a difference of 4 points in HAM-D represents an effect that can be accepted as unequivocally clinically relevant. This value fits very well to the difference between tDCS and placebo observed in a previously conducted trial where the standard deviation of the change in HAM-D was 6.3. For sample size calculation, it is supposed that $\Delta_{EP,A} = (\mu_E - \mu_P)_A = 4$ and $\sigma = 6$. Furthermore, as commonly done in non-inferiority settings, we assume equal efficacy of the two active treatments, i.e. $\Delta_{ER,A} = (\mu_E - \mu_R)_A = 0$, from which $\Delta_{RP,A} = 4$ results. A non-inferiority margin of $\delta = 2$ is used defining the maximally acceptable amount of inferiority of E versus R. The experimentwise type I error rate is to be controlled at one-sided level $\alpha/2 = 0.025$, and a power of $1 - \beta = 0.80$ is aspired to reject the null-hypotheses concerning the comparisons between E and P as well as E and R (global null-hypothesis H_0) or between E and P, E and R, as well as R and P (global null-hypothesis $\tilde{H}_0$), respectively. For balanced groups, i.e. $r_E = r_R = 1$, the required sample size to reject H_0 amounts to $426 = 3 \times 142$ and the same sample size is necessary for rejection of $\tilde{H}_0$. By applying a grid search [see Stucke and Kieser (2012) for a detailed description] under the restrictions $r_E \geq 1$, $r_R \geq 1$, and $r_E \geq r_R$, the required total sample size can be minimized. For the rejection of H_0, the optimal allocation is given by $r_E = n_E/n_P = 3.53$ and $r_R = n_R/n_P = 3.33$ resulting in a total sample size of $n = 338$ ($n_P = 43, n_E = 152, n_R = 143$). When the three null-hypotheses for all pairwise comparisons shall be rejected with power 0.80, the minimum required total sample size is $n = 348$ ($n_P = 48, n_E = n_R = 150$), which is achieved for the allocation ratios $r_E = r_R = 3.125$. It may be sensible to allocate equal sample size to the two active groups, i.e. to set $r_E = r_R = r$. Then, the minimum total sample size for rejecting H_0 ($\tilde{H}_0$) with power 0.80 is achieved for $r = 3.17$ ($r = 3.125$) and amounts to $n = 338$ with $n_P = 46$ and $n_E = n_R = 146$ ($n = 348$ with $n_P = 48$ and $n_E = n_R = 150$). For this situation, Fig. 9.1 shows the total sample size in dependence of the allocation ratios.

Remarks

1. Simulation studies performed for a wide range of parameter scenarios showed that the power (and with it the required sample size) for the case that σ^2 is unknown is very close to the situation of known variance (Stucke and Kieser 2012).
2. Stucke and Kieser (2012) proposed a general approach for sample size calculation in the situations considered above including as special cases not only continuous but, for example, also binary and Poisson distributed outcomes.
3. When analyzing a three-arm non-inferiority trial with "gold standard" design, an alternative testing strategy may be considered which is based on ordering of the hypotheses (see Sect. 11.2.2b). The test procedure starts with testing H_0^{EP} first at

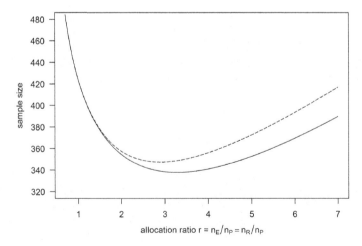

Fig. 9.1 Total sample size for rejecting the null-hypotheses for the comparisons E versus P and E versus R (solid line) and E versus P, E versus R, and R versus P (dashed line) depending on the allocation ratio $r = n_E/n_P = n_R/n_P$ (see Example 9.1)

one-sided level $\alpha/2$, continuing in case of rejection with testing H_0^{ER} at one-sided level $\alpha/2$ and then, in case of rejection, with testing H_0^{RP} at one-sided level $\alpha/2$. In contrast to the "all-or-none" approach considered above, this strategy may end with solely rejecting H_0^{EP} when assessing H_0 or with solely rejecting H_0^{EP} and H_0^{ER} when assessing $\tilde{H}_0$, respectively. Although not all study aims are reached in these cases, the result may nevertheless be of use. As the same critical values are applied, the methods for sample size calculation for this approach are equal to those for the intersection-union test described above.

9.3 Fraction Effect Approach

In the fraction effect approach, the acceptable amount of inferiority of the experimental treatment as compared to the reference is expressed as a fraction of the difference of effects between the reference and placebo. In the following, it will be described how this concept translates into an adequate formulation of the test problem, how this test problem is analyzed, and how the required sample size is calculated.

Test problem and test procedure

In the fraction effect approach, the non-inferiority margin is chosen as a fraction $f > 0$ of the (unknown) difference of the population means μ_R and μ_P, i.e.

$$\delta = f \cdot (\mu_R - \mu_P). \tag{9.8}$$

Therefore, the non-inferiority test problem (9.2b) reads

$$H_0^{ER}: \mu_E - \mu_R \leq -f \cdot (\mu_R - \mu_P)$$
$$\text{versus}$$
$$H_1^{ER}: \mu_E - \mu_R > -f \cdot (\mu_R - \mu_P),$$

which is equivalent to the following test problem for $\mu_R - \mu_P > 0$:

$$H_0^{ER}: \frac{\mu_E - \mu_P}{\mu_R - \mu_P} \leq \gamma$$
$$\text{versus} \tag{9.9}$$
$$H_1^{ER}: \frac{\mu_E - \mu_P}{\mu_R - \mu_P} > \gamma$$

with $\gamma = 1 - f$. Non-inferiority as expressed by test problem (9.9) means that the experimental treatment retains more than γ times 100% of the effect of the reference, each compared to placebo. For the assessment of non-inferiority within the fraction effect approach, additionally to the assessment of (9.9) superiority of R versus P has to be shown as the formulation of the non-inferiority hypothesis in (9.9) requires $\mu_R - \mu_P > 0$:

$$H_0^{RP}: \mu_R - \mu_P \leq 0$$
$$\text{versus} \qquad (9.10)$$
$$H_1^{RP}: \mu_R - \mu_P > 0.$$

Statistical tests for the assessment of the test problems as well as related methods for sample size calculation can easily be derived without major technical difficulties for the case of unknown σ^2, and therefore we consider in the following this situation. The null-hypothesis in (9.9) can be written as $H_0^{ER}: \mu_E - \gamma \cdot \mu_R - (1 - \gamma) \cdot \mu_P \leq 0$, and this formulation can be used to derive a test for the linear contrast $\mu_E - \gamma \cdot \mu_R - (1 - \gamma) \cdot \mu_P$ (Pigeot et al. 2003). The test statistic is given by

$$T = \sqrt{\frac{n_P \cdot n_E \cdot n_R}{n_P \cdot n_R + \gamma^2 \cdot n_P \cdot n_E + (1 - \gamma)^2 \cdot n_E \cdot n_R}} \cdot \frac{\bar{X}_E - \gamma \cdot \bar{X}_R - (1 - \gamma) \cdot \bar{X}_P}{S},$$
$$(9.11)$$

where $S^2 = \frac{1}{n-3} \cdot \sum_{i=P,E,R} (n_i - 1) \cdot S_i^2$ with $S_i^2 = \frac{1}{n_i-1} \cdot \sum_{j=1}^{n_i} (X_{ij} - \bar{X}_i)^2$, $\bar{X}_i = \frac{1}{n_i} \cdot \sum_{j=1}^{n_i} X_{ij}$, $i = P, E, R$, and $n = \sum_{i=P,E,R} n_i$. Under H_0^{ER}, T follows the central t-distribution with $n - 3$ degrees of freedom. Thus, non-inferiority is demonstrated at one-sided level $\alpha/2$ if $t > t_{1-\alpha/2, n_P+n_E+n_R-3}$. Of course, the null-hypothesis in (9.10) can be tested by the common two-sample t-test with critical value $t_{1-\alpha/2, n_P+n_R-2}$. As testing of H_0^{ER} makes only sense under the assumption $\mu_R - \mu_P > 0$, a hierarchical multiple test procedure is applied for the assessment of non-inferiority within the fraction effect approach (see Sect. 11.2.2b): Firstly, H_0^{RP} is tested at one-sided level $\alpha/2$. If H_0^{RP} can be rejected, H_0^{ER} is tested at one-sided level $\alpha/2$; otherwise, neither H_0^{RP} nor H_0^{ER} can be rejected. This test procedure controls the experimentwise type I error rate in the strong sense by $\alpha/2$.

Sample size calculation

(a) *Assessment of fraction effect retention*

For sample size calculation, an alternative $H_{1,A}^{ER}$ has to be specified by fixing a value $\theta_A = (\mu_E - \gamma \cdot \mu_R - (1 - \gamma) \cdot \mu_P)_A > 0$. Under $H_{1,A}^{ER}$, T follows the non-central t-distribution with $n - 3$ degrees of freedom and non-centrality parameter

$$nc = \sqrt{\frac{r_E \cdot r_R}{r_R + \gamma^2 \cdot r_E + (1 - \gamma)^2 \cdot r_E \cdot r_R}} \cdot n_P \cdot \frac{(\mu_E - \gamma \cdot \mu_R - (1 - \gamma) \cdot \mu_P)_A}{\sigma}$$
$$= \sqrt{\frac{r_E \cdot r_R}{r_R + \gamma^2 \cdot r_E + (1 - \gamma)^2 \cdot r_E \cdot r_R}} \cdot n_P \cdot \left(\left(\frac{\mu_E - \mu_P}{\mu_R - \mu_P} \right)_A - \gamma \right) \cdot \left(\frac{\Delta_{RP,A}}{\sigma} \right)$$
$$(9.12a, b)$$

with $\Delta_{RP,A} = (\mu_R - \mu_P)_A$. For the alternative of equal efficacy of E and R, which is commonly applied in non-inferiority trials, the non-centrality parameter is given by

$$nc = \sqrt{\frac{r_E \cdot r_R}{r_R + \gamma^2 \cdot r_E + (1 - \gamma)^2 \cdot r_E \cdot r_R}} \cdot n_P \cdot f \cdot \left(\frac{\Delta_{RP,A}}{\sigma}\right), \qquad (9.12c)$$

where $f = 1 - \gamma$. Representations (9.12b) and (9.12c) express the non-centrality parameter in terms of the fraction of retained effect of E versus R, both compared to P, and the difference between R and P and enable sample size calculation based on these quantities. Thus, for given allocation ratios $r_E = n_E/n_P$ and $r_R = n_R/n_P$, the required sample size to reject H_0 at one-sided level $\alpha/2$ with power $1 - \beta$ results from the basic inequality (2.7) by inserting the respective distribution functions. The sample size can therefore be obtained by determining the smallest integer n fulfilling the inequality

$$F_{T;n-3,0}^{-1}(1 - \alpha/2) \leq F_{T;n-3,nc}^{-1}(\beta), \qquad (9.13)$$

where $F_{T;n-3,0}$ and $F_{T;n-3,nc}$, respectively, denote the distribution functions of the central and non-central t-distribution with $n - 3$ degrees of freedom and $n = n_P + n_E + n_R = n_P \cdot (1 + r_E + r_R)$.

Approximating in (9.13) the distribution functions of the t-distribution by those of the standard normal distribution enables to apply Eq. (2.10), i.e. $nc = z_{1-\alpha} + z_{1-\beta}$, and to derive an explicit approximation formula:

$$n = (1 + r_E + r_R) \cdot \frac{\left(r_R + \gamma^2 \cdot r_E + (1 - \gamma)^2 \cdot r_E \cdot r_R\right)}{r_E \cdot r_R}$$

$$\cdot (z_{1-\alpha/2} + z_{1-\beta})^2 \cdot \left[\frac{\sigma}{\left(\left(\frac{\mu_E - \mu_P}{\mu_R - \mu_P}\right)_A - \gamma\right) \cdot \Delta_{RP,A}}\right]^2. \qquad (9.14)$$

By using this expression for the (approximate) total sample size for the allocation ratios, Pigeot et al. (2003) determined (approximate) optimal allocations for different situations. In practical applications, it is often sensible to allocate the same sample size to the two active treatments, i.e. to chose $r_E = r_R = r$. Then, the optimal allocation achieving the minimum total sample size is given for the allocation ratio

$$r = \frac{\sqrt{(1 + \gamma^2)/2}}{(1 - \gamma)}. \qquad (9.15)$$

Note that for $\gamma > 0.27$, which is usually the case, $r > 1$ holds true. Therefore, unbalanced designs with larger sample size allocated to the active treatment groups than to placebo lead to smaller total sample size as compared to allocating equal sample size to all groups. If no restrictions on the allocation ratios are imposed, optimal allocation is realized by choosing $r_E = \frac{1}{(1-\gamma)}$ and $r_R = \frac{\gamma}{(1-\gamma)}$. For $\gamma > 0.5$, r_E and r_R are again larger than 1 thus making allocation of more patients to the active groups than to placebo not only desirable from an ethical viewpoint but also attractive with respect to statistical efficiency.

(b) *Assessment of both superiority of R versus P and fraction effect retention*

Due to the hierarchical test procedure that includes a pre-test for superiority of R versus P, the power for demonstrating non-inferiority within the fraction effect approach may be smaller than the power for rejecting H_0^{ER}. If the sample size is used which is calculated for the assessment of the fraction effect retention test problem, the exact power for rejecting H_0^{RP} is equal to $1 - F_{T;n_P+n_R-2,nc}$, where $F_{T;n_P+n_R-2,nc}$ denotes the distribution function of the non-central t-distribution with $n_P + n_R - 2$ degrees of freedom and non-centrality parameter $nc = \sqrt{\frac{r_R}{(1+r_R)^2} \cdot n} \cdot \left(\frac{\Delta_{RP,A}}{\sigma}\right)$ with $n = n_P + n_R$ and n_P as well as n_R calculated from (9.13) (see Sect. 3.3). Using the approximation formula (9.14) and the approximation of the power of the t-test by that of the z-test, the following explicit formula for the power $1 - \beta_{RP}$ for rejecting H_0^{RP} can be derived:

$$
1 - \beta_{RP} = 1 - \Phi\left\{ z_{1-\alpha/2} - \sqrt{\frac{r_R}{(1+r_R)} \cdot \frac{r_R + \gamma^2 \cdot r_E + (1-\gamma)^2 \cdot r_E \cdot r_R}{r_E \cdot r_R}} \right.
$$
$$
\left. \cdot \frac{(z_{1-\alpha/2} + z_{1-\beta})}{\left[\left(\frac{\mu_E - \mu_P}{\mu_R - \mu_P}\right)_A - \gamma\right]} \right\}. \tag{9.16}
$$

where Φ denotes the distribution function of the standard normal distribution. Equation (9.16) can be used to evaluate a potential impact of the pre-test assessing H_0^{RP} on the power of the multiple test procedure if the sample size is calculated in order to reject H_0^{ER} with power $1 - \beta$. For the values $\alpha/2 = 0.025$, $1 - \beta = 0.80$ or 0.90 and the commonly used alternative $((\mu_E - \mu_P)/(\mu_R - \mu_P))_A = 1$, the power to reject H_0^{RP} according to Eq. (9.16) is at least 0.99 for $\gamma \geq 0.5$ and $r_E, r_R \leq 2$ as well as for $\gamma \geq 0.65$ and $r_E, r_R \leq 6$. In these situations, the power of the multiple test procedure is not noticeably affected by the pre-test of H_0^{RP} and thus the sample size calculated for the test problem H_0^{ER} versus H_1^{ER} is sufficient to assure also a power $1 - \beta$ for rejecting both null-hypotheses. For other scenarios, especially if it is expected that the experimental treatment is much more efficacious than the reference, simulations may be used to explore the performance characteristics of the multiple test procedure and to determine the required sample size (see Chap. 16). Alternatively, the power of the complete test procedure and the related required sample size can be calculated under the assumption of known variance by applying the following considerations. Denoting by Z_{ER} and Z_{RP} the test statistics for the assessment of H_0^{ER} and H_0^{RP}, respectively, where the estimators S of σ are replaced by σ, Zhou and Kundu (2016) showed that

$$
\text{Corr}(Z_{ER}, Z_{RP}) = \sqrt{\frac{r_E}{(r_R + \gamma^2 \cdot r_E + (1-\gamma)^2 \cdot r_E \cdot r_R) \cdot (1 + r_R)}} \tag{9.17}
$$
$$
\cdot ((1-\gamma) \cdot r_R - \gamma).
$$

The power for the complete multiple test procedure is given by

$$
\text{Pr}_{H_{1,A}}\left(Z_{ER} > z_{1-\alpha/2}, Z_{RP} > z_{1-\alpha/2}\right). \tag{9.18}
$$

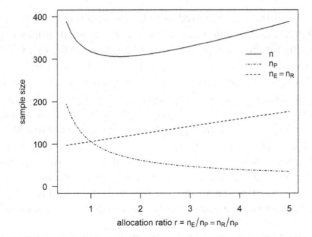

Fig. 9.2 Total sample size n (solid line), sample size for placebo group n_P (dot-dashed line), and sample size for active groups $n_E = n_R$ (dashed line) for rejecting the non-inferiority null-hypothesis formulated in terms of fraction effect depending on the allocation ratio $r = n_E/n_P = n_R/n_P$ (see Example 9.2)

Therefore, Eq. (9.18) can be evaluated according to the methods presented in Chap. 11 by numerical integration of the bivariate normal distribution with correlation given by (9.17). Based on this method, Zhou and Kundu (2016) determined the optimal allocation for specified overall power.

Example 9.2 The **ELECT-TDCS trial** (see Example 9.1) was planned according to the fraction effect approach (Brunoni et al. 2015). To declare transcranial direct-stimulation (E) non-inferior to drug treatment with escitalopram (R), an effect fraction of $f = 0.50$ as compared to placebo (P) had to be retained, i.e., the non-inferiority test problem (9.9) with $\gamma = 0.50$ had to be assessed. Sample size calculation assumed equal efficacy of E and R, i.e. $\mu_{E,A} = \mu_{R,A}$, and a standardized effect of R versus P of $\left(\Delta_{RP,A}/\sigma\right) = \left(\frac{4}{6}\right) = 0.67$ (cf. Example 9.1). The aspired power to reject H_0^{ER} with a test at one-sided level $\alpha/2 = 0.025$ was $1 - \beta = 0.80$. For balanced groups, i.e. $r_E = r_R = 1$, the exact required total sample size amounts to $321 (= 3 \times 107)$; the approximate sample size according to (9.14) is $3 \times 106 = 318$. If equal numbers of patients are to be allocated to the two active treatment groups but a different sample size is allowed for the placebo group, the optimal allocation ratio $r = n_E/n_P = n_R/n_P$ leading to the minimum exact total sample size is $r = \sqrt{2.5} = 1.58$ with related sample size $n = 308$ $(n_P = 74, n_E = n_R = 117)$; the same sample size results from the approximation formula (9.14). Based on these considerations, the sample size allocation $n_P : n_E : n_R = 2 : 3 : 3$ was used in the ELECT-TDCS trial resulting in an exact total required sample size of 312 $(n_P = 78, n_E = n_R = 117)$; the same sample size is obtained from the approximation formula (9.14).

Figure 9.2 shows the total required sample size resulting from the approximate formula (9.14) for equal sample sizes in the active treatment groups depending on the allocation ratio for the assumptions made when planning the ELECT-TDCS trial.

It can be seen that a considerable saving in sample size can be achieved by appropriate choice of the allocation ratio. If the restriction of equal sample size in the active groups is abandoned, the optimal allocation is by applying $r_E = n_E/n_P = 2$ and $r_R = n_R/n_P = 1$ which results in an exact total sample size of 288 ($n_P = n_R = 72, n_E = 144$; the approximate sample size obtained from (9.14) is $284 = 71 + 142 + 71$). Therefore, the improvement in efficiency as compared to the more restrictive condition $r_E = r_R$ amounts in a saving of 20 patients. For all sample sizes reported above, the (exact and approximate) power for rejecting H_0^{RP} amounts to more than 0.99. Therefore, the power of the multiple test procedure is hardly affected by the pre-test and is nearly equal to that for rejecting H_0^{ER}.

Remark Statistical tests and methods for sample size calculation for the assessment of non-inferiority according to the fraction effect retention approach are also available for binary endpoints with effect retention on the difference scale (Kieser and Friede 2007) as well as the effect measured by the relative risk and the odds ratio (Chowdhury et al. 2019), for censored time-to-event endpoints (Mielke et al. 2008; Kombrink et al. 2013), and for negative binomially distributed endpoints (Mütze et al. 2016).

Chapter 10
Comparison of Two Groups for Normally Distributed Outcomes and Test for Equivalence

10.1 Background and Notation

The aim of equivalence trials is to demonstrate that the difference between an experimental treatment (E) and a control (C) with respect to the endpoint of interest lies within a pre-defined interval, the so-called equivalence margin. Prominent examples of such trials are bioequivalence studies which aim to show that rate and extent of drug absorption of a generic medicinal product and the original formulation differ maximally by a pre-defined clinically irrelevant amount (Hauschke et al. 2007). In a general formulation and with the notation used in Chap. 8, the equivalence test problem is given by $H_0: \theta \leq \delta_L$ or $\theta \geq \delta_U$ versus $H_1: \theta \in]\delta_L, \delta_U[$, where δ_L and δ_U, $\delta_L < \delta_U$ denote the lower or upper boundary of the equivalence margin, respectively. The test problem for the assessment of equivalence is composed of two one-sided test problems, namely $H_0^L: \theta \leq \delta_L$ versus $H_1^L: \theta > \delta_L$ and $H_0^U: \theta \geq \delta_U$ versus $H_1^U: \theta < \delta_U$. These test problems constitute the intersection-union test problem $H_0: H_0^L \cup H_0^U$ versus $H_1: H_1^L \cap H_1^U$. Therefore, the null-hypothesis H_0 can be rejected at level α if both H_0^L and H_0^U can be rejected at level α (Berger 1982). By the way, this test procedure also controls the experimentwise type I error rate for the assessment of the two null-hypotheses H_0^L and H_0^U: As $\delta_L < \delta_U$, not both of the null hypotheses H_0^L and H_0^U can be true at the same time. Therefore, testing each of them at level α with a related test controls the multiple level α. The above-described approach to equivalence testing is denoted by the "two one-sided tests" (TOST) procedure. Alternative equivalence tests have been proposed (Wellek 2010), but we will in the following restrict to the TOST procedure. Equivalence trials with non-normally distributed outcomes are quite rare in practice. Therefore, the presentation is focussed to the case of outcomes which (potentially after an appropriate transformation) follow the normal distribution. For this situation, exact as well as approximate sample size calculation methods are given both for the setting that the treatment effect is expressed in terms of the difference and in terms of the ratio of

© Springer Nature Switzerland AG 2020

M. Kieser, *Methods and Applications of Sample Size Calculation and Recalculation in Clinical Trials*, Springer Series in Pharmaceutical Statistics, https://doi.org/10.1007/978-3-030-49528-2_10

means. However, the general ideas applied to construct these methods by appropriate utilization of techniques for the related non-inferiority test problems can be used to derive analogous procedures for other types of outcomes.

10.2   Difference of Means as Effect Measure

In this section, we consider the situation that the treatment effect is expressed in terms of the difference of means, i.e. $\theta = \mu_E - \mu_C$. We consider the case of a symmetric equivalence margin, i.e. $\delta_U = -\delta_L = \delta$. Then, the test problem for the assessment of equivalence in case of normally distributed data and the treatment effect expressed in terms of the difference of means is given by

$$H_0: \mu_E - \mu_C \leq -\delta \text{ or } \mu_E - \mu_C \geq \delta$$
$$\text{versus} \tag{10.1}$$
$$H_1: -\delta < \mu_E - \mu_C < \delta$$

with pre-specified $\delta > 0$. The two one-sided null-hypotheses $H_0^L: \mu_E - \mu_C \leq -\delta$ and $H_0^U: \mu_E - \mu_C \geq \delta$ constituting H_0 can be tested applying the shifted t-test statistics

$$T^L = \sqrt{\frac{n_C \cdot n_E}{n_C + n_E}} \cdot \frac{\bar{X}_E - \bar{X}_C + \delta}{S} \tag{10.2a}$$

and

$$T^U = \sqrt{\frac{n_C \cdot n_E}{n_C + n_E}} \cdot \frac{\bar{X}_E - \bar{X}_C - \delta}{S}, \tag{10.2b}$$

where S denotes the pooled standard deviation given in (3.8a) and (3.8b). The null-hypothesis H_0^L can be rejected at level α if $t^L > t_{1-\alpha,n_c+n_E-2}$, and H_0^U can be rejected if $t^U < -t_{1-\alpha,n_c+n_E-2}$. Hence, H_0 can be rejected at level α by the two one-sided t-tests procedure if both $t^L > t_{1-\alpha,n_c+n_E-2}$ and $t^U < -t_{1-\alpha,n_c+n_E-2}$. The random vector (T^L, T^U) has a bivariate t-distribution with $n_C + n_E - 2$ degrees of freedom, $\text{Corr}(T^L, T^U) = 1$, and non-centrality parameter components

$$nc_L = \sqrt{\frac{r}{(1+r)^2}} \cdot n \cdot \left(\frac{\mu_E - \mu_C + \delta}{\sigma}\right) \tag{10.3a}$$

$$nc_U = \sqrt{\frac{r}{(1+r)^2}} \cdot n \cdot \left(\frac{\mu_E - \mu_C - \delta}{\sigma}\right), \tag{10.3b}$$

where $r = n_E/n_C$ (Hauschke et al. 2007, p. 107ff). Therefore, for a specified alternative H_A defined by $\Delta_A = (\mu_E - \mu_C)_A \in \,]-\delta, \delta[$, allocation ratio r, sample size

n, and level α, the power to reject H_0 can be calculated by computing the probability that a random vector with the above-mentioned bivariate non-central t-distribution falls under H_A into the rejection region:

$$\text{Power} = \Pr(T^L > t_{1-\alpha, n_c + n_E - 2}, \, T^U < -t_{1-\alpha, n_c + n_E - 2})$$

$$= \int_{-\infty}^{-t_{1-\alpha, n-2}} \int_{t_{1-\alpha, n-2}}^{\infty} \text{bivt}(n - 2; nc_{L,A}, nc_{U,A}; x, y, 1) dx dy, \qquad (10.4)$$

where $\text{bivt}(df, nc_{L,A}, nc_{U,A}; x, y, 1)$ denotes the density of the bivariate t-distribution with df degrees of freedom, non-centrality parameter vector $(nc_{L,A}, nc_{U,A})$ defined by inserting the specified alternative Δ_A in (10.3a) and (10.3b), and correlation 1. The other way round, the exact sample size required to achieve a desired power $1 - \beta$ can be determined iteratively by equating the integral in (10.4) to $1 - \beta$ and resolving the equation for n (or more precisely: by determining the smallest integer n for which the integral in (10.4) is at least $1 - \beta$).

An approximate sample size formula can be derived based on the following considerations. For the alternative of exact equality of E and C, i.e. $\Delta_A = (\mu_E - \mu_C)_A = 0$, the probabilities for rejecting H_0^L and for rejecting H_0^U are equal due to the symmetry of the equivalence range. A conservative approach for assuring a power of $1 - \beta$ for rejecting both null-hypotheses is to calculate the sample size at a Bonferroni-adjusted type II error rate for each of the test problems. This sample size is well approximated by the Guenther-Schouten approach resulting in

$$n = \frac{(1+r)^2}{r} \cdot \left(z_{1-\alpha} + z_{1-\beta/2}\right)^2 \cdot \left(\frac{\sigma}{\delta}\right)^2 + \frac{(z_{1-\alpha})^2}{2} \quad \text{for } \Delta_A = 0. \qquad (10.5)$$

If $\Delta_A > 0$ ($\Delta_A < 0$), the power to reject H_0^L (H_0^U) is higher than to reject H_0^U (H_0^L). Therefore, if $|\Delta_A|$ is "not too near" to 0, rejection of H_0^L (H_0^U) is "included" if the sample size is chosen such that the power for rejecting H_0^U (H_0^L) is high enough. For the common power values, the approximate sample size can therefore be determined by assuring a power of $1 - \beta$ for the respective one-sided test problem. This leads to

$$n = \frac{(1+r)^2}{r} \cdot \left(z_{1-\alpha} + z_{1-\beta}\right)^2 \cdot \frac{\sigma^2}{(\delta - |\Delta_A|)^2} + \frac{(z_{1-\alpha})^2}{2} \quad \text{for } \Delta_A \neq 0 \qquad (10.6)$$

with $\Delta_A = (\mu_E - \mu_C)_A$ and $r = n_E/n_C$, i.e. $n_C = \left(\frac{1}{1+r}\right) \cdot n$ and $n_E = \left(\frac{r}{1+r}\right) \cdot n$.

Remark The most frequent application of equivalence trials is to demonstrate bioequivalence. Here, the test problem for concentration-related pharmacokinetic characteristics is on the original scale given by

$$H_0: \frac{\mu_E}{\mu_C} \leq 1/\delta \quad \text{or} \quad \frac{\mu_E}{\mu_C} \geq \delta$$
$$\text{versus} \tag{10.7}$$
$$H_1: 1/\delta < \frac{\mu_E}{\mu_C} < \delta$$

with pre-specified $\delta > 1$. Based on pharmacokinetic relationships, a lognormal distribution is commonly assumed for these characteristics. Then, on the logarithmic scale, test problem (10.7) transforms into the following test problem for normally distributed outcomes which is formulated in terms of the difference of means:

$$H_0: \ln \mu_E - \ln \mu_C \leq -\ln \delta \quad \text{or} \quad \ln \mu_E - \ln \mu_C \geq \ln \delta$$
$$\text{versus} \tag{10.8}$$
$$H_1: -\ln \delta < \ln \mu_E - \ln \mu_C < \ln \delta.$$

Therefore, the above-described exact and approximate methods for sample size calculation can also be applied to this setting by employing the appropriate quantities.

Bioequivalence trials are frequently conducted in a two-way crossover design, where each study participant receives sequentially E and C (interrupted by a washout phase) and where the sequence is commonly allocated by randomization. The methods for sample size calculation presented above can then also be applied by—in addition to taking into account the log-transformation—replacing "σ^2" by "$\sigma^2/2$". In the literature on bioequivalence trials, specification of the variability is often made in terms of the coefficient of variation CV instead of σ^2, where the relationship $CV = \sqrt{\exp(\sigma^2) - 1}$ is applied.

Example 10.1 Dutasteride at a daily dose of 0.5 mg is approved for the treatment and prevention of progression of benign prostatic hyperplasia as well as of androgenetic alopecia mono- and combination therapy. The open-label, randomized, single dose, two-way cross-over **bioequivalence trial** conducted by Fossler et al. (2015) was performed to demonstrate the bioequivalence of five 0.1 mg dutasteride capsules to one 0.5 mg dutasteride capsule in healthy adult male subjects. Each of the healthy male subjects participated in both treatment periods which were separated by a washout phase of 28 days. The sequences were balanced and patients were randomly allocated to one of them. Blood samples for pharmacokinetic analysis were collected up to 72 h post-dose. Primary endpoints were the maximum concentration (Cmax) and the area under the serum concentration-time curve to the last quantifiable concentration (AUC(0–t)). For both endpoints, the equivalence range on the original scale was defined as the commonly applied interval $[1/\delta, \delta] = [0.80, 1.25]$. Analysis was planned to be performed by application of the two one-sided t-tests procedure at level $\alpha = 0.05$ for each endpoint. Sample size calculation was based on the largest CV for Cmax and AUC(0–t) observed in previous trials which amounts to 0.248 and which corresponds to $\sigma = 0.2442$. The sample size should ensure a power of $1 - \beta = 0.90$ for each of the endpoints under the assumption of perfect equality, i.e. for $(\mu_E/\mu_C)_A = 1$. Exact and approximate sample size calculations both resulted in a total sample size of $n = 28$. A total of 36 subjects were enrolled in the study to

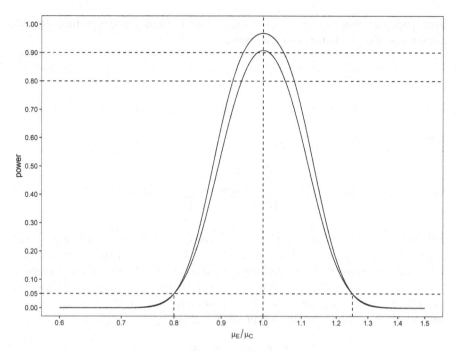

Fig. 10.1 Power for demonstrating bioequivalence with a total sample size of $n = 28$ (lower line) or $n = 36$ (upper line) depending on the ratio μ_E/μ_C. The equivalence range is [0.80, 1.25], the two one-sided t-tests procedure at level $\alpha = 0.05$ is applied, and $\sigma = 0.2442$ is assumed for power calculation (see Example 10.1)

ensure that 28 subjects completed dosing and critical assessments for both treatment periods. In fact, all 36 subjects could be evaluated for the primary pharmacokinetic endpoints. For this sample size, the exact power at $\mu_E/\mu_C = 1$ amounts to 0.969, and the power is at least 0.90 within the interval [0.951, 1.052]. Figure 10.1 shows the exact power for demonstrating bioequivalence for a total sample size of 28 and 36, respectively, for the planning scenario of the dutasteride study over a range of values for $\mu_E/\mu_C = 1$.

Remark Tang (2018) derived exact sample size calculation methods for the assessment of the equivalence test problem (10.1) with unstratified or stratified ANCOVA.

10.3 Ratio of Means as Effect Measure

As already described in Sect. 8.2.2, non-inferiority or equivalence margins, respectively, are frequently expressed as portion of the (unknown) population mean of the control. In equivalence trials, the lower and upper margins on the difference scale $\gamma_L, \gamma_U > 0$, respectively, can then be expressed as $\gamma_L = f_L \cdot \mu_C$ and $\gamma_U = f_U \cdot \mu_C$

with specified values f_L and f_U, $0 < f_L, f_U < 1$. Then, the test problem for the assessment of equivalence can be written as

$$H_0: \mu_E - \mu_C \leq -f_L \cdot \mu_C \text{ or } \mu_E - \mu_C \geq f_U \cdot \mu_C$$
$$\text{versus}$$
$$H_1: -f_L \cdot \mu_C < \mu_E - \mu_C < f_U \cdot \mu_C,$$

which is equivalent to

$$H_0: \tfrac{\mu_E}{\mu_C} \leq \delta_L \text{ or } \tfrac{\mu_E}{\mu_C} \geq \delta_U$$
$$\text{versus} \tag{10.9}$$
$$H_1: \delta_L < \tfrac{\mu_E}{\mu_C} < \delta_U$$

with $\delta_L = 1 - f_L$ and $\delta_U = f_U + 1$. We consider in the following equivalence ranges $]\delta_L, \delta_U[$ with $\delta_L = 1/\delta_U$, where $\delta_U < 1$. This is reasonable as then the assessment of equivalence is invariant when exchanging the population means in nominator and denominator. Methods for sample size calculation in the general situation $\delta_L < 1 < \delta_U$ can be found in Hauschke et al. (1999). For equivalence ranges with $\delta_L = 1/\delta_U = \delta$, the test problem is given by

$$H_0: \tfrac{\mu_E}{\mu_C} \leq \delta \text{ or } \tfrac{\mu_E}{\mu_C} \geq \tfrac{1}{\delta}$$
$$\text{versus} \tag{10.10}$$
$$H_1: \delta < \tfrac{\mu_E}{\mu_C} < \tfrac{1}{\delta}$$

with pre-specified $\delta < 1$. The null-hypothesis H_0 is composed of the two one-sided null-hypotheses $H_0^L: \tfrac{\mu_E}{\mu_C} \leq \delta$ and $H_0^U: \tfrac{\mu_E}{\mu_C} \geq \tfrac{1}{\delta}$ which can be tested by using the following test statistics:

$$T^L = \sqrt{\frac{n_C \cdot n_E}{n_C + \delta^2 \cdot n_E}} \cdot \frac{\bar{X}_E - \delta \cdot \bar{X}_C}{S} \tag{10.11a}$$

and

$$T^U = \sqrt{\frac{n_C \cdot n_E}{n_C + (1/\delta^2) \cdot n_E}} \cdot \frac{\bar{X}_E - (1/\delta) \cdot \bar{X}_C}{S}. \tag{10.11b}$$

The null-hypothesis H_0^L can be rejected at level α if $t^L > t_{1-\alpha,n_C+n_E-2}$, and H_0^U can be rejected at level α if $t^U < -t_{1-\alpha,n_C+n_E-2}$. Therefore, the two one-sided tests procedure rejects H_0 at level α if both conditions are met. The random vector (T^L, T^U) follows the bivariate t-distribution with $n_C + n_E - 2$ degrees of freedom, $\text{Corr}(T^L, T^U) = \frac{1+r}{\sqrt{(1+\delta^2 \cdot r) \cdot (1+(1/\delta^2) \cdot r)}}$, and components of the non-centrality parameter vector given by

$$nc_L = \sqrt{\frac{r}{(1+r)\cdot\left(1+\delta^2\cdot r\right)}}\cdot n\cdot\frac{\mu_E - \delta\cdot\mu_C}{\sigma} \tag{10.12a}$$

$$= \sqrt{\frac{r}{(1+r)\cdot\left(1+\delta^2\cdot r\right)}}\cdot n\cdot\frac{\mu_E/\mu_C - \delta}{CV} \tag{10.12b}$$

$$nc_U = \sqrt{\frac{r}{(1+r)\cdot\left(1+(1/\delta^2)\cdot r\right)}}\cdot n\cdot\left(\frac{\mu_E - (1/\delta)\cdot\mu_C}{\sigma}\right) \tag{10.12c}$$

$$= \sqrt{\frac{r}{(1+r)\cdot\left(1+(1/\delta^2)\cdot r\right)}}\cdot n\cdot\left(\frac{(\mu_E/\mu_C)-(1/\delta)}{CV}\right) \tag{10.12d}$$

As for the equivalence test problem formulated in terms of the difference of means, the power can be calculated by numerical integration of the density of the respective bivariate t-distribution over the rejection region. For this, the total sample size n, the alternative (defined by values $\mu_{E,A}$ and $\mu_{C,A}$ or a value $\theta_A = (\mu_E/\mu_C)_A$, respectively, depending on which version of the expression for the non-centrality parameter is applied), and the assumed variability (σ or CV, respectively) are to be specified. Inversely, the exact total sample size can be determined by fixing the desired power $1 - \beta$ and solving the power equation for n.

Based on the same considerations as for the difference in means, the following approximate sample size formula can be derived. For allocation ratios $r = n_E/n_C \geq 1$, the formula is given by the following expressions:

$$n = \begin{cases} \frac{(1+r)\cdot(1+(1/\delta^2)\cdot r)}{r}\cdot\left(z_{1-\alpha}+z_{1-\beta/2}\right)^2\cdot\left(\frac{CV}{1-\frac{1}{\delta}}\right)^2 + \frac{(z_{1-\alpha})^2}{2} & \text{for } \theta_A = 1 \\[2ex] \frac{(1+r)\cdot(1+\delta^2\cdot r)}{r}\cdot\left(z_{1-\alpha}+z_{1-\beta}\right)^2\cdot\left(\frac{CV}{\theta_A-\delta}\right)^2 + \frac{(z_{1-\alpha})^2}{2} & \text{for } \theta_A < 1 \\[2ex] \frac{(1+r)\cdot(1+(1/\delta^2)\cdot r)}{r}\cdot\left(z_{1-\alpha}+z_{1-\beta}\right)^2\cdot\left(\frac{CV}{\frac{1}{\delta}-\theta_A}\right)^2 + \frac{(z_{1-\alpha})^2}{2} & \text{for } \theta_A > 1 \end{cases}$$
$$\tag{10.13a–c}$$

with $\theta_A = \left(\frac{\mu_E}{\mu_C}\right)_A$ and $r = n_E/n_C$, i.e. sample sizes per group of $n_C = \left(\frac{1}{1+r}\right)\cdot n$ and $n_E = \left(\frac{r}{1+r}\right)\cdot n$. For $r = n_E/n_C < 1$, $(1/\delta^2)$ and $\frac{1}{\delta}$ have to be replaced by δ^2 and δ, respectively, in (10.13a).

If the standard deviation instead of the coefficient of variation is to be used for sample size calculation, Eqs. (10.13a) to (10.13c) can easily be adapted by employing the non-centrality parameters (10.12a) and (10.12c) instead of (10.12b) and (10.12d).

Remark The related methodology for exact sample size calculation for equivalence assessment in case that the trial is performed in a two-period cross-over design is provided by Hauschke et al. (1999).

Example 10.2 When the equivalence trial by Van der Linden et al. (2013) was initiated, human serum albumin (HSA) was the most commonly used colloid in **pediatric surgery**. In adult surgery, however, artificial colloids such as hydroxyethyl starch

(HES) were applied increasingly for volume replacement. The two-center, randomized, controlled, double-blind trial in children undergoing elective open-heart surgery aimed at demonstrating equivalence between HSA and HES. The primary endpoint was the total volume of colloid solution required in the intraoperative volume period, and the treatment group difference was expressed in terms of the ratio of means. The equivalence range agreed with the US Food and Drug Administration was defined by the interval $]\delta, 1/\delta[=]0.55, 1.82[$. For sample size calculation, equality of HSA and HES was assumed, i.e. $\theta_A = 1$, and a coefficient of variation of $CV = 0.363$ was supposed. Balanced allocation was to be applied ($r = 1$) and, untypically for equivalence assessment, a type I error rate of $\alpha = 0.025$ was used. If the primary outcome is assumed as normally distributed, the correlation between the test statistics T^L and T^U amounts to $\mathrm{Corr}(T^L, T^U) = \frac{2\cdot\delta}{1+\delta^2} = 0.845$, and a total sample size of $2 \times 13 = 26$ is required according to exact calculations to assure the desired power of $1 - \beta = 0.90$ under these assumptions. The approximation formula (10.13a) results in $2 \times 12 = 24$ patients. In order to get more safety information, $2 \times 30 = 60$ children were planned to be included. For this sample size, the exact power for demonstrating equivalence under the above-described assumptions amounts to 0.9999, and for values of θ_A within the interval $[0.800, 1.346]$ the exact power is at least 0.90.

Chapter 11
Multiple Comparisons

11.1 Background and Notation

Up to now, we considered the situation that a single test problem is assessed in confirmatory analysis. There are, however, numerous scenarios where it is necessary to perform multiple confirmatory tests within a clinical trial. Multiplicity may, for example, arise when multiple endpoints are defined as primary outcomes in a two-group trial, when more than two groups are included in the trial which are to be compared with respect to a single primary endpoint, or when a combination of both occurs. Formally, the situation is described by k test problems

$$
\begin{array}{c}
H_{0,i} \\
\text{versus} \\
H_{1,i}, \quad i = 1, \ldots, k, \quad k > 1
\end{array}
\tag{11.1}
$$

to be assessed in confirmatory analysis. The commonly applied concept of type I error rate control in confirmatory clinical trials, which is also required by regulatory agencies (CHMP 2016), is that of strong control of the familywise error rate: A multiple test procedure for the assessment of (11.1) controls the familywise type I error rate in the strong sense by α, if the probability of erroneously rejecting any true null-hypothesis is not larger than α, regardless of which and how many of the null-hypotheses are true (Dmitrienko et al. 2010). In the following, we consider only multiple test procedures that fulfill this requirement.

There exists a universe of multiple test procedures [see, for example, the textbooks by Hochberg and Tamhane (1987), Westfall and Young (1993), and Dmitrienko et al. (2010), and the review by Dmitrienko et al. (2013)]. In this book, we can only consider some specific multiple test procedures and related methods for sample size calculation. In the following, the significance levels to be applied for testing the individual null-hypotheses $H_{0,i}$ within a multiple test procedure are denoted by local levels. We distinguish procedures with pre-fixed and data-dependent local levels. In

© Springer Nature Switzerland AG 2020
M. Kieser, *Methods and Applications of Sample Size Calculation and Recalculation in Clinical Trials*, Springer Series in Pharmaceutical Statistics,
https://doi.org/10.1007/978-3-030-49528-2_11

procedures with pre-fixed local levels, each null-hypothesis is tested at the respective level which is already known in the planning stage. In contrast, for procedures with data-dependent local levels it is not known beforehand which levels will be applied to the null-hypotheses $H_{0,i}$ in the analysis. A popular and generally applicable procedure with pre-fixed local levels controlling the familywise error rate in the strong sense is the Bonferroni procedure where each null-hypothesis $H_{0,i}$, $i = 1, \ldots, k$, is tested at level α/k. The Bonferroni procedure is an example of a single-step procedure where all individual p-values are compared simultaneously to their respective critical level. In contrast, the Bonferroni-Holm procedure (Holm 1979) is a stepwise procedure with data-dependent local levels that also controls the familywise error rate in the strong sense for arbitrary sets of test problems (11.1). Here, the observed p-values are ordered and denoted by $p_{(1)} \leq p_{(2)} \leq \ldots \leq p_{(k)}$, and the respective null-hypotheses are indicated by $H_{0,(i)}$. The procedure starts comparing $p_{(1)}$ with α/k: If $p_{(1)} \geq \alpha/k$, all $H_{0,i}$ cannot be rejected and the procedure stops; if $p_{(1)} < \alpha/k$, $H_{0,(1)}$ can be rejected and the procedure continues comparing $p_{(2)}$ with $\alpha/(k-1)$: If $p_{(2)} \geq \alpha/(k-1)$, $H_{0,(i)}$ cannot be rejected for all $i = 2, \ldots, k$, and the procedure stops; if $p_{(2)} < \alpha/(k-1)$, $H_{0,(2)}$ can be rejected and the procedure continues. In the mth step, $m = 1, \ldots, k$, $p_{(m)}$ is compared to $\alpha/(k-m+1)$ and the same decision rules as described above are applied. The Bonferroni-Holm procedure rejects at least as much null-hypotheses as the Bonferroni procedure. However, sample size calculation is more challenging as, in contrast to the Bonferroni approach, it is not known in the planning stage which local level is to be applied in the analysis for a specific null-hypothesis. The gernerally applicable methods for sample size calculation presented in the following Sect. 11.2 are based on pre-fixed levels. Power and sample size considerations for stepwise procedures with data-dependent local levels can, for example, be found in Bretz et al. (2011).

The definition of power we used for the case of testing a single null-hypothesis in the confirmatory analysis can be extended in various ways when assessing multiple test problems. Which concept to use is defined by the primary objective of the clinical trial. In this chapter, we consider the following two concepts: If the aim of the trial is to reject all false null-hypotheses, the related power is denoted as conjunctive power (Senn and Bretz 2007) or complete power (Westfall et al. 2011). If the aim of the trial is to reject at least one false null-hypothesis, the related power is denoted as disjunctive power (Senn and Bretz 2007) or minimal power (Westfall et al. 2011).

In the next section, generally applicable methods for sample size calculation for these power concepts are presented. We assume that strong control of the experimentwise type I error rate is assured by application of a multiple test procedure with pre-fixed local levels α_i, $i = 1, \ldots, k$. No specific assumptions are made, which is both a strength and a weakness: As we will see, these methods can be applied in any multiple comparison situation. However, as crude (Bonferroni) adjustments are employed that do not incorporate any information on the dependence structure of the involved test statistics, they may provide unnecessarily large sample sizes. In

Sects. 11.3 and 11.4, examples of test procedures and related sample size calculation methods are presented for multiple endpoints and for the comparison of more than two groups. These methods are tailored to specific distributions of the applied test statistics and the correlations between them. They exploit the characteristics of the specific situation and may therefore be more efficient than the generally applicable procedures presented in the following Sect. 11.2.

11.2 Generally Applicable Sample Size Calculation Methods and Applications

11.2.1 Methods

We assume in the following that a multiple test procedure with pre-fixed local levels $\alpha_i, i = 1, \ldots, k$, is applied to assure control of the experimentwise type I error rate in the strong sense. Furthermore, for sample size calculation, all null-hypotheses $H_{0,i}$, $i = 1, \ldots, k$, are supposed to be false; otherwise one would hardly include the related test problem in the confirmatory analysis. The following general approaches can then be pursued to assure a specified conjunctive or disjunctive power, respectively.

(a) *Study aim: rejection of all null-hypotheses (conjunctive power)*

A simple method to assure a conjunctive power of at least $1 - \beta$ is by guaranteeing that the power for each test problem $H_{0,i}$ versus $H_{1,i}, i = 1, \ldots, k$, amounts to at least $1 - \beta/k$ under the specified alternative $H_{A,i}$; this corresponds to a Bonferroni adjustment of the type II error rate. Validity of this approach can be seen as follows. Under the specified alternative hypotheses, the following inequality holds true:

$$\text{Pr(reject all null-hypotheses)} = 1 - \text{Pr(do not reject at least one null-hypothesis)}$$

$$\geq 1 - \sum_{i=1}^{k} \text{Pr}\big(\text{do not reject } H_{0,i}\big)$$

$$\geq 1 - k \cdot (\beta/k) = 1 - \beta.$$

Therefore, calculating for each test problem $H_{0,i}$ versus $H_{1,i}$ the required sample size to reject $H_{0,i}$ at level α_i with power $1 - \beta/k$ and taking the maximum of these sample sizes assures a conjunctive power of $1 - \beta$.

(b) *Study aim: rejection of at least one null-hypothesis (disjunctive power)*

Obviously, a disjunctive power of at least $1 - \beta$ can be assured if the power for each test problem $H_{0,i}$ versus $H_{1,i}, i = 1, \ldots, k$, amounts to at least $1 - \beta$ under the specified alternative $H_{A,i}$. Then, $\text{Pr(reject at least one null-hypothesis)} \geq \text{Pr}\big(\text{reject } H_{0,i}\big) \geq 1 - \beta$ for all $i = 1, \ldots, k$. Therefore, a disjunctive power of $1 - \beta$ can be assured by calculating for each test problem $H_{0,i}$ versus $H_{1,i}$ the required sample size to reject $H_{0,i}$ at level α_i with power $1 - \beta$ and taking the maximum of these sample sizes.

11.2.2   Applications

(a)   *Equal importance of test problems*

Frequently, there is no ordering in the importance of the test problems to be assessed in the confirmatory analysis but all hypotheses are of equal relevance. Then the following two situations can be distinguished.

(a1)   *Study is successful if and only if all null-hypotheses can be rejected ("all or none" success criterion)*

There are situations where a clinical trial is successful only if all null-hypotheses assessed in confirmatory analysis can be rejected. In the literature, this is called the "all or none" success criterion. The related test problem is denoted as intersection-union test problem and is given by

$$H_0: \bigcup_{i=1}^{k} H_{0,i}$$
$$\text{versus} \tag{11.3}$$
$$H_1: \bigcap_{i=1}^{k} H_{1,i}.$$

The following intersection-union test procedure assures strong control of the familywise error rate α: "Reject H_0 if all $H_{0,i}, i = 1, \ldots, k$, can be rejected at level α, accept H_0 otherwise" (Berger 1982). Hence, the local type I error rate to be applied for testing the single null-hypotheses has not to be adjusted when assessing the "all or none" success criterion with the intersection-union test. The probability for a successful trial is equal to the conjunctive power and therefore the disjunctive power is not relevant in this situation. According to Sect. 11.2.1a, the following method can be applied for sample size calculation to assure a conjunctive power of $1 - \beta$: For each test problem $H_{0,i}$ versus $H_{1,i}$, the required sample size is calculated for the local level α and a power of $1 - \beta / k$ and the maximum of the resulting sample sizes is taken.

(a2)   *Study is successful if at least one null-hypothesis can be rejected ("at least one" success criterion)*

It is often sufficient for trial success that at least one of the null-hypotheses is rejected in the confirmatory analysis. Then, the simplest multiple test procedure with pre-fixed local levels controlling the familywise error rate in the strong sense is the Bonferroni procedure where each null-hypothesis $H_{0,i}, i = 1, \ldots, k$, is tested at level $\alpha_i = \alpha / k$. The power concept that corresponds to the "at least one" success criterion is that of the disjunctive power. A value of $1 - \beta$ for the disjunctive power can be assured by calculating for each test problem $H_{0,i}$ versus $H_{1,i}$ the required sample size for the local level $\alpha_i = \alpha/k$ and a power of $1 - \beta$ and taking the maximum of the resulting sample sizes.

Even if rejection of one null-hypothesis is sufficient to make the trial successful, it may nevertheless be desirable to reject not only one but all null-hypotheses. For example, such a result may allow to make more far-reaching claims for the investigated therapy. Then, a conjunctive power of $1 - \beta$ is aspired which can be assured by calculating for each test problem $H_{0,i}$ versus $H_{1,i}$ the required sample size for the local level $\alpha_i = \alpha/k$ and a power of $1 - \beta/k$ and taking the maximum of the resulting sample sizes.

Note that the Bonferroni procedure can be generalized in that the local levels α_i need not necessarily be equal: The familywise error rate is controlled by α in the strong sense as long as $\sum_{i=1}^{k} \alpha_i \leq \alpha$. The related methods for sample size calculation are then analogous to those described above.

(b) *Different importance of test problems*

We now assume that the test problems $H_{0,i}$ versus $H_{1,i}$, $i = 1, \ldots, k$, can in the planning phase be ordered according to their importance such that rejection of $H_{0,l}$ is only of interest if all $H_{0,i}$ with $i = 1, \ldots, l, l \leq k$, can be rejected. For example, when comparing two groups with respect to a single endpoint measured at k time points between start of the intervention and treatment end, group differences before end of therapy are only of interest in case that there is an effect at the final assessment at the end of therapy. In such cases, the following test procedure for *a priori* ordered hypotheses controls the type I error rate in the strong sense (Maurer et al. 1995).

Step 1 $H_{0,1}$ is tested at level α:

- If $H_{0,1}$ cannot be rejected, all $H_{0,i}$, $i = 1, \ldots, k$, cannot be rejected and the procedure stops.
- If $H_{0,1}$ can be rejected, the procedure continues with step 2.

Step 2 $H_{0,2}$ is tested at level α:

- If $H_{0,2}$ cannot be rejected, all $H_{0,i}$, $i = 2, \ldots, k$, cannot be rejected and the procedure stops.
- If $H_{0,2}$ can be rejected, the procedure continues with step 3.

Step m ($m = 1, \ldots, k$) $H_{0,m}$ is tested at level α:

- If $H_{0,m}$ cannot be rejected, all $H_{0,i}$, $i = m, \ldots, k$, cannot be rejected and the procedure stops.
- If $H_{0,m}$ can be rejected and $m < k$, the procedure continues with step $m + 1$; if $m = k$, the procedure stops.

Therefore, to control the experimentwise type I error rate in the strong sense, the local type I error rates have not to be adjusted when this test procedure is applied.

According to 11.2.1a, a conjunctive power of $1 - \beta$ can be assured by calculating for each test problem $H_{0,i}$ versus $H_{1,i}$ the required sample size for the local level α and a power of $1 - \beta/k$ and taking the maximum of the resulting sample sizes.

Remark While the aims and decision criteria of the intersection-union test and the multiple test procedure for *a priori* ordered hypotheses differ, the methods for sample size calculation to assure a conjunctive power of $1 - \beta$ coincide. Furthermore, due to the testing strategy, in case of ordered test problems the disjunctive power equals to the power for rejecting the null-hypothesis $H_{0,1}$ relating to the most important test problem and therefore the sample size can be calculated accordingly.

The general methods described above may overestimate the required sample size in specific situations and may be improved by utilizing information about the distribution and the correlation of the involved test statistics. In the next two sections, we will demonstrate this approach for normally distributed outcomes and the cases of multiple endpoints and multi-armed clinical trials.

11.3 Multiple Endpoints

11.3.1 Background and Notation

In many indications, the disease has multiple facets and therefore the efficacy of therapies cannot be captured by a single endpoint. Instead, implementation of multiple primary endpoints is required. There exist methodological regulatory guidelines (CHMP 2016; FDA 2017) as well as indication-specific guidelines, for example in the field of Alzheimer's disease (CHMP 2008; FDA 2013), for dealing with this type of multiplicity. Both the "all or none" and the "at least one" success criterion described in the preceding Sect. 11.2.2 are relevant in practical applications. For both situations, sample size calculation methods for normally distributed endpoints are presented in this section. Besides providing insight in the methodology for an important example for the occurrence of multiplicity in clinical trials, considering the multiple endpoint situation will also illustrate the principle of sample size calculation tailored to specific multiple comparison procedures.

In the following, we assume two groups to be compared with respect to $k > 1$ normally distributed primary endpoints. To be concrete, we deal with independent multivariate normally distributed random vectors $(X_{C1l}, \ldots, X_{Ckl})$, $l = 1, \ldots, n_C$, and $(X_{E1m}, \ldots, X_{Ekm})$, $m = 1, \ldots, n_E$, with expectation $\boldsymbol{\mu}_C = (\mu_{C1}, \ldots, \mu_{Ck})$ and $\boldsymbol{\mu}_E = (\mu_{E1}, \ldots, \mu_{Ek})$ and common and known covariance matrix $\boldsymbol{\Sigma}$. The single test problems to be assessed are given by

$$H_{0,i}: \mu_{Ei} - \mu_{Ci} = 0$$
$$\text{versus}$$
$$H_{1,i}: \mu_{Ei} - \mu_{Ci} \neq 0, \quad i = 1, \ldots, k.$$

As in the related single primary endpoint situation, the z-test statistic to be used for the assessment of $H_{0,i}$ is given by $Z_i = \sqrt{\frac{r}{(1+r)^2} \cdot n} \cdot \frac{\bar{X}_{Ei} - \bar{X}_{Ci}}{\sigma_i}, i = 1, \ldots, k,$ where $\bar{X}_{Ci} = \frac{1}{n_C} \cdot \sum_{l=1}^{n_C} X_{Cil}, \bar{X}_{Ei} = \frac{1}{n_E} \cdot \sum_{m=1}^{n_E} X_{Eim}; \sigma_i^2$ is the common and known variance of X_{Cil} and $X_{Eim}, l = 1, \ldots, n_C, m = 1, \ldots, n_E, r = n_E/n_C$, and $n = n_E + n_C$. The vector of test statistics $\mathbf{Z} = (Z_1, \ldots, Z_k)$ follows the k-variate standardized normal distribution with components of the expectation being equal to $\sqrt{\frac{r}{(1+r)^2} \cdot n} \cdot \left(\frac{\Delta_i}{\sigma_i}\right), i = 1, \ldots, k$, with $\Delta_i = \mu_{Ei} - \mu_{Ci}$, and off-diagonal elements of the covariance matrix ρ_{ij} being equal to the correlation between endpoints i and $j, i \neq j$. The sample size is to be calculated for a specified alternative $\mathbf{\Delta}_A = (\Delta_{1,A}, \ldots, \Delta_{k,A})$ with $\Delta_{i,A} = (\mu_{Ei} - \mu_{Ci})_A, i = 1, \ldots, k$. We assume $\Delta_{i,A} \neq 0$ for all $i = 1, \ldots, k$, as one would not include endpoints in the confirmatory analysis for which no treatment group difference is expected. Furthermore, for simplicity of notation, the endpoints are assumed to be defined such that $\Delta_{i,A} > 0$ for all $i = 1, \ldots, k$. Under $\mathbf{\Delta}_A$, the vector of centered test statistics $\mathbf{Z}^* = (Z_1^*, \ldots, Z_k^*)$ with $Z_i^* = Z_i - \sqrt{\frac{r}{(1+r)^2} \cdot n} \cdot \left(\frac{\Delta_{i,A}}{\sigma_i}\right), i = 1, \ldots, k$, follows the k-variate standardized normal distribution with expectation zero and off-diagonal elements of the covariance matrix ρ_{ij} being equal to the correlation between endpoints i and $j, i \neq j$. This result will be used below to derive methods for sample size calculation for the "all or none" and the "at least one" success criterion.

11.3.2 Methods

(a) "All or none" success criterion

In the situation that the trial is successful only if the null-hypotheses for all primary endpoints can be rejected, the primary outcomes are denoted as co-primary endpoints. There are a number of disease areas where the full picture of the treatment effect cannot be captured by a single outcome and where multiple co-primary endpoints are to be used in clinical trials (see Remark 3 below). According to the intersection-union principle (see Sect. 11.2.2a1), the familywise type I error rate can be controlled in the strong sense by α if the intersection of the rejection regions of the univariate two-sided level-α tests is used as rejection region for the assessment of $H_0: \bigcup_{i=1}^{k} H_{0,i}$, i.e. $\bigcap_{i=1}^{k} \{|Z_i| > z_{1-\alpha/2}\}$. Therefore, the conjunctive power to reject H_0 with the intersection-union test at two-sided level α under the specified alternative $H_{1,A}$ defined by $\mathbf{\Delta}_A$ is given by

$$\mathrm{Pr}_{H_{1,A}} \left(\bigcap_{i=1}^{k} \{|Z_i| > z_{1-\alpha/2}\} \right) \approx \mathrm{Pr}_{H_{1,A}} \left(\bigcap_{i=1}^{k} \{Z_i > z_{1-\alpha/2}\} \right)$$

$$= \left(\mathrm{Pr}_{H_{1,A}} \left\{ \bigcap_{i=1}^{k} Z_i - \sqrt{\frac{r}{(1+r)^2} \cdot n} \cdot \left(\frac{\Delta_{i,A}}{\sigma_i}\right) > z_{1-\alpha/2} \right. \right.$$

$$-\sqrt{\frac{r}{(1+r)^2} \cdot n \cdot \left(\frac{\Delta_{i,A}}{\sigma_i}\right)}\})$$

$$= \mathrm{Pr}_{H_{1,A}}\left(\bigcap_{i=1}^{k}\{Z_i^* > c_i^*\}\right),$$

with $\mathbf{Z}^* = (Z_1^*, \ldots, Z_k^*)$ as defined above and $c_i^* = z_{1-\alpha/2} - \sqrt{\frac{r}{(1+r)^2} \cdot n} \cdot \left(\frac{\Delta_{i,A}}{\sigma_i}\right)$.
Therefore, under a specified alternative $\boldsymbol{\Delta}_A$ and for given variances of endpoints $\sigma_i^2, i = 1, \ldots, k$, correlations between endpoints $\rho_{ij}, i, j = 1, \ldots, k, i \neq j$, and sample size n, the power can be computed numerically by evaluating the distribution function of the k-variate standardized normal distribution. The other way round, the required sample size can be calculated by searching for the smallest integer assuring that the conjunctive power is at least $1 - \beta$, i.e. such that the following inequality is fulfilled:

$$\mathrm{Pr}_{H_{1,A}}\left(\bigcap_{i=1}^{k}\{Z_i^* > c_i^*\}\right) \geq 1 - \beta. \tag{11.4}$$

As a simple grid search may require considerable computation time, Sugimoto et al. (2012) and Hamasaki et al. (2013) proposed to apply the Newton-Raphson algorithm or a linear interpolation algorithm, respectively, to reduce the computational burden.

If the standardized effects are equal for all endpoints, i.e. $(\Delta_{i,A}/\sigma_i) = (\Delta/\sigma)_A$ for all $i = 1, \ldots, k$, an explicit formula that does not require integration can be derived. Then, $c_i^* = z_{1-\alpha/2} - \sqrt{\frac{r}{(1+r)^2} \cdot n} \cdot \left(\frac{\Delta}{\sigma}\right)_A = c^*$ for all $i = 1, \ldots, k$, and inequality (11.4) is fulfilled if

$$-c^* \geq z_{k,\rho_{ij};1-\beta}, \tag{11.5}$$

where $z_{k,\rho_{ij};\beta}$ denotes the one-sided equicoordinate (i.e. the quantiles in each dimension coincide) β-quantile of the k-variate standard normal distribution with ρ_{ij} as matrix of correlation coefficients. Therefore, resolving the equation

$$z_{1-\alpha/2} - \sqrt{\frac{r}{(1+r)^2} \cdot n} \cdot \left(\frac{\Delta}{\sigma}\right)_A = -z_{k,\rho_{ij};1-\beta}$$

for n leads to the following formula for the required total sample size:

$$n = \frac{(1+r)^2}{r} \cdot \left(z_{1-\alpha/2} + z_{k,\rho_{ij};1-\beta}\right)^2 \cdot \left(\frac{\sigma}{\Delta}\right)_A^2. \tag{11.6}$$

To give an impression of how the required sample size changes in dependence of the number of endpoints and the correlation, Table 11.1, third column, shows the total sample size required according to formula (11.6) to obtain a conjuctive power of $1 - \beta = 0.90$ for the situation of $k = 1, 2, \ldots, 5$ endpoints with equal standardized

Table 11.1 Total sample size to achieve a conjunctive or disjunctive power of $1 - \beta = 0.90$, respectively, for k endpoints with equal standardized effects $(\Delta/\sigma)_A = 0.2$ and equal correlation ρ between endpoints ($\alpha/2 = 0.025$, one-sided, $r = n_E/n_C = 1$)

k	ρ	Conjunctive power $1 - \beta = 0.90$	Disjunctive power $1 - \beta = 0.90$
1		1052	1052
2	0	1292	740
	0.3	1274	838
	0.5	1252	910
	0.8	1194	1044
3	0	1428	618
	0.3	1400	752
	0.5	1366	854
	0.8	1270	1050
4	0	1524	548
	0.3	1488	706
	0.5	1444	826
	0.8	1324	1058
5	0	1598	504
	0.3	1556	674
	0.5	1504	806
	0.8	1362	1066

effects $(\Delta/\sigma)_A = 0.2$ and with equal correlation $\rho_{ij} = \rho$ between the endpoints, if the sample size is allocated in a balanced way to the groups and if a one-sided level of $\alpha/2 = 0.025$ is applied. It can be seen that for constant correlation the required sample size increases with increasing number of endpoints while it decreases with increasing correlation. The latter is due to the fact that it becomes more easier to obtain significant results for all endpoints in case they are highly correlated.

Remark The above-described methods assume that the variance inserted in the test statistic is known. Sozu et al. (2011) showed that the related sample sizes are very close to those required in case of unknown variance. For the situations considered by Sozu et al. (2011), just one more subject has to be added in each group in case that the variance is unknown for the analysis. This reflects the results described in Sect. 3.3 for a single primary endpoint.

When determining the sample size by evaluating Eq. (11.4) or using Eq. (11.6), assumptions on the correlations between the endpoints have to be made. On the one hand, if the assumptions are true, this may add precision to the calculations as compared to the crude approach described in Sect. 11.2.2a1. On the other hand, wrong assumptions may lead to inaccurate results even when using the refined approach thus defeating the purpose of applying the complex method: If too large (small) correlations are employed in sample size calculations, the resulting sample size is smaller (larger) than actually required. However, depending on the scenario, the difference may be negligible and a simple approach may be sufficient. Therefore, the question arises in which situations application of the crude procedure presented in Sect. 11.2.2a1 is sufficient and when it is advisable to incorporate the correlations between the endpoints in sample size calculation. Sozu et al. (2011) showed that the influence of the correlation is strong if the standardized treatment group differences $\Delta_{i,A}/\sigma_i$ are approximately equal for the endpoints. Then, application of the above-described methods that exploit the correlation is sensible. However, if the standardized difference is much smaller for one endpoint than for the others, the sample size required to assure rejection of the related null-hypothesis with sufficient power leads to an extremely high power for the other endpoints. Therefore, in this case the sample size can be determined by just calculating the sample size required to reject the null-hypothesis for the endpoint with the smallest standardized effect at two-sided level α with power $1 - \beta$.

For the intermediate situation where the standardized effect sizes differ but do not vary extremely, Varga et al. (2017) proposed a simple procedure for sample size calculation. It is based on the assumption of independent endpoints which is commonly not strictly true. Nevertheless, it provides quite accurate sample sizes in the situation the method is tailored for. The basic idea is to assign higher individual power values to endpoints with higher anticipated standardized treatment group differences such that overall the conjunctive power is $1 - \beta$. If the individual power for endpoint i is denoted by $1 - \beta_i$ and the endpoints are independent, the conjunctive power is

$$\prod_{i=1}^{k}(1 - \beta_i). \tag{11.7}$$

The method proposed by Varga et al. (2017) consists of the following steps:

Step 1 Assign equal individual power $(1 - \beta)^{1/k}$ to each endpoint and calculate the corresponding total sample sizes required for rejecting the individual null-hypotheses $H_{0,i}$ at level α with power $(1 - \beta)^{1/k}$.

Step 2 Calculate for each endpoint the individual power obtained for the smallest sample size obtained in step 1 (i.e. those for the endpoint with the largest standardized treatment group difference) and calculate the conjunctive power based on (11.7).
        If the conjunctive power is at least $1 - \beta$: stop
        Else: go to step 3.

Step 3 Increase the smallest total sample size obtained in step 1 by the smallest amount possible for the allocation ratio at hand (for example by 2 for $r = 1$ or by 3 for $r = 2$) and repeat steps 2 and 3 until the aspired conjunctive power is reached.

In the following, application of the various methods for calculating the required sample size for the "all or none" success criterion is illustrated by an example.

Example 11.1 The European Medicines Agency (EMA) and the US Food and Drug Administration (FDA) require in their guidelines on the investigation of therapies for the treatment of **Alzheimer's disease** that efficacy is to be demonstrated on both a cognitive and a functional assessment scale (CHMP 2008; FDA 2013). Therefore, in these clinical trials a co-primary endpoints approach is pursued (see Sect. 11.3.2a). Commonly used scales for the assessment of these domains are the change in Alzheimer's Disease Assessment Scale cognitive subscale (ADAS-cog; Rosen et al. 1984) and the change in Alzheimer's Disease Cooperative Study Activities of Daily Living scale (ADCS-ADL; Galasko et al. 1997), respectively. Generally, there is a moderate to strong relationship between these two endpoints (Liu-Seifert et al. 2016).

Green et al. (2009) performed a multicenter, randomized, double-blind, placebo-controlled clinical trial investigating the efficacy and tolerability of the compound tarenflurbil in patients with mild Alzheimer's disease. The changes between baseline and 18 months in ADAS-cog and ADCS-ADL were used as co-primary endpoints. These scales are ordinal but considered as (approximately) normally distributed for sample size calculation. The assumed treatment group differences (standard deviations) for the change in ADAS-cog and ADCS-ADL, respectively, were 2 (10) and 2.6 (13), and thus the anticipated standardized effect amounts to $(\Delta_{1,A}/\sigma_1) = (\Delta_{2,A}/\sigma_2) = (\Delta/\sigma)_A = 0.2$ for both endpoints. Balanced sample size was calculated for the quite unusual individual power values $1 - \beta_1 = 1 - \beta_2 = 0.98$. For a one-sided type I error rate of $\alpha/2 = 0.025$, this results in a total sample size of $1612 = 2 \times 806$ (based on Eq. (3.5) for the z-test). In the article, the authors mentioned that "because the co-primary endpoints were expected to be correlated, the joint power would have been in excess of $0.96 \ (= 0.98^2)$ for detecting treatment differences" (Green et al. 2009, p. 2559).

In the following, we illustrate the impact of taking the correlation between the endpoints into account for the more convenient conjunctive power value of $1 - \beta = 0.90$. For the generally applicable approach based on Bonferroni adjustment of the type II error rate, the individual power values amount to $1 - \beta_1 = 1 - \beta_2 = 0.95$ and the related total sample size based on Eq. (3.5) is 1300. Note that in case of equal standardized effect sizes for both endpoints, the procedure proposed by Varga et al. (2017) results in the same sample size as obtained when assuming independent endpoints, i.e., when calculating the sample size for individual power values of $1 - \beta_1 = 1 - \beta_2 = \sqrt{0.90} = 0.9487$. This approach leads to a total sample size of 1292. Furthermore, evaluation of inequality (11.4) by numerical integration and computation of the sample size according to formula (11.6) lead to the same results for equal standardized effect sizes for the endpoints. Employing the correlation in sample size calculation according to (11.4) or (11.6) and assuming correlations of 0.3, 0.5, and 0.8, respectively, results in total sample sizes of 1274, 1252, and 1194 which are by about 1%, 3%, and 8% smaller than the sample size of 1292 required when the endpoints are independent. The quantiles of the bivariate standard normal distribution to be used for the calculations based on formula (11.6) are $z_{2,0.3;0.90} = 1.6069$, $z_{2,0.5;0.90} = 1.5770$, and $z_{2,0.8;0.90} = 1.4931$, respectively.

Let us now assume for illustrative purposes that a standardized treatment group difference of $\Delta_{1,A}/\sigma_1 = 0.3$ is anticipated for the change in ADAS-cog and of $\Delta_{2,A}/\sigma_2 = 0.2$ for the change in ADCS-ADL. Then, the sample size resulting from the approaches based on Bonferroni adjustment of the type II error rate and on the assumption of independence of the endpoints and equal allocation of the power result in the same sample size as for equal standardized effects $\Delta_{1,A}/\sigma_1 = \Delta_{2,A}/\sigma_2 = 0.2$, i.e. 1300 and 1292. The reason for this is that the sample size necessary for the endpoint with the smaller effect is needed to assure the aspired conjunctive power, and this sample size leads to an individual power for the other endpoint of more than 0.999. The approach of Varga et al. (2017) allocates higher power to the endpoint with the larger effect $(1 - \beta_1 = 0.9982, 1 - \beta_2 = 0.9019)$ while assuring a conjunctive power of $1 - \beta = (1 - \beta_1) \cdot (1 - \beta_2) = 0.90$ under the assumption of independent endpoints. The resulting total sample size amounts to 1058. If a correlation of 0.3, 0.5, and 0.8, respectively, between the endpoints is employed in the calculations based on the evaluation of (11.4), a total sample size of 1056, 1054, and 1052 results (Hamasaki et al. 2018, Table 2, p. 35). Note that in the situation of endpoints with unequal effect sizes the reduction in sample size as compared to assuming independent endpoints is negligible and considerably smaller as compared to the situation of equal effect sizes of the endpoints.

Finally, we assume standardized treatment group differences of $\Delta_{1,A}/\sigma_1 = 0.4$ and $\Delta_{2,A}/\sigma_2 = 0.2$. Again, the first two approaches lead to the same sample sizes as in the above scenarios because they apply the same power and because the sample size is defined by the one required for the endpoint with the smaller standardized effect. The method of Varga et al. (2017) allocates power values of close to 1 to the first endpoint and $1 - \beta_2 = 0.9003$ to the second resulting in a total sample size of 1052. Employing correlations in sample size calculation and evaluating (11.4) leads to the same sample size of 1052 for all non-negative correlations (Hamasaki et al. 2018, Table 2, p. 35). The total sample size of 1052 is just the number of subjects required to assure an individual power of 0.90 for rejecting the null-hypothesis corresponding to the endpoint with the smaller standardized effect size at two-sided level $\alpha = 0.05$. This sample size also provides a conjunctive power of 0.90 as the individual power for the other endpoint with a much larger effect is nearly 1. The results for this example are summarized in Table 11.2.

(b) *"At least one" success criterion*

In case of the "at least one" success criterion, the local type I error rate needs to be adjusted in order to control the familywise error rate in the strong sense. Numerous multiple test procedures exist dealing with this requirement. Here, we base our considerations on the generally applicable single-step Bonferroni approach where each individual test is performed at local level α/k. This multiple test procedure is conservative, especially if the test statistics are highly positively correlated (Dmitrienko et al. 2010; Sozu et al. 2015). If the correlation is known, this conservatism can be removed by adjusting the critical boundaries (James 1991; Bang et al. 2005). However, the actual type I error rate of this approach is sensitive to

Table 11.2 Total sample size for Example 11.1 when applying different methods for sample size calculation ($\alpha/2 = 0.025$, one-sided, conjunctive power $1 - \beta = 0.90$, $r = n_E/n_C = 1$)

Standardized difference	Bonferroni adjustment of type II error rate	Method by Varga et al. (2017)	Calculation acc. to formula (11.4) assuming correlation ρ			
			0.0	0.3	0.5	0.8
$\Delta_{1,A}/\sigma_1 = 0.2$, $\Delta_{2,A}/\sigma_2 = 0.2$	1300	1292	1292	1274	1252	1194
$\Delta_{1,A}/\sigma_1 = 0.3$, $\Delta_{2,A}/\sigma_2 = 0.2$	1300	1058	1292	1056	1054	1052
$\Delta_{1,A}/\sigma_1 = 0.4$, $\Delta_{2,A}/\sigma_2 = 0.2$	1300	1052	1292	1052	1052	1052

misspecifications of the correlations and this method is therefore not considered here.

Using the notations and assumptions introduced above, the disjunctive power to reject at least one individual null-hypothesis under a specified alternative hypothesis $H_{1,A}$ defined by $\boldsymbol{\Delta}_A$ when using the Bonferroni adjustment is given by

$$\Pr\nolimits_{H_{1,A}} \left(\bigcup_{i=1}^{k} \{|Z_i| \geq z_{1-\alpha/2k}\} \right) \approx \Pr\nolimits_{H_{1,A}} \left(\bigcup_{i=1}^{k} \{Z_i \geq z_{1-\alpha/2k}\} \right)$$

$$= 1 - \Pr\nolimits_{H_{1,A}} \left(\bigcap_{i=1}^{k} \{Z_i < z_{1-\alpha/2k}\} \right)$$

$$= 1 - \Pr\nolimits_{H_{1,A}} \left(\bigcap_{i=1}^{k} \left\{ Z_i - \sqrt{\frac{r}{(1+r)^2} \cdot n} \cdot \left(\frac{\Delta_{i,A}}{\sigma_i}\right) \right. \right.$$

$$\left. \left. < z_{1-\alpha/2k} - \sqrt{\frac{r}{(1+r)^2} \cdot n} \cdot \left(\frac{\Delta_{i,A}}{\sigma_i}\right) \right\} \right)$$

$$= 1 - \Pr\nolimits_{H_{1,A}} \left(\bigcap_{i=1}^{k} \{Z_i^* < c_i^*\} \right), \tag{11.8}$$

where $\mathbf{Z}^* = \left(Z_1^*, \ldots, Z_k^*\right)$ is as defined in Sect. 11.3.1 and follows under $H_{1,A}$ the k-variate standardized normally distribution with expectation vector zero and off-diagonal elements of the covariance matrix ρ_{ij} being equal to the correlation between endpoints i and j, $i \neq j$, and where $c_i^* = z_{1-\alpha/2k} - \sqrt{\frac{r}{(1+r)^2} \cdot n} \cdot \left(\frac{\Delta_{i,A}}{\sigma_i}\right)$. The sample size required to assure the aspired disjunctive power $1 - \beta$ can thus be obtained by using numerical integration to determine the smallest integer for which the expression (11.8) is at least $1 - \beta$, i.e. for which

$$\Pr\nolimits_{H_{1,A}} \left(\bigcap_{i=1}^{k} \{Z_i^* < c_i^*\} \right) \leq \beta. \tag{11.9}$$

Again, as for the "all or none" success criterion, an explicit formula for the total required sample size can be derived for the case that the standardized effects are equal for all endpoints, i.e. $(\Delta_{i,A}/\sigma_i) = (\Delta/\sigma)_A$ for all $i = 1, \ldots, k$. Then, $c_i^* = z_{1-\alpha/2k} - \sqrt{\frac{r}{(1+r)^2} \cdot n} \cdot \left(\frac{\Delta}{\sigma}\right)_A = c^*$ for all $i = 1, \ldots, k$, and inequality (11.9) is fulfilled if $c^* \leq z_{k,\rho_{ij};\beta}$. Therefore, the following formula provides the total sample size required to assure a disjunctive power of $1-\beta$ in case of equal effect sizes $(\Delta/\sigma)_A$ for all endpoints when the Bonferroni test procedure is applied in the analysis:

$$ n = \frac{(1+r)^2}{r} \cdot \left(z_{1-\alpha/2k} - z_{k,\rho_{ij};\beta}\right)^2 \cdot \left(\frac{\sigma}{\Delta}\right)_A^2. \tag{11.10} $$

Table 11.1, fourth column, shows the total required sample size calculated according to formula (11.10) for a desired disjunctive power of $1 - \beta = 0.90$ and the case of $k = 2$ to 5 endpoints with equal standardized effects $(\Delta/\sigma)_A = 0.2$ and equal correlation between the endpoints, if equal sample size allocation and a one-sided level of $\alpha/2 = 0.025$ is applied. The pattern observed for the conjunctive power is reversed to that seen for the disjunctive power (third column). Now, for fixed correlation, the required sample size decreases with increasing number of endpoints. Furthermore, in contrast to the conjunctive power, the sample size required for the disjunctive power is decreasing with increasing correlation between the endpoints when keeping the number of endpoints fixed. This is due to the fact that for increasing correlation it becomes more difficult that at least one of the test statistics falls above the critical value. As for the conjunctive power, the impact is generally the highest for similar effect sizes of the endpoints. Further investigations on the behaviour of the conjunctive and disjunctive power in case of multiple endpoints were presented by Senn and Bretz (2007).

Example 11.2 For illustrative purposes, we again consider the **Alzheimer's disease trial** by Green et al. (2009) and now assume that sample size is calculated to assure a disjunctive power of $1 - \beta = 0.90$ when applying the Bonferroni procedure at one-sided familywise type I error rate $\alpha/2 = 0.025$. For equal standardized treatment group differences for both endpoints $\Delta_{1,A}/\sigma_1 = \Delta_{2,A}/\sigma_2 = 0.2$, the total required balanced sample size calculated by evaluating inequality (11.9) by numerical integration or by applying formula (11.10) amounts to 740, 838, 910, and 1044 for correlations $0.0, 0.3, 0.5$, and 0.8, respectively. The related quantiles of the bivariate standard normal distribution to be imputed in Eq. (11.10) are $z_{2,0.0;0.10} = -0.4783$, $z_{2,0.3;0.10} = -0.6501$, $z_{2,0.5;0.10} = -0.7721$, and $z_{2,0.8;0.10} = -0.9869$, respectively. Figure 11.1 shows the total required sample size to achieve a conjunctive power or disjunctive power, respectively, of $1 - \beta = 0.90$ for equal standardized effect sizes $\Delta_{1,A}/\sigma_1 = \Delta_{2,A}/\sigma_2 = 0.2$ depending on the correlation between the two endpoints. Note that here and elsewhere in the book, if not stated otherwise, figures showing sample sizes in dependence of determining quantities do not restrict to integer values for the sample size but are based on unrounded values to avoid "jumps" of the graphs. Therefore, minor differences to numbers given in the text may occur.

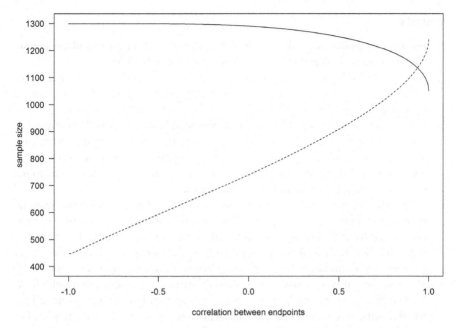

Fig. 11.1 Total sample size to achieve a conjunctive power (solid line) or disjunctive power (dahed line), respectively, of $1 - \beta = 0.90$ in case of two endpoints with equal standardized differences $\Delta_{1,A}/\sigma_1 = \Delta_{2,A}/\sigma_2 = 0.2$ depending on the correlation between the endpoints ($\alpha/2 = 0.025$, one-sided; see Example 11.2)

For standardized effects of $\Delta_{1,A}/\sigma_1 = 0.2$ and $\Delta_{2,A}/\sigma_2 = 0.3$, the corresponding total sample sizes for the correlations $0.0, 0.3, 0.5$, and 0.8, respectively, are 444, 488, 516, and 548, and for $\Delta_{1,A}/\sigma_1 = 0.2$ and $\Delta_{2,A}/\sigma_2 = 0.4$, the numbers are 278, 296, 304, and 312 (Sozu et al. 2015, Table 5.1, p. 64). The more the standardized treatment group differences differ between the endpoints, the more the burden of providing at least one rejection is shifted to the endpoint with the largest effect size. As a consequence, for such scenarios the sample size calculated for this endpoint for a significance level α/k and an individual power $1 - \beta$ is a good approximation to the sample size actually required in case of positive correlations. For example, for $\Delta_A/\sigma = 0.4$, the total sample size to achieve a power of 0.90 at one-sided level $0.025/2 = 0.0125$ is 312 which is quite close to the numbers reported above for the setting $\Delta_{1,A}/\sigma_1 = 0.2$ and $\Delta_{2,A}/\sigma_2 = 0.4$ and correlations ≥ 0.3.

Remarks

1. For a single endpoint, i.e. $k = 1$, the situation is that of the z-test and the sample size formulas (11.6) and (11.10) coincide with formula (3.5).
2. Sozu et al. (2015) give a comprehensive overview on sample size calculation for clinical trials with continuous or binary multiple endpoints including software programs.
3. Offen et al. (2007) considered medical and statistical issues with multiple co-primary endpoints. Among other things, they presented examples for disease areas where regulatory agencies require to apply two or more co-primary endpoints, and they give estimates for the correlations between the endpoints that may be helpful for sample size calculation.
4. In specific settings, multiple endpoints can be combined to a single outcome by building a composite endpoint. Here, the overall impact of an intervention is captured by including several events of interest in one variable. For example, in the REPRISE III trial comparing two transcatheter aortic valve replacement devices (see Example 8.3), the primary efficacy endpoint was a composite of all-cause mortality, disabling stroke, and moderate or greater paravalvular leak. By combining several endpoints to a single one, multiplicity aspects are eliminated. Furthermore, combining components with small effect sizes may lead to a composite with a larger effect and thus the required sample size may become smaller. Methods as well as R code for sample size calculation for binary and time-to-event composite endpoints are presented in the monograph of Rauch et al. (2018).

11.4   More Than Two Groups

11.4.1   Background and Notation

In Chap. 7, methods for sample size calculation were presented for the comparison of more than two groups. The resulting sample sizes assure a pre-specified power for the rejection of the global null-hypothesis stating that there is no difference between any of the multiple treatment groups. However, one is usually not only interested in whether there is any difference but which of the treatment groups differ. This question can be addressed under control of the familywise error rate in the strong sense by applying an appropriate multiple test procedure. As for multiple endpoints, many multiple test procedures are available when analyzing multi-armed trials. Here, we restrict again to single-step procedures where all involved null-hypotheses are simultaneously tested at a level which is known already in the planning stage. This makes sample size calculation much simpler than for stepwise procedures with data-dependent local levels.

Basically, two types of questions can be distinguished which may be addressed when performing multi-armed trials. In so-called many-to-one comparisons, k of the $k + 1$ included treatments are compared to a common control. For example, in placebo-controlled dose-finding studies, it is often the main interest to demonstrate efficacy of the active groups by comparing each of them to placebo. If, in contrast, not only comparisons to a defined treatment group but between all arms are of interest, this is denoted as the pairwise comparisons approach leading to $k \cdot (k + 1)/2$ null-hypotheses to be assessed. In the following, we illustrate for the case of many-to-one comparisons and a normally distributed endpoint how sample size calculation can be improved upon the general applicable methods by exploiting the specific characteristics of the involved test statistics. Hints to methods for pairwise comparisons will be given at the end of this section. Furthermore, sample size calculation procedures for the important application of pairwise comparisons within the so-called "gold standard" non-inferiority design are presented in Chap. 9.

We consider the many-to-one comparisons situation with a normally distributed outcome and assume independent random samples $X_{ij}, i = 0, \ldots, k, j = 1, \ldots, n_i$, from normal distributions $N(\mu_i, \sigma^2), i = 0, \ldots, k$, with common and known variance σ^2. The aim is to compare $\mu_1, \ldots, \mu_k$ against μ_0. If higher values of the outcome correspond to favorable results, the related k one-sided test problems are given by

$$H_{0,i}: \mu_i - \mu_0 \leq 0$$
$$\text{versus} \tag{11.11a}$$
$$H_{1,i}: \mu_i - \mu_0 > 0, \quad i = 1, \ldots, k,$$

and the analogous two-sided test problems are

$$H_{0,i}: \mu_i - \mu_0 = 0$$
$$\text{versus} \tag{11.11b}$$
$$H_{1,i}: \mu_i - \mu_0 \neq 0, \quad i = 1, \ldots, k.$$

We restrict in the following to the more convenient one-sided approach. The z-test statistic can be used for the assessment of the null-hypotheses which is given by

$$Z_{i,0} = \sqrt{\frac{r_i}{1 + r_i}} \cdot n_0 \cdot \frac{\bar{X}_i - \bar{X}_0}{\sigma}, \quad i = 1, \ldots, k,$$

where

$$\bar{X}_i = \frac{1}{n_i} \cdot \sum_{j=1}^{n_i} X_{ij} \text{ and } r_i = n_i/n_0, i = 0, \ldots, k.$$

In contrast to the multiple endpoint situation, the correlation between the test statistics is known here: It can easily be seen that $\text{Corr}(Z_{i,0}, Z_{j,0}) = \rho_{ij} = \sqrt{\frac{r_i \cdot r_j}{(1 + r_i) \cdot (1 + r_j)}}$, $i, j = 1, \ldots, k, i \neq j$. Hence, if the allocation ratio relative to the control group is constant, i.e. $r_i = n_i/n_0 = r$ for $i = 1, \ldots, k$, the correlations are equal to $r/(r + 1)$

which equals to 0.5 in case of balanced groups. Knowledge of the correlations enables to improve the Bonferroni procedure for the many-to-one comparison situation.

In the next section, the most popular single-step test procedure for many-to-one comparisons, the Dunnett test, and related methods for sample size calculation are presented.

11.4.2 Dunnett Test

Test procedure
When testing the k one-sided null-hypotheses $H_{0,i}$ in (11.11a), the familywise type I error rate is controlled in the strong sense at level $\alpha/2$ by the following multiple test procedure proposed by Dunnett (1955): "Reject all null-hypotheses $H_{0,i}$ with $z_{i,0} > c$, accept all $H_{0,i}$ with $z_{i,0} \leq c$", where the critical value c for the test statistics is determined such that

$$\Pr_{H_0:\bigcap_{i=1}^{k} H_{0,i}} \left(\bigcup_{i=1}^{k} \{Z_{i,0} > c\} \right) = \alpha/2$$

or equivalently

$$\Pr_{H_0:\bigcap_{i=1}^{k} H_{0,i}} \left(\bigcap_{i=1}^{k} \{Z_{i,0} \leq c\} \right) = 1 - \alpha/2. \tag{11.12}$$

Validity of Eq. (11.12) does not only guarantee control of the type I error rate under the global null-hypothesis H_0 but assures control of the familywise type I error rate in the strong sense, i.e. control is guaranteed independent of which of the $H_{0,i}$ are true or false. This can be seen as follows. Let us assume that all $H_{0,i}$ are true for $i \leq t$ and false for $i > t, t < k$. Then a type I error occurs if $z_{i,0} > c$ for at least one $i \leq t$. Therefore, the probability that a type I error occurs is given by

$$\Pr_{\bigcap_{i=1}^{t} H_{0,i}} \left(\bigcup_{i=1}^{t} \{Z_{i,0} > c\} \right) \leq \Pr_{\bigcap_{i=1}^{k} H_{0,i}} \left(\bigcup_{i=1}^{k} \{Z_{i,0} > c\} \right) = \alpha/2.$$

Under the global null-hypothesis $H_0: \bigcap_{i=1}^{k} H_{0,i}$, $\mathbf{Z} = (Z_{1,0}, \ldots, Z_{k,0})$ follows the k-variate standardized normal distribution with expectation vector zero and off-diagonal elements of the covariance matrix being equal to the correlations $\rho_{ij}, i \neq j$. Therefore, it follows from (11.12) that the critical value c for the one-sided Dunnett test at level $\alpha/2$ is given by the one-sided equicoordinate (i.e. the quantiles in each dimension coincide) $(1 - \alpha/2)$-quantile of the k-variate standard normal distribution with ρ_{ij} as matrix of correlation coefficients; analogously to Sect. 11.3, this quantile is in the following denoted by $z_{k,\rho_{ij};1-\alpha/2}$. The critical value for the test statistics of the Dunnett procedure is smaller than that for the Bonferroni procedure and therefore it is uniformly more powerful. For example, for $k = 2$ and balanced groups (i.e. $\rho_{ij} = \rho = 0.5$), the critical value of the Dunnett procedure at one-sided level $\alpha/2 = 0.025$ amounts to $z_{2,0.5;0.975} = 2.2121$ as compared to the Bonferroni boundary $z_{0.9875} = 2.2414$. For $k = 3$ and 4, respectively, the corresponding critical values are $z_{3,0.5;0.975} = 2.3490$ versus $z_{0.9917} = 2.3940$ and $z_{4,0.5;0.975} = 2.4418$ versus $z_{0.99375} = 2.4977$.

In the following, we present methods for sample size calculation when the many-to-one comparisons are performed with the Dunnett test.

Sample size calculation

For sample size calculation, an alternative $H_{1,A}$ has to be specified by fixing $\mathbf{\Delta}_A = (\Delta_{1,A}, \ldots, \Delta_{k,A})$ with $\Delta_{i,A} = (\mu_i - \mu_0)_A$, $i = 1, \ldots, k$. Without loss of generality, we assume $\Delta_{i,A} > 0$ for all $i = 1, \ldots, k$ as one would hardly include an experimental treatment in a superiority trial which is assumed to be not better than the control. Under $\mathbf{\Delta}_A$, the vector of centered test statistics $\mathbf{Z}^* = (Z_1^*, \ldots, Z_k^*)$ with $Z_{i,0}^* = Z_{i,0} - \sqrt{\frac{r_i}{1+r_i}} \cdot n_0 \cdot \left(\frac{\Delta_{i,A}}{\sigma}\right)$, $i = 1, \ldots, k$, follows the k-variate standardized normal distribution with expectation zero and off-diagonal elements of the covariance matrix ρ_{ij} being equal to the correlation between the test statistics $Z_{i,0}$ and $Z_{j,0}$, $i \neq j$, i.e. $\rho_{ij} = \sqrt{\frac{r_i \cdot r_j}{(1+r_i) \cdot (1+r_j)}}$. Analogously to the case of multiple endpoints considered in Sect. 11.3, the conjunctive or disjunctive power, and the other way round the related sample sizes, can be determined by numerical integration of the multivariate standard normal distribution, now employing the decision boundaries of the Dunnett test. Further details are given in the next paragraphs.

(a) *Study aim: rejection of all null-hypotheses (conjunctive power)*

The conjunctive power under an alternative $H_{1,A}$ specified by $\mathbf{\Delta}_A$ when performing the k many-to-one comparisons with the Dunnett test at level $\alpha/2$ is given by

$$\mathrm{Pr}_{H_{1,A}}\left(\bigcap_{i=1}^{k} Z_{i,0} > z_{k,\rho_{ij};1-\alpha/2}\right)$$

$$= \left(\mathrm{Pr}_{H_{1,A}}\left\{\bigcap_{i=1}^{k} Z_{i,0} - \sqrt{\frac{r_i}{1+r_i}} \cdot n_0 \cdot \left(\frac{\Delta_{i,A}}{\sigma}\right) > z_{k,\rho_{ij};1-\alpha/2} \right.\right.$$

$$\left.\left. - \sqrt{\frac{r_i}{1+r_i}} \cdot n_0 \cdot \left(\frac{\Delta_{i,A}}{\sigma}\right)\right\}\right)$$

$$= \mathrm{Pr}_{H_{1,A}}\left(\bigcap_{i=1}^{k} \{Z_{i,0}^* > c_i^*\}\right).$$

Here, $\mathbf{Z}^* = (Z_{1,0}^*, \ldots, Z_{k,0}^*)$ is defined as above and follows under the alternative the k-variate standard normal distribution with expectation vector zero and off-diagonal elements of the covariance matrix being equal to $\rho_{ij} = \sqrt{\frac{r_i \cdot r_j}{(1+r_i) \cdot (1+r_j)}}$ and $c_i^* = z_{k,\rho_{ij};1-\alpha/2} - \sqrt{\frac{r_i}{1+r_i}} \cdot n_0 \cdot \left(\frac{\Delta_{i,A}}{\sigma}\right)$. The required total sample size to achieve a power of $1 - \beta$ is therefore given by the smallest integer n for which

$$\text{Pr}_{H_{1,A}}\left(\bigcap_{i=1}^{k}\{Z_{i,0}^* \geq c_i^*\}\right) > 1 - \beta, \tag{11.13}$$

which can determined by numerical integration.

For equal allocation ratios $r_i = n_i/n_0 = r$ and equal differences between experimental treatments and control $\Delta_{i,A} = \Delta_A$ for all $i = 1, \ldots, k$, we have $\rho = r/(r + 1)$ and $c_i^* = z_{k,\rho;1-\alpha/2} - \sqrt{\frac{r}{(1+r)\cdot(1+r\cdot k)}} \cdot n \cdot \left(\frac{\Delta_A}{\sigma}\right) = c^*$. Therefore, the left part in (11.13) is equal to the right one if $-c^* = z_{k,\rho;1-\beta}$. Solving this equation results in the following explicit formula for the total required sample size

$$n = \frac{(1 + r) \cdot (1 + r \cdot k)}{r} \cdot \left(z_{k,\rho;1-\alpha/2} + z_{k,\rho;1-\beta}\right)^2 \cdot \left(\frac{\sigma}{\Delta_A}\right)^2, \tag{11.14}$$

where $\rho = r/(r + 1)$ with $r = n_i/n_0$ for all $i = 1, \ldots, k$.

(b) *Study aim: rejection of at least one null-hypothesis (disjunctive power)*

The disjunctive power under an alternative $H_{1,A}$ specified by Δ_A when performing the k many-to-one comparisons with the Dunnett test at level $\alpha/2$ is given by

$$1 - \text{Pr}_{H_{1,A}}\left(\bigcap_{i=1}^{k}\{Z_{i,0} \leq z_{k,\rho_{ij};1-\alpha/2}\}\right)$$
$$= 1 - \text{Pr}_{H_{1,A}}\left(\bigcap_{i=1}^{k}\{Z_{i,0}^* \leq c_i^*\}\right),$$

where $Z_{i,0}^*$ and c_i^* are defined as above for the conjunctive power and where $\mathbf{Z}^*$ follows under $H_{1,A}$ the same multivariate normal distribution as specified there. The total required sample size can be obtained analogously as above by determining the smallest integer n for which

$$1 - \text{Pr}_{H_{1,A}}\left(\bigcap_{i=1}^{k}\{Z_{i,0}^* \leq c_i^*\}\right) \geq 1 - \beta$$

i.e.

$$\text{Pr}_{H_{1,A}}\left(\bigcap_{i=1}^{k}\{Z_{i,0}^* \leq c_i^*\}\right) \leq \beta. \tag{11.15}$$

Inequality (11.15) can be solved by numerical integration.

Again, for equal allocation ratios $r_i = n_i/n_0 = r$ and equal differences between experimental treatments and control $\Delta_{i,A} = \Delta_A$ for all $i = 1, \ldots, k$, an explicit formula for the required sample size can be derived. We then have $\rho = r/(r+1)$ and $c_i^* = z_{k,\rho;1-\alpha/2} - \sqrt{\frac{r}{(1+r)\cdot(1+r\cdot k)}} \cdot n \cdot \left(\frac{\Delta_A}{\sigma}\right) = c^*$. Therefore, inequality (11.15) is fulfilled if $c^* = z_{k,\rho;\beta}$ and solving this equation results in the following formula for the total required sample size

$$n = \frac{(1+r)\cdot(1+r\cdot k)}{r} \cdot \left(z_{k,\rho;1-\alpha/2} - z_{k,\rho;\beta}\right)^2 \cdot \left(\frac{\sigma}{\Delta_A}\right)^2, \qquad (11.16)$$

where $\rho = r/(r+1)$ with $r = n_i/n_0$ for all $i = 1, \ldots, k$.

Table 11.3 considers the situation that $k = 1$ to 5 groups are compared to the control. Shown are the total sample sizes which are required according to formulas (11.14) or (11.16), respectively, to obtain a conjunctive or disjunctive power of $1 - \beta = 0.90$ for equal allocation ratios $r = n_i/n_0$ in case of equal standardized differences $(\Delta/\sigma)_A = 0.5$ between the groups $i, i = 1, \ldots, k$, and the control.

Table 11.3 Total sample size to achieve a conjunctive or disjunctive power of $1 - \beta = 0.90$, respectively, for comparing k groups to a control with equal standardized difference between the k groups and the control $(\Delta/\sigma)_A = 0.5$ and equal allocation ratios $r = n_i/n_0, i = 1, \ldots, k$ ($\alpha/2 = 0.025$, one-sided)

k	r	Conjunctive power $1 - \beta = 0.90$			Disjunctive power $1 - \beta = 0.90$		
		n	n_0	n_i	n	n_0	n_i
1	1	170	85	85	170	85	85
	2	191	127	64	191	127	64
2	$1/k = 0.5$	352	176	88	203	101	51
	$1/\sqrt{k} = 0.707$	341	141	100	202	84	59
	1	345	115	115	216	72	72
	2	418	84	167	285	57	114
3	$1/k = 0.333$	557	278	93	233	116	39
	$1/\sqrt{k} = 0.577$	511	187	108	232	85	49
	1	536	134	134	268	67	67
	2	669	96	191	382	55	109
4	$1/k = 0.25$	776	388	97	261	129	33
	$1/\sqrt{k} = 0.5$	689	229	115	258	86	43
	1	735	147	147	320	64	64
	2	936	104	208	477	53	106
5	$1/k = 0.2$	1008	503	101	287	142	29
	$1/\sqrt{k} = 0.447$	861	266	119	282	87	39
	1	942	157	157	372	62	62
	2	1216	111	221	572	52	104

Note that when calculating the sample sizes, the unrounded n is used to calculate $n_0 = \left(\frac{1}{1+k\cdot r}\right) \cdot n$ and $n_i = \left(\frac{r}{1+r}\right) \cdot n$, and the returned values are both rounded up to the next integer. The resulting allocation ratio might then be slightly different to the specified allocation ratio.

Fixing the number of groups and the allocation ratio, the required sample size is smaller for the disjunctive power than for the conjunctive power as the underlying power concept constitutes a lower hurdle for success. For fixed number of groups, the sample size shows for both the conjunctive and the disjunctive power a U-shape with smallest sample size for the square-root allocation $n_i/n_0 = 1/\sqrt{k}$. For this allocation ratio, the sample size of the control group and, to a less extent, of the groups, $i, i = 1, \ldots k$ increase with increasing k when considering the conjunctive power. In contrast, for the disjunctive power, n_0 hardly changes while n_i decreases when increasing the number of groups.

Example 11.3 The **ChroPac trial** (see Sect. 1.2.1) compared two surgical treatments in patients with chronic pancreatitis. Let us assume for illustrative purposes that within the study two experimental interventions are compared to a common control and that a clinically relevant difference of $\Delta_{1,A} = \Delta_{2,A} = \Delta_A = 10$ between the experimental groups and the control is assumed for the primary outcome, the average physical functioning score 24 months after surgery. A common standard deviation of $\sigma = 20$ in all three groups is stipulated, and a conjunctive or disjunctive power, respectively, of $1 - \beta = 0.90$ is desired.

For balanced groups, i.e. $r_i = n_i/n_0 = 1$ for all $i = 1, 2$, the critical value for the test statistics when applying the single-step Dunnett procedure at one-sided level $\alpha/2 = 0.025$ is $z_{2,\rho_{bal};0.975} = 2.2121$ with $\rho_{bal} = 0.5$. The square-root allocation rule assigns equal sample size $n_i, i = 1, \ldots, k$, to the treatments and $n_0 = \sqrt{k} \cdot n_i$ to the control, i.e. $r = 1/\sqrt{k}$. In the situation at hand ($k = 2$), the ratio n_i/n_0 amounts to $r = 0.707$ for this allocation rule and the critical value is given by $z_{2,\rho_{sq};0.975} = 2.2206$ with $\rho_{sq} = 1/(1 + \sqrt{2}) = 0.414$. For sample size calculation, the quantiles $z_{2,\rho_{bal};0.90} = 1.5770$ and $z_{2,\rho_{bal};0.10} = -0.7721$ (balanced groups) or $z_{2,\rho_{sq};0.90} = 1.5915$ and $z_{2,\rho_{sq};0.10} = -0.7186$ (square-root allocation), respectively, are additionally required. Then, for balanced allocation total sample sizes of $n = 345 (= 3 \times 115)$ according to (11.14) or $n = 216 (= 3 \times 72)$ according to (11.16), respectively, are required to assure a conjunctive or disjunctive power of 0.90. For the square-root allocation, the corresponding sample sizes are $n = 341 (= 141 + 2 \times 100)$ for the conjunctive and $n = 202 (= 84 + 2 \times 59)$ for the disjunctive power, which are smaller than for balanced groups. Figure 11.2 shows the (unrounded) total sample sizes to achieve a conjunctive or disjunctive power of 0.90 under the above assumptions which are obtained from Eqs. (11.14) and (11.16), respectively, depending on the allocation ratio.

For the conjunctive power, the minimum value amounts to 338.46 which is achieved for $r = 0.756$, and the sample size is at most 339 within the range $r \in [0.695, 0.821]$. If the total sample size shall be minimized, an allocation ratio within this range can be chosen for which the sample sizes (rounded up to integers) sum up to 339. This holds, for example, true for the allocation ratios $r = 0.702$

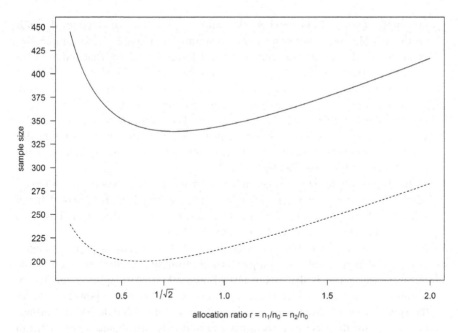

Fig. 11.2 Total sample size to achieve a conjunctive power (solid line) or disjunctive power (dahed line), respectively, of $1 - \beta = 0.90$ for the Dunnett test comparing $k = 2$ groups to a common control with equal standardized differences between the k groups and the control $\Delta_{1,A}/\sigma = \Delta_{2,A}/\sigma = 0.5$ depending on the allocation ratio ($\alpha/2 = 0.025$, one-sided; see Example 11.3)

(leading to $339 = 141 + 2 \times 99$) or $r = 0.719$ ($339 = 139 + 2 \times 100$). The curve for the disjunctive power is also quite flat in the neighborhood of the minimum 199.88 which is attained for $r = 0.593$. The total sample sizes resulting from Eq. (11.16) are at most 200 for $r \in [0.566, 0.623]$. The minimum sample size can, for example, be obtained by applying the allocation ratios $r = 0.587$ (resulting in the partition $200 = 92 + 2 \times 54$) or $r = 0.611$ (leading to $200 = 90 + 2 \times 55$).

If the generally applicable methods based on Bonferroni adjustment of the type I and type II error rate, respectively, are applied (see Sect. 11.2) and if sample size formula (3.5) for the z-test is used, the resulting total sample size for balanced groups amounts to $363 = 3 \times 121$ or $300 = 3 \times 100$, respectively, for a conjunctive or disjunctive power of 0.90. The imputed quantiles are then $z_{1-0.025/2} = 2.2414$ and $z_{1-0.10/2} = 1.6449$ (conjunctive power) or $z_{1-0.025/2} = 2.2414$ and $z_{1-0.10} = 1.2816$ (disjunctive power), respectively.

Remarks

1. For $k = 1$, the situation corresponds to that of the z-test, and accordingly formulas (11.13) and (11.16) are identical to sample size formula (3.5).
2. The Dunnett test procedure and related methods for sample size calculation can also be derived for the case of unknown variance. Then, the multivariate t-distribution (instead of the multivariate standard normal distribution) is involved

in sample size calculation, where the number of degrees of freedom depends on the sample size. Therefore, iterative solution is required. The sample size obtained for the known variance case is a lower bound for that required for unknown variance. In fact, for the situations they considered, Horn and Vollandt (1998) found that both sample sizes were often equal and rarely differed by more than one per group in case of equally balanced sample size.

3. For the Dunnett procedure, the square-root allocation rule (i.e. equal n_i, $i = 1, \ldots, k$, and $n_0 = \sqrt{k} \cdot n_i$) is approximately equal to the optimal sample size allocation minimizing the required total sample size for fixed conjunctive power [see Dunnett (1955) and Tamhane (1987)].

4. Methods for sample size calculation for a Dunnett-type test procedure for binary outcomes and for nonparametric analysis can be found in Horn and Vollandt (2000). Related procedures for time-to-event outcomes and analysis based on the log-rank test were derived by Jung et al. (2008).

5. *Pairwise comparisons*: For normally distributed outcomes, a popular single-step procedure for pairwise comparisons controlling the familywise error rate in the strong sense is the Tukey test (Tukey 1953). An explicit sample size formula for this approach assuring a pre-specified conjunctive power exists for the special case of three arms, i.e. $k = 2$ (Horn and Vollandt 2000). Furthermore, also for the case of three arms and normally distributed endpoints, Liu (1996) provided methods for sample size calculation for the step-down procedure of Newman (1939) and Keuls (1952) and the step-up procedure of Welsch (1977). The overviews by Horn and Vollandt (2000, 2001) give useful collections of sample size formulas for pairwise comparisons including also binary outcomes and nonparametric approaches. For those situations for which an analytical determination of the required sample size is intractable or impossible, the simulation-based approach described in Chap. 16 is an attractive option.

Chapter 12
Assessment of Safety

12.1 Background and Notation

When investigating a new treatment, both efficacy and safety are of interest. In most clinical trials in phase II and III, the primary endpoint measures efficacy, and therefore sample size calculation is based on this outcome. However, there are also situations were the primary objective in these trials concerns safety. Furthermore, even if sample size calculation is based on an efficacy endpoint, one is commonly interested which extent of information can be gained on safety aspects with the number of patients included in the study. Phase IV clinical trials, also denoted as post-marketing surveillance trials, are performed after a treatment is approved and when it is on the market. Here, primary focus is usually on the investigation of safety aspects of a new treatment. Due to limitations of clinical trials conducted before approval with respect to, for example, sample size, included patient population, and follow-up time, the information on the safety profile of a novel therapy is often uncomplete when it enters the market. Aims of phase IV trials are, amongst others, to identify side effects not observed in earlier trials or to estimate the risk of a known adverse event with higher precision. Adequate sample size is required to achieve these tasks. An overview by Zhang et al. (2016) showed, however, that too small sample sizes are a major concern in phase IV trials evaluating drug safety.

In this section, we consider a binary endpoint measuring an undesirable incident, such as an adverse event. For example, in oncology trials the Common Terminology Criteria of Adverse Events (CTCAE) is frequently used for the reporting of adverse events. It provides a systematic classification of adverse events together with a grading (National Institute of Health 2017). We assume in the following that the sum of observed events X follows the binomial distribution $Bin(n, p)$. The setting of a single group, commonly the patients receiving the experimental treatment, is considered throughout this chapter. The situation of sample size calculation for the comparison of two or more groups with respect to a binary outcome can be addressed

© Springer Nature Switzerland AG 2020

M. Kieser, *Methods and Applications of Sample Size Calculation and Recalculation in Clinical Trials*, Springer Series in Pharmaceutical Statistics,
https://doi.org/10.1007/978-3-030-49528-2_12

with the methods presented in Chap. 5 or Sect. 7.4, respectively. In the following sections, three approaches to the analysis of such outcomes and related methods for sample size or power calculation are presented.

12.2   Testing Hypotheses on the Event Probability

In this section, we present an exact and an approximate test for the assessment of the two-sided test problem

$$H_0: p = p_0$$
$$\text{versus} \tag{12.1a}$$
$$H_1: p \neq p_0$$

or the related one-sided problem

$$H_0: p \geq p_0$$
$$\text{versus} \tag{12.1b}$$
$$H_1: p < p_0$$

together with the associated methods for sample size calculation. Here, $p_0, 0 < p_0 < 1$, denotes a specified event probability. The considered tests are dual to the confidence intervals presented in the next section in that they lead to the same decisions about rejecting or accepting H_0 (see first paragraph in Sect. 12.3).

12.2.1   Exact Binomial Test

The exact binomial test is the most commonly applied procedure for the assessment of (12.1a) and (12.1b), respectively. Let denote $k_{L,\alpha/2}$ and $k_{U,\alpha/2}$ the largest or smallest integer, respectively, for which

$$\Pr\left(X \leq k_{L,\alpha/2}\right) = \sum_{i=1}^{k_{L,\alpha/2}} \binom{n}{i} \cdot p_0^i \cdot (1 - p_0)^{n-i} \leq \alpha/2$$

and

$$\Pr\left(X \geq k_{U,\alpha/2}\right) = \sum_{i=k_{U,\alpha/2}}^{n} \binom{n}{i} \cdot p_0^i \cdot (1 - p_0)^{n-i} \leq \alpha/2.$$

Then the set of integers $\{k: k \leq k_{L,\alpha/2} \text{ or } k \geq k_{U,\alpha/2}\}$ defines the rejection region for a two-sided test of H_0: $p = p_0$ at level α, and the set $\{k: k \leq k_{L,\alpha/2}\}$ defines a one-sided level-$\alpha/2$ test for the assessment of (12.1b). For given sample size n and specified alternative $p_A \neq p_0$, the power can be calculated by summing up the probabilities under the alternative of those observations that lead to rejection of the null-hypothesis. For the two-sided case, the power is thus given by

$$\text{Power}(n, p_A) = \sum_{i=1}^{k_{L,\alpha/2}} \binom{n}{i} \cdot p_A^i \cdot (1 - p_A)^{n-i} + \sum_{i=k_{U,\alpha/2}}^{n} \binom{n}{i} \cdot p_A^i \cdot (1 - p_A)^{n-i},$$

$$(12.2)$$

and for the one-sided test at level $\alpha/2$ the power is just the first summand of (12.2). The required sample size can be determined by setting the left side of (12.2) equal to the desired power $1 - \beta$ and resolving the equation for n.

12.2.2 Approximate Score Test

The test statistic of the score test is given by $Z = \sqrt{n} \cdot \frac{\hat{p} - p_0}{\sqrt{p_0 \cdot (1 - p_0)}}$. This test statistic is related with that of the z-test because it is assumed that the variance of the nominator is known. Under the null-hypothesis, Z is approximately standard normally distributed. Under an alternative specified by a value $p_A \neq p_0$, it holds true that $\hat{p} - p_0 \sim N\left(p_A - p_0, \frac{p_A \cdot (1 - p_A)}{n}\right)$. Using the notations $\sigma_0^2 = \frac{p_0 \cdot (1 - p_0)}{n}$, $\sigma_A^2 = \frac{p_A \cdot (1 - p_A)}{n}$, and $\Delta_A = (p_A - p_0)$, we can apply the method for sample size determination described in Sect. 5.2.1. An approximate sample size formula can be obtained by resolving Eq. (5.7), i.e. $\Delta_A = \sigma_0 \cdot z_{1-\alpha/2} + \sigma_A \cdot z_{1-\beta}$, for n. In the current situation, the equation is given by

$$(p_A - p_0) = \sqrt{\frac{p_0 \cdot (1 - p_0)}{n}} \cdot z_{1-\alpha/2} + \sqrt{\frac{p_A \cdot (1 - p_A)}{n}} \cdot z_{1-\beta},$$

which results in the following sample size formula:

$$n = \frac{\left(z_{1-\alpha/2} \cdot \sqrt{p_0 \cdot (1 - p_0)} + z_{1-\beta} \cdot \sqrt{p_A \cdot (1 - p_A)}\right)^2}{(p_A - p_0)^2}.$$

$$(12.3)$$

Table 12.1 shows the sample size required to reject $H_0 : p \geq p_0$ at one-sided level $\alpha/2 = 0.025$ with the exact binomial test and the score test, respectively, for some null- and alternative hypotheses defined by p_0 and $p_A < p_0$, respectively.

Table 12.1 Sample size for exact binomial test (n_{bin}) and score test (n_{score}) to reject the null-hypothesis H_0: $p \geq p_0$ at one-sided level $\alpha/2 = 0.025$ with power $1 - \beta = 0.80$ for the alternative defined by p_A

p_0	p_A	n_{bin}	n_{score}
0.1	0.05	231	239
0.2	0.15	466	471
0.2	0.1	107	108
0.3	0.25	633	638
0.3	0.2	155	153
0.3	0.1	36	34
0.4	0.35	744	742
0.4	0.3	181	182
0.4	0.2	45	43
0.5	0.45	786	783
0.5	0.4	199	194
0.5	0.3	49	47
0.6	0.55	767	761
0.6	0.5	195	191
0.6	0.4	49	48
0.7	0.65	689	676
0.7	0.6	183	172
0.7	0.5	47	44
0.8	0.75	540	528
0.8	0.7	144	137
0.8	0.6	41	36
0.9	0.85	341	316
0.9	0.8	94	86
0.9	0.7	29	24

Note that the sample sizes for the parameter constellations (p_0, p_A) and $(1 - p_0, 1 - p_A)$ are equal

12.3 Estimating the Event Probability with Specified Precision

Point estimates and confidence intervals are commonly used to quantify the probability of a defined undesirable event. The precision of the estimate can be judged by the width of a two-sided $(1 - \alpha)$ confidence interval. In the following, both an exact confidence interval that guarantees control of the confidence level $1 - \alpha$ and an approximate confidence interval are presented with associated methods for sample size calculation. As strict control of the coverage probability is frequently associated with an undershoot of the confidence level below the nominal value $1 - \alpha$, the price is an increased width of the confidence interval. Therefore, the approximate methods are preferred as long as a potential violation of the confidence level is negligible. The most appropriate approach should be chosen based on a careful evaluation for the situation at hand. For methodology and results of such investigations see, for example, Fagerland et al. 2017, page 42ff.

The two methods for calculating confidence intervals which we consider here can be obtained by inverting tests for hypotheses on a binomial probability. Therefore, the decisions made by applying the two-sided test at level α to the null-hypothesis $H_0 : p = p_0$ or by looking whether or not p_0 is contained in the related two-sided $(1 - \alpha)$ confidence interval are equivalent: H_0 is rejected at two-sided level α by the test if and only if the two-sided $(1 - \alpha)$ confidence interval excludes the value p_0.

12.3.1 Exact Clopper-Pearson Confidence Interval

The most common exact confidence interval for a binomial probability is that proposed by Clopper and Pearson (1934). In the sense mentioned in the previous paragraph, it is the counterpart of the exact binomial test (see Sect. 12.2.1). The two-sided exact Clopper-Pearson confidence interval at confidence level $1 - \alpha$ is obtained by inverting two one-sided exact binomial tests at level $\alpha/2$. An explicit representation of the lower (LCL_{CP}) and upper (UCL_{CP}) confidence limit of the two-sided $(1 - \alpha)$ Clopper-Pearson confidence interval is given by

$$\text{LCL}_{CP} = \text{Beta}(\alpha/2; k, n - k + 1)$$
$$\text{UCL}_{CP} = \text{Beta}(1 - \alpha/2; k + 1, n - k),$$

where $\text{Beta}(\gamma; a, b)$ is the γ-quantile of the beta distribution with parameters a and b, n is the number of patients, and k is the number of observed events. Therefore, the width of the confidence interval is given by

$$W_{CP}(n, k) = \text{Beta}(1 - \alpha/2; k + 1, n - k) - \text{Beta}(\alpha/2; k, n - k + 1)$$

and for event probability p the expected width is

$$EW_{CP}(p, n) = E(W_{CP}(n, k)) = \sum_{k=0}^{n} \Pr(X = k) \cdot W_{CP}(n, k)$$
$$= \sum_{k=0}^{n} \binom{n}{k} \cdot p^k \cdot (1 - p)^{n-k} \cdot W_{CP}(n, k). \quad (12.4)$$

For specified p and n, formula (12.4) allows to calculate the precision of the estimate of p in terms of the expected width of the $(1 - \alpha)$ Clopper-Pearson confidence interval. The other way round, if a defined expected width is desired, the required sample size can be calculated by resolving (12.4) for n.

12.3.2 Approximate Wilson Score Confidence Interval

The Wilson score interval (Wilson 1927) is obtained by inverting the approximate score test (see Sect. 12.2.2). The lower and upper confidence limits of the two sided $(1 - \alpha)$ Wilson score confidence interval can be expressed as

$$\text{LCL}_{\text{Wilson}} = \frac{2 \cdot n \cdot \hat{p} + z^2_{1-\alpha/2} - z_{1-\alpha/2} \cdot \sqrt{z^2_{1-\alpha/2} + 4 \cdot n \cdot \hat{p} \cdot (1 - \hat{p})}}{2 \cdot \left(n + z^2_{1-\alpha/2}\right)}$$

$$\text{UCL}_{\text{Wilson}} = \frac{2 \cdot n \cdot \hat{p} + z^2_{1-\alpha/2} + z_{1-\alpha/2} \cdot \sqrt{z^2_{1-\alpha/2} + 4 \cdot n \cdot \hat{p} \cdot (1 - \hat{p})}}{2 \cdot \left(n + z^2_{1-\alpha/2}\right)},$$

where n denotes the number of patients, k the number of observed events, and $\hat{p} = k/n$ the ML-estimator of p. The width of the two-sided $(1 - \alpha)$ Wilson score confidence interval is thus given by

$$W_{\text{Wilson}}(n, k) = \frac{z_{1-\alpha/2} \cdot \sqrt{z^2_{1-\alpha/2} + 4 \cdot n \cdot \hat{p}(1 - \hat{p})}}{n + z^2_{1-\alpha/2}}$$

and the expected width for event probability p and sample size n can be calculated according to Eq. (12.4) by just replacing $W_{\text{CP}}(n, k)$ by $W_{\text{Wilson}}(n, k)$. To assure that the expected width is not larger than a pre-specified amount, the required sample size n can be determined by solving Eq. (12.4) (with W_{CP} replaced by W_{Wilson}) for n. Table 12.2 shows the expected width of the 95% Clopper-Pearson and Wilson score confidence intervals for some scenarios. Throughout, the expected width of the Wilson score confidence interval is smaller than that of the Clopper-Pearson interval, but the difference is decreasing with increasing sample size. Figure 12.1 shows the required sample size to achieve a pre-specified expected width of the Clopper-Pearson or Wilson score confidence interval, respectively. Consistent with the results shown in Table 12.2, the sample size required for the Wilson score interval is always smaller than for the Clopper-Pearson confidence interval.

Example 12.1 Vemurafenib is an orally available BRAF kinase inhibitor for which efficacy was demonstrated in patients with metastatic melanoma that has a BRAF^{V600} mutation. An open-label, multicentre **safety study including patients with advanced metastatic melanoma** with BRAF^{V600} mutations was performed in 44 countries. The aim of this trial was to investigate whether the safety results

Table 12.2 Expected width of two-sided 95% Clopper-Pearson (EW_{CP}) and Wilson score (EW_{Wilson}) confidence interval for event probability p and sample size n

p	n	EW_{CP}	EW_{Wilson}
0.1	50	0.1805	0.1665
0.2		0.2335	0.2149
0.3		0.2645	0.2438
0.4		0.2814	0.2595
0.5		0.2868	0.2646
0.1	100	0.1256	0.1177
0.2		0.1640	0.1543
0.3		0.1864	0.1759
0.4		0.1985	0.1876
0.5		0.2024	0.1914
0.1	250	0.0779	0.0744
0.2		0.1025	0.0985
0.3		0.1168	0.1126
0.4		0.1246	0.1203
0.5		0.1271	0.1228

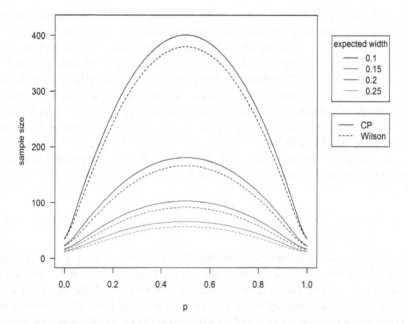

Fig. 12.1 Sample size to achieve a specified expected width of the two-sided 95% Clopper-Pearson (CP) or Wilson score (Wilson) confidence interval for event probability p

from registration trials can be extrapolated to clinical practice and to determine the frequencies of rare adverse events (Larkin et al. 2014). The sample size should allow to estimate an adverse event incidence of $p = 0.01$ with a 95% confidence interval of width 0.007. With the sample size of 3300 patients planned for this study, the expected width of the 95% Clopper-Pearson confidence interval amounts for $p = 0.01$ to 0.0071 and that of the Wilson score confidence interval to 0.0069.

Remark As a result of comprehensive evaluations of its characteristics, Fagerland et al. (2017) recommended the Wilson score test and the associated confidence interval as methods to be applied when testing hypotheses or calculating confidence intervals for a binomial probability. Especially, these methods perform better than the more widespread Wald test or Wald confidence interval, where for calculation of the test statistic the variance of $\hat{p}$ is estimated by employing the ML-estimate of p instead of p_0 in the variance expression.

12.4   Observing at Least One Event with Specified Probability

Due to the limited number of patients under exposure, only the most frequent adverse events are identified when a new drug is registered and comes into market. This is one of the reasons for the necessity of post-marketing surveillance studies. Here, treatments are monitored systematically for a large number of patients under conditions that are used in practice. The choice of the sample size of such studies may be guided by the following considerations. These thoughts give also insight in the occurrence probabilities of risks that may be detected in studies for which the sample size is determined based on other considerations as, for example, to achieve the desired power for the proof of efficacy.

If an event of interest occurs with probability p and if n patients are under exposure, then the probability to observe at least one event is given by

$$\Pr(X \geq 1) = 1 - \Pr(X = 0)$$
$$= 1 - (1 - p)^n. \tag{12.5}$$

The other way round, if the desired probability to observe at least one event which occurs with probability p is specified, the required sample size can be obtained by resolving (12.5) for n: The sample size assuring that this probability is at least $1 - \beta$ is (with eventual up-rounding to the next integer) given by $n = \frac{\ln \beta}{\ln(1-p)}$. If the study is conducted with sample size n and no adverse event of a specific kind is observed, then it can be concluded that this is either due to chance (for risk probability p, this chance is less or equal to β), or the risk of such an event is smaller than p. The following Table 12.3 gives some numerical examples for this approach. Especially, it can be seen that the chance to detect risks with small occurrence probability is

Table 12.3 **a** Sample size n to observe an event with occurrence probability p at least once with probability 0.80 (0.90). **b** Probability to observe an event with occurrence probability p at least once ($\Pr(X \geq 1)$) for sample size n

a

p	n
0.01	161 (230)
0.005	322 (460)
0.001	1609 (2302)
0.0005	3219 (4605)
0.0001	16,094 (23,025)

b

p	n	$\Pr(X \geq 1)$
0.01	100	0.6340
0.005		0.3942
0.001		0.0952
0.0005		0.0488
0.0001		0.0100
0.01	500	0.9934
0.005		0.9184
0.001		0.3936
0.0005		0.2212
0.0001		0.0488
0.01	1000	>0.9999
0.005		0.9933
0.001		0.6323
0.0005		0.3935
0.0001		0.0952

rather low with sample sizes commonly used in phase I–III trials thus demonstrating the need of post-marketing surveillance studies. Note that Table 12.3a shows that a sample size of about $1.6/p$ or $2.3/p$ is required to observe an event with occurrence probability p at least once with probability 0.80 or 0.90, respectively. For a detection probability of 0.95, the required sample size is about $3/p$ as already mentioned by Eypasch et al. (1995).

There are obvious relationships between some numbers depicted in Table 12.3. The connection between sample sizes shown in Table 12.3a follows from the approximation $\ln(1 - k \cdot p) \approx k \cdot \ln(1 - p)$ that holds true for "small" values of p and appropriate k. From this, it follows that $n(k \cdot p) \approx (1/k) \cdot n(p)$. Furthermore, Taylor expansion and further approximations result in $(1 - p)^k \approx 1 - k \cdot p + (k \cdot p)^2 - (k \cdot p)^3$ for "small" values of p and appropriate k. Therefore $\Pr(X \geq 1)(n, p) = 1 - (1 - p)^k \approx k \cdot p - (k \cdot p)^2 + (k \cdot p)^3 \approx \Pr(X \geq 1)(n/k, k \cdot p)$.

Example 12.2 In the **York acupuncture safety study**, the type and frequency of adverse events after acupuncture were investigated (MacPherson et al. 2001). The study was performed as a prospective survey. All professional acupuncturists who were member of the British Acupuncture Council and who were practicing in the United Kingdom were invited to record details of adverse events occurring after treatment during a four week period. The required sample size was calculated to assure a probability of 0.95 to observe at least one of those adverse events with occurrence probability $p = 1/10,000 = 0.0001$. This results in a sample size of $n = \ln(0.05)/\ln(0.9999) = 29,956$, which was rounded up to 30,000. Note that the result of this calculation fits well to the rule of thumb mentioned above: For detecting an adverse event with occurrence probability $p = 1/m$ with probability 0.95, a sample size of about $3 \cdot m$ is required.

Chapter 13
Cluster-Randomized Trials

13.1 Background and Notation

There is consensus that randomized trials are the gold standard for the evaluation of new therapies or health interventions. In some situations, it is preferable or necessary not to randomize individual patients but groups of patients, for example clinics, care homes, or households. These groups are in the following denoted as "clusters". There are a number of textbooks (for example Donner and Klar 2000; Hayes and Moulton 2009; Eldridge and Kerry 2012; Campbell and Walters 2014) and overviews (for example Turner et al. 2017a, b) on the rationale, design, analysis, and reporting of cluster-randomized trials and we restrict in the following to methods for sample size calculation for some basic scenarios.

The fact that patients are not randomized individually but in clusters has to be taken into account in planning and analysis as individuals of the same cluster commonly share characteristics and therefore the outcome of subjects within a cluster tend to be more similar than those from other clusters. For this reason, the extent of information obtained from the data of a patient is smaller than in case of a trial with individual randomization and this has, of course, to be taken into account in sample size calculation. Furthermore, two approaches for the analysis of cluster-randomized trials can be distinguished and the chosen method has to be adequately incorporated in sample size calculation. In cluster-level analyses (also denoted as aggregated analyses), summary measures are calculated for each cluster condensing the data to be included in the analysis to a single observation per cluster. Individual-level analyses (also referred to as non-aggregated analyses) use the individual outcomes of the subjects in the analysis. While the conduct of cluster-level analyses is straightforward, covariates on an individual level cannot be incorporated. This, however, can be done in individual-level analyses by using regression models. In the following, sample size calculation methods for both analysis approaches will be presented. We will exemplify the principle of sample size calculation for cluster-randomized trials and its differences to individually randomized trials in detail for normally distributed

© Springer Nature Switzerland AG 2020   167
M. Kieser, *Methods and Applications of Sample Size Calculation and Recalculation in Clinical Trials*, Springer Series in Pharmaceutical Statistics,
https://doi.org/10.1007/978-3-030-49528-2_13

outcomes and will then give a condensed overview of related methods for other scale
levels of the outcomes.

Throughout Sects. 13.2.1 and 13.2.2, the sample size per cluster is assumed to
be equal for all clusters and is denoted by m. Furthermore, the number of clusters
allocated to the experimental (E) and the control (C) intervention, respectively, are
denoted by K_E and K_C, the total number of clusters by $K = K_E + K_C$, the cluster
allocation ratio by $r = K_E/K_C$, and the total sample size by $n = K \cdot m$.

13.2   Normally Distributed Outcomes

We consider the same situation as in Chap. 3, namely that E and C are compared
with respect to a normally distributed outcome. The two- or one-sided test problem,
respectively, to be assessed is

$$
\begin{aligned}
H_0&: \mu_E - \mu_C = 0 \\
&\text{versus} \\
H_1&: \mu_E - \mu_C \neq 0
\end{aligned}
\tag{13.6a}
$$

or

$$
\begin{aligned}
H_0&: \mu_E - \mu_C \leq 0 \\
&\text{versus} \\
H_1&: \mu_E - \mu_C > 0
\end{aligned}
\tag{13.6b}
$$

implying that higher values of the outcome refer to better results. The following
model for the random variables $X_{ijk}, i = C, E, j = 1, \ldots, m, k = 1, \ldots, K_i$, is
assumed:

$$
X_{ijk} = \mu_i + Y_{ik} + \varepsilon_{ijk},
\tag{13.7}
$$

where $Y_{ik} \sim N\left(0, \sigma_b^2\right)$, $\varepsilon_{ijk} \sim N\left(0, \sigma_w^2\right)$ with unknown σ_b^2 and σ_w^2, and where Y_{ik} and
ε_{ijk} are independent. This means that the cluster effects Y_{ik} vary randomly within
each group between the clusters with variance σ_b^2 and expectation zero, while ε_{ijk}
vary within the clusters with variance σ_w^2 and expectation zero. In cluster-randomized
trials, the intra-cluster correlation coefficient (ICC) quantifies the extent of the within-
cluster dependence of the outcomes of subjects from the same cluster. It is given by the
proportion of the total variance $\sigma^2 = \sigma_b^2 + \sigma_w^2$ accounted for by the between-cluster
variation, i.e.

$$
\begin{aligned}
\text{ICC} &= \frac{\sigma_b^2}{\sigma_b^2 + \sigma_w^2} \\
&= \frac{\sigma_b^2}{\sigma^2}
\end{aligned}
\tag{13.8}
$$

with $\sigma^2 = \sigma_b^2 + \sigma_w^2$. Another important measure is the so-called design effect (DE) defined by

$$DE = 1 + (m - 1) \cdot ICC. \tag{13.9}$$

As $ICC \geq 0$ and $m \geq 1$, $DE \geq 1$ holds true.

13.2.1 Cluster-Level Analysis

In the cluster-level analysis, the information each cluster contributes is the mean $\bar{X}_{ik} = \frac{1}{m} \cdot \sum_{j=1}^{m} X_{ijk}$. Obviously, $E(\bar{X}_{ik}) = \mu_i$ and $Var(\bar{X}_{ik}) = \sigma_b^2 + \frac{\sigma_w^2}{m} = \sigma_b^2 + \left(\frac{\sigma^2 - \sigma_b^2}{m}\right) = \sigma^2 \cdot \frac{DE}{m}$. Thus, for the group means calculated from these summary measures it holds true that $\bar{X}_i = \frac{1}{K_i} \cdot \sum_{k=1}^{K_i} \bar{X}_{ik} \sim N\left(\mu_i, \frac{1}{K} \cdot \frac{\sigma^2}{m} \cdot DE\right)$. The following t-test statistic can be used for the assessment of the test problems (13.6a) and (13.6b):

$$T = \sqrt{\frac{K_C \cdot K_E}{K_C + K_E}} \cdot \frac{\bar{X}_E - \bar{X}_C}{S} \tag{13.10}$$

with pooled standard deviation calculated according to (3.8a) and (3.8b) taking into account that the included observations are the cluster means $\bar{X}_{ik}$. Following the same arguments as in Sect. 3.3, the distribution of T under the null- and the alternative hypothesis can be derived. For $\mu_E - \mu_C = 0$, T follows the central t-distribution with $K - 2$ degrees of freedom. Under an alternative specified by $\Delta_A = (\mu_E - \mu_C)_A \neq 0$, T follows the non-central t-distribution with $K - 2$ degrees of freedom and non-centrality parameter $nc = \sqrt{\frac{r}{(1+r)^2} \cdot K} \cdot \left(\frac{\Delta_A}{\sigma \cdot \sqrt{\frac{DE}{m}}}\right)$, where $r = K_E / K_C$. Therefore, the exact number of required clusters, each with sample size m, can be obtained as in Sect. 3.3 by iteratively searching for the smallest integer K for which

$$F_{T,K-2,0}^{-1}(1 - \alpha/2) \leq F_{T,K-2,nc}^{-1}(\beta), \tag{13.11}$$

where $F_{T,K-2,0}$ and $F_{T,K-2,nc}$ denote the cumulative distribution functions of the central and non-central t-distribution, respectively, with $K - 2$ degrees of freedom. The smallest integer K resolving (13.11) results in a total sample size of $n = K \cdot m$, group-wise number of clusters $K_E = \frac{r}{(1+r)} \cdot K$ and $K_C = \frac{1}{1+r} \cdot K$, respectively, and related sample sizes $n_E = K_E \cdot m$ and $n_C = K_C \cdot m$. These figures assure a power of $1 - \beta$ for rejecting H_0 at two-sided level α or one-sided level $\alpha/2$, respectively, if the alternative specified by Δ_A holds true.

The approximate sample size formula (3.5) for the z-test can be adapted to the current situation by just replacing "σ^2" by "$\sigma^2 \cdot \frac{DE}{m}$". This results in the approximation formula

$$K = \frac{DE}{m} \cdot \frac{(1+r)^2}{r} \cdot \left(z_{1-\alpha/2} + z_{1-\beta}\right)^2 \cdot \left(\frac{\sigma}{\Delta_A}\right)^2. \tag{13.12}$$

As $m = n/K$, it follows that

$$n = DE \cdot \frac{(1+r)^2}{r} \cdot \left(z_{1-\alpha/2} + z_{1-\beta}\right)^2 \cdot \left(\frac{\sigma}{\Delta_A}\right)^2, \tag{13.13}$$

i.e., $n = DE \cdot n_{indiv}$, where n_{indiv} denotes the sample size required in case of randomizing the subjects individually. Hence, the design effect is (approximately) just the factor by which the sample size is increased when applying cluster randomization instead of allocating the individuals randomly to the groups. It should be noted that the approximate sample size formula should, if at all, be even more carefully applied in cluster-randomized trials analyzed on a cluster-level than when randomizing patients individually as the number of involved clusters is usually considerably smaller than the number of patients commonly included in individually randomized studies (for an illustration see Example 13.1).

Example 13.1 The **ChroPac trial** compared two surgical interventions in patients with chronic pancreatitis (see Sect. 1.2.1). Primary endpoint was the mean quality of life within 24 months after surgery measured by the physical functioning score of the EORTC QLQ-C30 quastionnaire. Patients were randomized individually applying balanced allocation. Sample size calculation for the two-sided t-test at level $\alpha = 0.05$ and desired power $1-\beta = 0.90$ assuming a clinically relevant difference of $\Delta_A = 10$ and a standard deviation of $\sigma = 20$ resulted in a required total sample size of $n = 172$ (see Example 3.2). For illustrative purposes, we assume that the ChroPac trial is performed as a cluster-randomized trial with cluster allocation ratio $r = 1$. Cook et al. (2012) calculated ICCs for various outcomes from a database of surgical trials. Based on their results for quality of life outcomes, assuming ICC $= 0.01$ seems to be reasonable. For cluster size $m = 6$ (12, 24), the design effect amounts to DE $= 1.05$ (1.11, 1.23) and the exact required number of clusters according to (13.11) is $K = 32$ (18, 12) with related total sample size $n = 192$ (216, 288). Note that while for $m = 6$ and 12 the actual power is close to the desired value 0.90 (actual power 0.905), it overshoots the aspired power for $m = 24$ (actual power 0.93). The reason for this is that only multiples of m are possible as sample sizes and therefore this issue may become even more relevant for larger cluster sizes. Applying the approximate formulas (13.12) and (13.13), respectively, results in $K = 30$ (16, 10) and $n = 180$ (192, 240). This illustrates the remark made above stating that the approximate sample size formula may be quite inaccurate due to the commonly moderate number of clusters.

13.2.2 Individual-Level Analysis

A common method for individual-level analysis of cluster-randomized trials is by using Generalized Estimating Equations (GEE). This approach can account for the hierarchical structure imposed by the clustered data and can also include covariates on a cluster- or individual level. A description of GEE analysis is beyond the scope of this book and can, for example, be found in Hardin and Hilbe (2012). Shih (1997) showed that for individual-level analysis of clustered data with GEE and equal cluster size m the same approximation formula for the total sample size (13.13) as for cluster-level analysis holds true. Especially, the required sample size is again increased by the factor DE as compared to non-clustered data.

13.2.3 Dealing with Unequal Cluster Size

The sample size formulas given above assumed equal cluster size m. However, frequently the sample size per cluster is variable. As recommended by Eldridge et al. (2006), this should be taken into account in sample size calculation if the ratio of the standard deviation of the cluster size $SD(m_k)$ and the mean cluster size $\bar{m}$ exceeds 0.23. Two kinds of analytical approaches can be distinguished which require knowledge of different quantities (Rutterford et al. 2015) and which are described below.

Sample size for each cluster is known

For a cluster-level analysis which weights the cluster mean of cluster k_i with weight $m_{k_i}/(1 + (m_{k_i} - 1) \cdot \text{ICC})$ ("minimum variance weights"; Kerry and Bland 2001) or individual-level analysis with GEE and exchangeable correlation structure (see Manatunga et al. 2001), the design effect is given by

$$\text{DE} = \frac{K \cdot \bar{m}}{\sum_{i=E,C} \sum_{k_i=1}^{K_i} \frac{m_{k_i}}{1+(m_{k_i}-1)\cdot\text{ICC}}}, \tag{13.14}$$

where m_{k_i} is the sample size of cluster $k_i, i = C, E, k_i = 1, \ldots, K_i$, and $\bar{m} = \frac{1}{K} \cdot \sum_{i=E,C} \sum_{k_i=1}^{K_i} m_{k_i}$. For equal cluster size, i.e. $m_{k_i} = m$, Eq. (13.14) equals (13.9).

Mean and standard deviation of cluster sample sizes are known

If not all cluster sample sizes are known but only the mean and standard deviation of their distribution, the following design effect can be used to accommodate varying cluster sample sizes when a cluster-level analysis is performed with cluster means weighted by the cluster size:

$$\text{DE} = 1 + ((CV^2 + 1) \cdot \bar{m} - 1) \cdot \text{ICC}, \tag{13.15}$$

where again $\bar{m} = \frac{1}{K} \cdot \sum_{i=E,C} \sum_{k_i=1}^{K_i} m_{k_i}$ and $CV = \frac{\sigma_m}{\mu_m}$ with μ_m and σ_m denoting the mean and standard deviation of the distribution of the cluster sizes (Eldridge et al. 2006). As individual-level analyses are more efficient than cluster-level analyses using the cluster sizes as weights (Lake et al. 2002), application of the design effect given in (13.15) provides a conservative approach for these analyses.

Remark Shih and Lee (2018) proposed a simulation-based approach for cluster-randomized trials with unequal cluster size that can be applied for different types of outcomes (see Chap. 16 for a general framework for simulation-based sample size determination). Here, prior information about the cluster sizes and the ICC can be incorporated by specifying probability functions for these quantities.

13.3 Other Scale Levels of the Outcome

Multiplying the sample size required for an individually randomized trial by the "standard" design effect $DE = 1 + (m - 1) \cdot ICC$ is an appropriate approach for cluster-randomized trials with constant cluster size not only for the unadjusted analysis of continuous outcomes but also for a number of other situations. For *analysis of covariance (ANCOVA)*, the sample size required for a cluster-randomized trial can be obtained by inflating the sample size resulting from the approximation formula (3.17) (Sect. 3.4) by DE (Teerenstra et al. 2012). For *ordered categorical data*, the sample size provided by (4.8) (Sect. 4.3.2) multiplied by DE gives adequate sample size for cluster-randomized trials analyzed on an individual level by use of a mixed model (Campbell and Walters 2014; for mixed model analysis see, for example, Stroup 2013). For *binary outcomes*, multiplication of the approximate sample size provided by formula (5.8) (Sect. 5.2.1) with DE results in appropriate sample sizes for analysis on cluster level (Donner et al. 1981) and for analysis on individual level with GEE (Shih 1997). Furthermore, the standard design effect DE can be used for *time-to-event data* to inflate the sample size obtained by Schoenfeld's approach (see Sect. 6.3.1) if the analysis is performed on cluster level with a weighted log-rank test (Gangnon and Kosorok 2004).

Remarks

1. If several observation units are captured for each patient, the same cluster structure as considered above holds true and the aforementioned methods can be applied accordingly. Examples for this situation are dentistry studies where the condition of several teeth is measured for all trial participants or ophthalmology trials where observations are taken from both eyes of each subject.

2. When calculating the sample size for cluster-randomized trials, information on the ICC is required. Appropriate values can be obtained from systematic investigations or reviews of cluster-randomized trials [see, for example, Adams et al. (2004), Thompson et al. (2012) and the systematic reviews reported in van Breukelen and Candel (2012)].

3. Further methods and aspects of sample size calculation for cluster-randomized trials can be found in the comprehensive review by Rutterford et al. (2015).

Chapter 14
Multi-Regional Trials

14.1 Background and Notation

Ideally, beneficial drugs are made available to patients globally at the same time. However, regional differences may lead to differences in drug effect and therefore results observed in one region may not necessarily be extrapolated to another. The ICH E5(R1) Guideline *Ethnic Factors in the Acceptability of Foreign Clinical Data* (1998) distinguishes extrinsic factors (such as medical practice and health-care system but also conventions in designing and conducting clinical trials) and intrinsic factors (such as genetic factors and ethnicity). Performing separate drug development programs in regions for which different drug effects may occur requires huge resources. Therefore, in the past drugs have first been approved in those regions where the clinical trial data have been captured leading to a time lag in approval for the other regions. For example, a document prepared by the Japanese Ministry of Health, Labour and Welfare (MHLW), which is the regulatory authority in Japan, states an aggravating delay of several years of new drug approval in Japan compared to other countries (MHLW 2007). To reduce this delay, the concept of multi-regional trials was discussed in the *Questions & Answers* document referring to the ICH E9 Guideline (ICH 2006). A multi-regional clinical trial (also denoted as global clinical trial) is conducted in several regions using a common protocol. The aims of a multi-regional trial are twofold: (a) to demonstrate efficacy for the entire study population across all involved regions; (b) to demonstrate consistency of results between regions. If part (b) is accomplished, i.e., if the assessment of consistent results between regions is successful, it is established that the drug effect is not sensitive to ethnic factors. Then, the global results obtained in part (a) can be transferred to the local regions. Two concepts are commonly used for the assessment of consistency which were provided in the document *Basic Principles on Global Trials* of the Japanese Ministry of Health, Labour and Welfare (2007), and which are commonly denoted as Method 1 and 2. Method 1 refers to the situation of proving consistency of the results obtained for a specific pre-defined region with the global results. This is done by demonstrating

© Springer Nature Switzerland AG 2020

M. Kieser, *Methods and Applications of Sample Size Calculation and Recalculation in Clinical Trials*, Springer Series in Pharmaceutical Statistics,
https://doi.org/10.1007/978-3-030-49528-2_14

that a pre-specified fraction of the observed global effect is retained for this region. Method 2 demands to establish a similar tendency for the effect estimates of all involved regions.

In the following, we consider two-arm trials with a normally distributed outcome comparing an experimental treatment (E) to placebo (C). The variance σ^2 is assumed to be known and equal for the two groups. Let K be the number of regions, n the total sample size, w_j, $j = 1, \ldots, K$, the proportion of patients in the jth region out of the total sample size, i.e. $\sum_{j=1}^{K} w_j = 1$, and n_j, $j = 1, \ldots, K$, the sample size in region j, i.e. $n_j = w_j \cdot n$. The allocation ratio is denoted by $r = n_E/n_C$, where n_E and n_C are the total sample sizes in group E and C, respectively, and n_{ij} is the sample size in group i and region j, $i = C, E$, $j = 1, \ldots, K$. We denote by $X_{ijk} \sim N(\mu_i, \sigma^2)$ the response of patient k in group i and region j, $i = C, E$, $j = 1, \ldots, K$, $k = 1, \ldots, n_{ij}$. Then, $\widehat{\Delta}_{all} = \frac{1}{n_E} \cdot \sum_{j=1}^{K} \sum_{k=1}^{n_{Ej}} X_{Ejk} - \frac{1}{n_C} \cdot \sum_{j=1}^{K} \sum_{k=1}^{n_{Cj}} X_{Cjk}$ denotes the observed overall mean difference from all regions and $\widehat{\Delta}_j = \overline{X}_{Ej} - \overline{X}_{Cj}$ the observed mean difference in region j, where $\overline{X}_{ij} = \frac{1}{n_{ij}} \cdot \sum_{k=1}^{n_{ij}} X_{ijk}$, $i = C, E$, $j \in \{1, \ldots, K\}$.

According to the above-mentioned aims of multi-regional trials, the challenges for sample size calculations are (a) to assure a defined power $1 - \beta$ for demonstrating superiority of E versus C in the entire study population across all regions at one-sided level $\alpha/2$ and (b) to assure a defined probability $1 - \beta'$ for demonstrating consistency of results across regions. According to (3.5), the total required sample size to fulfill requirement (a) for a specified alternative $\Delta_A = (\mu_{E,A} - \mu_{C,A}) > 0$ is approximately given by

$$n = \frac{(1+r)^2}{r} \cdot \left(z_{1-\alpha/2} + z_{1-\beta}\right)^2 \cdot \left(\frac{\sigma}{\Delta_A}\right)^2. \tag{14.1}$$

To fulfill part (b), the two consistency requirements mentioned above can be formulated as follows.

Method 1: For one defined region s, $s \in \{1, \ldots, K\}$ and a pre-specified value γ, $0 < \gamma < 1$, it is to be demonstrated that

$$\frac{\widehat{\Delta}_s}{\widehat{\Delta}_{all}} > \gamma. \tag{14.2}$$

Note that when applying Method 1, the *Basic Principles on Global Clinical Trials* document (MHLW 2007) recommends using values of 0.5 or more for the effect fraction γ to be retained and a probability of $1 - \beta' = 0.80$ or higher for demonstrating the consistency condition.

Method 2: It is to be demonstrated that

$$\widehat{\Delta}_j > 0 \quad \text{for all } j = 1, \ldots, K. \tag{14.3}$$

Two approaches for the probability for establishing consistency can be distinguished: the unconditional probability and the probability conditional on a statistically significant overall treatment effect. In practice, the latter one is more relevant as extrapolation of global results to regions does only matter in case that a significant effect can be shown in the entire study population. In the following, it will be shown how to calculate the sample size to fulfill requirement (b) for Method 1 and 2 as well as for both the unconditional and the conditional approach. Throughout, it is assumed that the treatment effect is the same in all regions, i.e. $\Delta_j = \Delta_{all}$ for all $j = 1, \dots, K$.

14.2 Demonstrating Consistency of Global Results and Results for a Specified Region

In this section, we consider the consistency criterion defined by Method 1. We assume that the trial is performed in two regions and that the global results shall be transferred to the pre-specified region $s \in \{1, 2\}$ with sample size $n_s = w_s \cdot n$.

Unconditional probability
It can be shown (Ikeda and Bretz 2010; Quan et al. 2010) that the unconditional probability to demonstrate (14.2) can be approximated by the following expression:

$$p_{uncond} = \Pr\left(\frac{\widehat{\Delta}_s}{\widehat{\Delta}_{all}} > \gamma\right)$$

$$\approx \Phi\left(\frac{\Delta_A \cdot (1 - \gamma) \cdot \sqrt{w_s \cdot n \cdot r}}{\sigma \cdot (1 + r) \cdot \sqrt{1 - w_s \cdot \gamma \cdot (2 - \gamma)}}\right). \tag{14.4}$$

If the total sample size n is calculated according to (14.1)

$$p_{uncond} \approx \Phi\left(\frac{(1 - \gamma) \cdot (z_{1-\alpha/2} + z_{1-\beta}) \cdot \sqrt{w_s}}{\sqrt{1 - w_s \cdot \gamma \cdot (2 - \gamma)}}\right). \tag{14.5}$$

Therefore, the fraction of the sample size to be recruited in the specified region in order to fulfill the condition $p_{uncond} \geq 1 - \beta'$ can be obtained by resolving the expression (14.5) for w_s:

$$w_s = \frac{z_{1-\beta'}^2}{(1 - \gamma)^2 \cdot (z_{1-\alpha/2} + z_{1-\beta})^2 + z_{1-\beta'}^2 \cdot \gamma \cdot (2 - \gamma)}. \tag{14.6}$$

Note that the sample size portion of the specified region does neither depend on the quantities Δ_A and σ assumed for sample size calculation tailored to demonstrate a global effect nor on the allocation ratio r. Figure 14.1 shows the portion of the total sample size required for the specified region depending on the effect fraction γ for

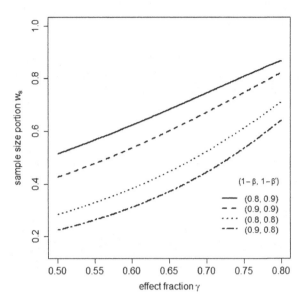

Fig. 14.1 Portion of the total sample size for the specified region w_s to assure an unconditional probability $1 - \beta'$ for demonstrating consistency of global results and results for a specified region (Method 1) depending on the effect fraction γ. The total sample size is calculated to achieve the power $1 - \beta$ for demonstrating a significant overall result at one-sided level $\alpha/2 = 0.025$

various combinations of the power $1 - \beta$ to demonstrate a significant overall result at one-sided level $\alpha/2 = 0.025$ and the unconditional probability $1 - \beta'$. Keeping the other quantities fixed, w_s increases with increasing γ and $1 - \beta'$, but decreases with increasing $1 - \beta$.

Conditional probability

The probability of obtaining consistent results between the overall study population and the specified region conditional on demonstrating significant superiority of E to C in all patients is given by

$$
\begin{aligned}
p_{cond} &= \Pr\left(\frac{\widehat{\Delta}_s}{\widehat{\Delta}_{all}} > \gamma, \; \sqrt{\frac{r}{(1+r)^2} \cdot n} \cdot \left(\frac{\widehat{\Delta}_{all}}{\sigma}\right) > z_{1-\alpha/2}\right) \\
&\approx \frac{\Pr\left(\widehat{\Delta}_s - \gamma \cdot \widehat{\Delta}_{all} > 0, \; \widehat{\Delta}_{all} - z_{1-\alpha/2} \cdot \sigma \cdot \sqrt{\frac{(1+r)^2}{r \cdot n}} > 0\right)}{\Pr\left(\widehat{\Delta}_{all} - z_{1-\alpha/2} \cdot \sigma \cdot \sqrt{\frac{(1+r)^2}{r \cdot n}} > 0\right)} \\
&= \frac{p_{joint}}{p_{sig}}.
\end{aligned}
\tag{14.7}
$$

Note that the second line is an approximation as there is a non-zero probability (which, however, is very small for situations typically met in clinical trials) for the event $\widehat{\Delta}_{all} \leq 0$ even if $\Delta_A > 0$. It can be shown (Quan et al. 2010) that

$$p_{joint} = \Pr(Z_1 > c_1, Z_2 > c_2)$$

where $\mathbf{Z} = (Z_1, Z_2)$ follows the two-dimensional standard normal distribution with correlation $\rho = \frac{(1-\gamma)\cdot\sqrt{n_s}}{\sqrt{n+\gamma\cdot(\gamma-2)\cdot n_s}} = \frac{(1-\gamma)\cdot\sqrt{w_s}}{\sqrt{1+\gamma\cdot(\gamma-2)\cdot w_s}}$ and where $c_1 = -\rho\cdot\sqrt{\frac{r}{(1+r)^2}}\cdot n\cdot\left(\frac{\Delta_A}{\sigma}\right)$ and $c_2 = z_{1-\alpha/2} - \sqrt{\frac{r}{(1+r)^2}}\cdot n\cdot\left(\frac{\Delta_A}{\sigma}\right)$. Plugging in (14.1) for the total sample size results in the simplified expressions $c_1 = -\rho\cdot\left(z_{1-\alpha/2} + z_{1-\beta}\right)$ and $c_2 = -z_{1-\beta}$. Therefore, p_{joint} can be calculated by integration over the density of a two-dimensional standard normal distribution. Furthermore,

$$p_{sig} = \Phi\left(\sqrt{\frac{r}{(1+r)^2}}\cdot n\cdot\left(\frac{\Delta_A}{\sigma}\right) - z_{1-\alpha/2}\right), \tag{14.8}$$

which is equal to $1 - \beta$ if the sample size is calculated with Eq. (14.1). As a consequence, the conditional probability p_{cond} can be calculated for given ingredients. Figure 14.2 shows the conditional probability for obtaining consistent results between the overall study population and the specified region depending on the sample size portion of the specified region w_s for various combinations of power $1 - \beta$ and effect fraction γ. If all other quantities are fixed, the conditional probability increases with increasing values of w_s and $1 - \beta$ and with decreasing effect fraction γ.

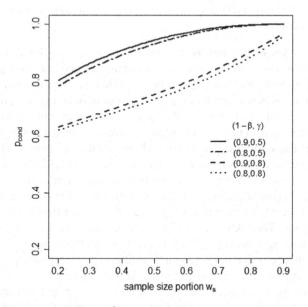

Fig. 14.2 Conditional probability $1 - \beta'$ for demonstrating consistency of global results and results for a specified region (Method 1) depending on the sample size portion of the specified region w_s. The total sample size is calculated to achieve the power $1 - \beta$ for demonstrating a significant overall result at one-sided level $\alpha/2 = 0.025$, and γ denotes the effect fraction for the consistency assessment

Table 14.1 Smallest sample size portion for the specified region w_s assuring an unconditional or conditional power $1 - \beta'$, respectively, for demonstrating consistency of global results and results for a specified region (Method 1). The total sample size is calculated to achieve the power $1 - \beta$ for demonstrating a significant overall result at one-sided level $\alpha/2 = 0.025$

	γ	$1 - \beta$			
		0.80		0.90	
		$1 - \beta'$		$1 - \beta'$	
		0.80	0.90	0.80	0.90
Unconditional	0.5	0.284	0.514	0.224	0.426
	0.8	0.713	0.869	0.644	0.822
Conditional	0.5	0.230	0.422	0.201	0.382
	0.8	0.651	0.821	0.611	0.794

Table 14.1 gives the smallest sample size portion required for the specified region to assure an unconditional or conditional power $1-\beta'$, respectively, for demonstrating consistency according to Method 1 if the total sample size is calculated to assure a power $1 - \beta$ for demonstrating a significant overall result at one-sided level $\alpha/2 = 0.025$. All other quantities fixed, the sample size portion is smaller for $1 - \beta = 0.90$ than for 0.80 and larger for $1 - \beta' = 0.90$ than for 0.80. Furthermore, the sample size portion of the specified region is smaller for the conditional as compared to the unconditional approach.

Example 14.1 **Atopic dermatitis** is a chronic inflammatory skin disease presumably caused by a combination of genetic, environmental, and immunological factors. Symptoms include dry and scaly skin, redness, and itching. In a randomized, placebo-controlled trial, which was conducted across 55 sites in Australia, Canada, Germany, Japan, Poland, and the United States, the human monoclonal antibody Tralokinumab was investigated in patients with moderate to severe atopic dermatitis (Wollenberg et al. 2019a). The study was performed with three different doses of Tralokinumab and placebo and with two co-primary endpoints (see Sect. 11.3.2a for an explanation of this term). To simplify illustration, we assume that the trial includes two arms (experimental treatment and placebo) and that the change of the Eczema Area and Severity Index (EASI) between start and end of treatment (week 12) is the single primary endpoint. The EASI is a tool for measuring the severity of atopic dermatitis with scores ranging from 0 to 72 points (Tofte et al. 1998). Sample size calculation for the change of the EASI assumed a difference of $\Delta_A = 7$ points between the active treatment and placebo and a standard deviation of $\sigma = 10$ points. For balanced samples, a one-sided t-test at level $\alpha/2 = 0.025$ and a desired power of $1 - \beta = 0.80$, an approximate required total sample size of $2 \times 33 = 66$ patients is required according to formula (14.1) for demonstrating a significant overall result.

A cohort of Japanese patients was included to identify potential inter-ethnic differences. An unconditional probability $1 - \beta' = 0.80$ was desired to demonstrate that the observed effect in Japanese patients is at least half of that in the full study population, i.e. $\gamma = 0.50$ (Wollenberg et al. 2019b). Based on these specifications, the sample size portion of the specified region Japan amounts to $w_s = 0.284$ according to formula (14.6), i.e., a cohort of $2 \times 10 = 20$ Japanese patients are to be included. For this allocation of the sample size, the conditional probability of observing consistent results between the overall study population and the Japanese patients conditional on demonstrating significant superiority of E to C in all patients amounts to 0.832 when calculated according to (14.7).

Remark Quan et al. (2010) presented sample size calculation procedures for Method 1 for the case of time-to-event endpoints.

14.3 Demonstrating a Consistent Trend Across All Regions

In this section, we consider consistency assessment according to Method 2 and assume that the clinical trial is conducted in $K \geq 2$ regions.

Unconditional probability
The unconditional probability of observing a consistent trend in the estimated treatment effect across all K regions was investigated by Kawai et al. (2008). Obviously,
$\Pr\left(\widehat{\Delta}_j > 0\right) = \Phi\left(\sqrt{\frac{r}{(1+r)^2} \cdot n_j} \cdot \left(\frac{\Delta_A}{\sigma}\right)\right)$ for any $j = 1, \ldots, K$, and as the data from the regions are independent it follows

$$p_{uncond} = \Pr\left(\widehat{\Delta}_j > 0 \text{ for all } j = 1, \ldots K\right)$$
$$= \prod_{j=1}^{K} \Phi\left(\sqrt{\frac{r}{(1+r)^2} \cdot n_j} \cdot \left(\frac{\Delta_A}{\sigma}\right)\right). \tag{14.9}$$

If the sample size is calculated according to formula (14.1) to assure a power of $(1 - \beta)$ for demonstrating superiority of E versus C in the entire study population, $n_j = w_j \cdot \frac{(1+r)^2}{r} \cdot \left(z_{1-\alpha/2} + z_{1-\beta}\right)^2 \cdot \left(\frac{\sigma}{\Delta_A}\right)^2$. Therefore, expression (14.9) can be written as

$$p_{uncond} = \prod_{j=1}^{K} \Phi\left(\sqrt{w_j} \cdot \left(z_{1-\alpha/2} + z_{1-\beta}\right)\right). \tag{14.10}$$

As for Method 1, p_{uncond} does not depend on Δ_A, σ, or r. Therefore, if the total sample size is calculated for demonstrating overall efficacy, the unconditional probability for demonstrating a consistent trend across regions depends, for fixed number of

Table 14.2 Maximum unconditional probability for demonstrating a consistent trend across K regions (Method 2) which is achieved for equal sample size in all regions. The total sample size is calculated to achieve the power $1 - \beta$ for demonstrating a significant overall result at one-sided level $\alpha/2 = 0.025$

K	$1 - \beta$	
	0.80	0.90
2	0.953	0.978
3	0.850	0.911
4	0.714	0.806
5	0.574	0.682
10	0.125	0.191

regions, only on the partitioning of the total sample size among the regions. It can be shown (Li et al. 2007) that the maximum unconditional probability is achieved for equal sample size in all regions, i.e. for $w_i = 1/K$. Note that this matches with the assumption of equal treatment effects in all regions made throughout this chapter. For this partitioning, $p_{uncond} = \Phi\left(\frac{(z_{1-\alpha/2}+z_{1-\beta})}{\sqrt{K}}\right)^K$. Table 14.2 shows some numerical examples for this scenario.

It is not always possible to split the total sample size equally between regions. Therefore, the question arises how large the proportion of the smallest region has to be in order to assure an unconditional probability of at least $1 - \beta'$. Note that for any partition of the total sample size among the regions, the unconditional probability for demonstrating a consistent trend is at most as large as the figures given in Table 14.2 which considerably limits the attainable values for $1 - \beta'$. We assume without loss of generality that the proportions are ordered such that $w_1 \leq w_2 \leq \ldots \leq w_K$. For fixed w_1, $p_{uncond}(w_1, w_2, \ldots, w_K)$ as given in (14.10) is monotonically increasing with w_j for each $j = 2, \ldots, K$. Thus, to determine the minimum value for w_1 assuring an unconditional probability for demonstrating consistency, only the parameter constellation $(w_1, w_1, \ldots, 1 - (K - 1) \cdot w_1)$ has to be considered for which

$$p_{uncond} = \left(\Phi\left(\sqrt{w_1} \cdot \left(z_{1-\alpha/2} + z_{1-\beta}\right)\right)\right)^{K-1}$$
$$\cdot \Phi\left(\sqrt{1 - (K - 1) \cdot w_1} \cdot \left(z_{1-\alpha/2} + z_{1-\beta}\right)\right). \tag{14.11}$$

Note that (14.11) is defined for values $0 \leq w_1 \leq 1/(K - 1)$. However, as w_1 refers to the smallest region, it is sensible to restrict the considerations to values $0 \leq w_1 \leq 1/K$. Figure 14.3 shows for this range of values the unconditional probability for demonstrating a consistent trend across K regions depending on the sample size of the smallest region.

Conditional probability
Kawai et al. (2008) investigated characteristics of the conditional probability for the case $K = 3$ by simulations. However, a general explicit expression for this quantity

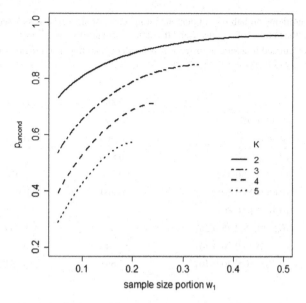

Fig. 14.3 Unconditional probability for demonstrating a consistent trend across K regions (Method 2) depending on the sample size portion of the smallest region w_1, i.e. $0 \leq w_1 \leq 1/K$. The total sample size is calculated to achieve the power $1 - \beta$ for demonstrating a significant overall result at one-sided level $\alpha/2 = 0.025$

can be derived analogously to Sect. 14.2:

$$
\begin{aligned}
p_{cond} &= \Pr\left(\widehat{\Delta}_j > 0 \text{ for all } j = 1, \ldots, K \middle| \sqrt{\frac{r}{(1+r)^2}} \cdot n \cdot \left(\frac{\widehat{\Delta}_{all}}{\sigma}\right) > z_{1-\alpha/2}\right) \\
&= \frac{\Pr\left(\widehat{\Delta}_j > 0 \text{ for all } j = 1, \ldots, K, \ \widehat{\Delta}_{all} - z_{1-\alpha/2} \cdot \sigma \cdot \sqrt{\frac{(1+r)^2}{r \cdot n}} > 0\right)}{\Pr\left(\widehat{\Delta}_{all} - z_{1-\alpha/2} \cdot \sigma \cdot \sqrt{\frac{(1+r)^2}{r \cdot n}} > 0\right)} \\
&= \frac{p_{joint}}{p_{sig}}. \tag{14.12}
\end{aligned}
$$

The probability to obtain a significant overall result is again given by (14.8). As different regions are disjunct, $\mathrm{Corr}\left(\widehat{\Delta}_j, \widehat{\Delta}_{j'}\right) = 0$ for $j \neq j'$, $j, j' \in \{1, \ldots, K\}$. Furthermore $\mathrm{Corr}\left(\widehat{\Delta}_j, \widehat{\Delta}_{all}\right) = \sqrt{n_j/n} = \sqrt{w_j}$ holds true. Therefore, the joint probability p_{joint} in the nominator of expression (14.12) is given by $p_{joint} = \Pr(Z_1 > c_1, Z_2 > c_2, \ldots, Z_K > c_K, Z_{all} > c_{all})$, where $\mathbf{Z} = (Z_1, \ldots, Z_K, Z_{all})$ follows the $(K+1)$-dimensional standard normal distribution with correlation matrix according to the expressions above and where $c_j = -\sqrt{\frac{r}{(1+r)^2} \cdot n_j} \cdot \left(\frac{\Delta_A}{\sigma}\right)$, $j =$

Table 14.3 Conditional probability for demonstrating a consistent trend across K regions (Method 2) depending on the number of regions K and the sample size portions w_j, $j = 1, \ldots, K$. The total sample size is calculated to assure a power $1 - \beta$ for demonstrating a significant overall result at one-sided level $\alpha/2 = 0.025$

K	$(w_1, \ldots, w_K)$	$1 - \beta$	
		0.80	0.90
2	(0.50,0.50)	0.992	0.995
	(0.30,0.70)	0.969	0.977
	(0.10,0.90)	0.846	0.865
3	(0.33,0.33,0.33)	0.931	0.950
	(0.25,0.25,0.50)	0.904	0.926
	(0.10,0.10,0.80)	0.712	0.746
4	(0.25,0.25,0.25,0.25)	0.818	0.859
	(0.10,0.20,0.30,0.40)	0.746	0.787
	(0.10,0.10,0.40,0.40)	0.691	0.731

$1, \ldots, K$, and $c_{all} = z_{1-\alpha/2} - \sqrt{\frac{r}{(1+r)^2} \cdot n} \cdot \left(\frac{\Delta_A}{\sigma}\right)$. Again, the expressions can be simplified by inserting the total sample size obtained from formula (14.1): $c_j = -\left(z_{1-\alpha/2} + z_{1-\beta}\right) \cdot \sqrt{w_j}$, $c_{all} = -z_{1-\beta}$, $p_{sig} = 1 - \beta$. Table 14.3 shows some numerical examples for the conditional probability of demonstrating consistency according to Method 2 for the situation that the total sample size is calculated for a power $1 - \beta = 0.80$ for demonstrating a global effect and for various combinations of the number of regions and the sample size allocation between regions. From its maximum in case of an equal split of the sample size over the regions, p_{cond} decreases the more unbalanced the allocation is.

Remarks

1. Several alternatives to the consistency criteria Method 1 and Method 2 were considered in the literature and related sample size calculation procedures were developed. Ikeda and Bretz (2010) suggested a modification of Method 1, which can be applied to various types of data, and compared its characteristics to those of Method 1. Further criteria were examined for continuous outcomes by Ko et al. (2010) and Chen et al. (2012) for two-armed trials and by Uesaka (2009) for multi-armed trials.
2. Huang et al. (2017) presented sample size calculation methods when assessing multiple co-primary endpoints (see Sect. 11.3.2a) in multi-regional trials.

Chapter 15
Integrated Planning of Phase II/III Drug Development Programs

15.1 Background and Notation

The conduct of drug development programs is extremely resource-intensive and time-consuming. These expenditures are at the same time risky as for some indication failure rates of 50% or more are reported for phase III trials and approval, respectively (Gan et al. 2012; Arrowsmith and Miller 2013). Phase II and III trials play a crucial role in this process. The aim of phase II is to provide a solid basis for a reliable decision whether to stop or continue drug development. If the results of phase II are promising enough for a go-decision, they provide the basis for planning the proceeding phase III trial whose goal is the ultimate proof of efficacy. Obviously, these requirements can be met the better, the higher the sample size in phase II is. On the other side, the available resources are always limited and, moreover, time constraints impose restrictions on the overall sample size of the phase II/III program. For these reasons, there is a strong link between phase II and III and an integrated planning is advantageous. The purpose of such an approach is to determine the sample size allocation between phase II and III as well as the go/no-go decisions in an optimal way. For this, a criterion has to be specified defining optimality. A useful approach is by considering a utility function that includes (fixed and variable per-patient) costs of the phase II/III program on the one side and expected revenues after successful launch of the drug on the market on the other. Determining the parameters such that they maximize the expected utility thus leads to the largest overall value. In the following, the principle of optimal planning of phase II/III programs in terms of sample size and go/no-go decision rule is illustrated by the example of trials with one normally distributed outcome. Note that the situation of separate phase II and phase III trials is considered here which is different to that of seamless phase II/III studies. The latter ones allow to integrate phase II and III within a single trial by implementing an adaptive two-stage design (see, for example, Bauer and Kieser 1999). Methods for sample size recalculation in adaptive two-stage design that use the interim results are presented in Part IIIB of this book.

© Springer Nature Switzerland AG 2020
M. Kieser, *Methods and Applications of Sample Size Calculation and Recalculation in Clinical Trials*, Springer Series in Pharmaceutical Statistics,
https://doi.org/10.1007/978-3-030-49528-2_15

In this chapter, we consider the scenario of a single phase II trial comparing an experimental treatment (E) with a control (C). In case of a go-decision, this trial is followed by a single phase III trial which also compares E to C and which is performed in the same population and with the same normally distributed endpoint. The total and group-wise sample sizes in phase II and III, respectively, are denoted by n_j, n_{Ej} and n_{Cj}, $j = 2, 3$, with allocation ratios $r_j = n_{Ej}/n_{Cj}$. We assume $X_{ijk} \sim N(\mu_i, \sigma^2)$, $i = C, E$, $j = 2, 3, k = 1, \ldots, n_{ij}$, with known variance σ^2 and denote the true treatment group difference by $\Delta = \mu_E - \mu_C$, where positive values of Δ correspond to higher efficacy of E.

15.2 Optimizing Phase II/III Programs

In the above-described scenario, we assume that the decision of whether drug development is continued after phase II is based on a threshold value κ for the treatment effect $\widehat{\Delta}_2$ observed in phase II. If the decision is to continue, the total sample size of the phase III trial is calculated based on the estimate $\widehat{\Delta}_2$, i.e.

$$n_3\left(\widehat{\Delta}_2\right) = \frac{(1+r_3)^2}{r_3} \cdot \left(z_{1-\alpha/2} + z_{1-\beta}\right)^2 \cdot \left(\frac{\sigma}{\widehat{\Delta}_2}\right)^2. \tag{15.1}$$

In the phase III trial, the test problem $H_0: \Delta \leq 0$ versus $H_1: \Delta > 0$ is assessed at one-sided level $\alpha/2$ using the test statistic of the z-test $Z = \sqrt{\frac{n_{C3} \cdot n_{E3}}{n_{C3} + n_{E3}}} \cdot \frac{\overline{X}_{E3} - \overline{X}_{C3}}{\sigma}$ which is standard normally distributed under H_0 (see Sect. 3.2). When optimizing phase II/III programs, the following quantities are considered:

probability of a go-decision after phase II: $p_{go} = \int_\kappa^\infty f\left(\widehat{\Delta}_2\right) \cdot d\widehat{\Delta}_2$, where $f\left(\widehat{\Delta}_2\right)$ denotes the density of the distribution of $\widehat{\Delta}_2 \sim N\left(\Delta, \frac{n_{C2} + n_{E2}}{n_{C2} \cdot n_{E2}} \cdot \sigma^2\right)$.

probability of a successful program: $\text{POSP} = p_{go} \cdot \Pr\left(Z \geq z_{1-\alpha/2}|go\right) = \int_\kappa^\infty \int_{z_{1-\alpha/2}}^\infty f(z) \cdot f\left(\widehat{\Delta}_2\right) dz d\widehat{\Delta}_2$, where $f(z)$ denotes the density of the distribution of $Z \sim N\left(\sqrt{\frac{r_3}{(1+r_3)^2} \cdot n_3} \cdot \frac{\Delta_A}{\sigma}, 1\right)$ with specified alternative $\Delta_A = (\mu_E - \mu_C)_A$ (see Sect. 3.2).

utility function u: difference between costs c for conducting the program and the prospective gain g after program success (Kirchner et al. 2016). The costs of phase II and III are composed of fixed costs c_{02} and c_{03}, respectively, and per-patient costs c_2 and c_3. As costs for phase III only incur in case of a go-decision after phase II, the costs can be expressed as

$$c(n_2, \kappa) = c_{02} + n_2 \cdot c_2 + c_{03} \cdot I_{\{go\}} + n_3 \cdot c_3 \cdot I_{\{go\}}, \tag{15.2}$$

where $I_{\{\cdot\}}$ denotes the indicator function. The gain achieved after bringing the new treatment on the market depends on the efficacy results observed in the phase III trial. There are various ways how to model this relationship. For illustration, we consider a framework inspired by the one used by the German Institute for Quality and Efficiency in Health Care (IQWIG 2017) whose assessment results are, besides others, the basis for negotiations on the reimbursement of drugs. Here, three categories of small, medium, and large treatment effects are defined by threshold values s, m and l, respectively, with $0 \leq s < m < l$. The effect of the experimental treatment is then classified as small, medium or large, respectively, if the lower boundary of the two-sided $(1 - \alpha)$ confidence interval calculated in the phase III trial falls above s, m or l. If the associated events are denoted by E_s, E_m, and E_l, respectively, and if the values for the benefit dedicated to these efficacy categories are named b_s, b_m, and b_l, the gain can be written as

$$g(n_2, \kappa) = I_{\text{go}} \cdot \left(b_s \cdot I_{E_s} + b_m \cdot I_{E_m} + b_l \cdot I_{E_l} \right). \tag{15.3}$$

expected utility:

$$E[u(n_2, \kappa)] = -E[c(n_2, \kappa)] + E[g(n_2, \kappa)], \tag{15.4}$$

where the costs $c(n_2, \kappa)$ and the gain $g(n_2, \kappa)$ are given by (15.2) and (15.3), respectively, and where the expectation is taken over $\widehat{\Delta}_2$. The expected costs are given by

$$E[c(n_2, \kappa)] = c_{02} + n_2 \cdot c_2 + c_{03} \cdot p_{\text{go}} + c_3 \cdot p_{\text{go}} \cdot E[n_3 \,|\, \text{go}].$$

Furthermore,

$$E[n_3 \,|\, \text{go}] = \frac{1}{p_{\text{go}}} \cdot \int_\kappa^\infty n_3\left(\widehat{\Delta}_2\right) \cdot f\left(\widehat{\Delta}_2\right) d\widehat{\Delta}_2, \tag{15.5}$$

where $n_3\left(\widehat{\Delta}_2\right)$ is given by (15.1).

The expected gain is given by

$$E[g(n_2, \kappa)] = E\left[I_{\text{go}} \cdot \left(b_s \cdot I_{E_s} + b_m \cdot I_{E_m} + b_l \cdot I_{E_l} \right)\right]$$
$$= p_{\text{go}} \cdot \left(b_s \cdot E[I_{E_s} | \text{go}] + b_m \cdot E[I_{E_m} | \text{go}] + b_l \cdot E[I_{E_l} | \text{go}] \right).$$

The expectations of the events to observe a result which is classified in a specific effect category can be determined as follows. The lower boundary of the two-sided $(1 - \alpha)$ confidence interval obtained in the phase III trial is given by $LB = \widehat{\Delta}_3 - z_{1-\alpha/2} \cdot \sqrt{\frac{n_{C3} + n_{E3}}{n_{C3} \cdot n_{E3}}} \cdot \sigma$, where $\widehat{\Delta}_3$ is the effect observed in phase III. Thus, the condition $a < LB \leq b$ is equivalent to $Z \in \left] \sqrt{\frac{n_{C3} \cdot n_{E3}}{n_{C3} + n_{E3}}} \cdot \frac{a}{\sigma} + z_{1-\alpha/2}, \sqrt{\frac{n_{C3} \cdot n_{E3}}{n_{C3} + n_{E3}}} \cdot \frac{b}{\sigma} + z_{1-\alpha/2} \right]$. Therefore,

$$E\left[I_{E_s} | \text{go}\right] = \frac{1}{p_{go}} \cdot \int_{\kappa}^{\infty} \int_{z_{1-\alpha/2}}^{u_m} f(z) \cdot f\left(\widehat{\Delta}_2\right) dz d\widehat{\Delta}_2 \qquad (15.6a)$$

$$E\left[I_{E_m} | \text{go}\right] = \frac{1}{p_{go}} \cdot \int_{\kappa}^{\infty} \int_{u_m}^{u_l} f(z) \cdot f\left(\widehat{\Delta}_2\right) dz d\widehat{\Delta}_2 \qquad (15.6b)$$

$$E\left[I_{E_l} | \text{go}\right] = \frac{1}{p_{go}} \cdot \int_{\kappa}^{\infty} \int_{u_l}^{\infty} f(z) \cdot f\left(\widehat{\Delta}_2\right) dz d\widehat{\Delta}_2, \qquad (15.6c)$$

where $u_i = \sqrt{\frac{r_3}{(1+r_3)^2} \cdot n_3} \cdot \frac{i}{\sigma} + z_{1-\alpha/2}$, $i = m, l$, and where it is assumed that a significant result is sufficient to be classified as a small effect (i.e. $s = 0$). Putting the above expressions together, the expected utility given in (15.4) can be calculated for any combination (n_2, κ). Maximization leads to the optimal design parameters $\left(n_2^*, \kappa^*\right)$.

Example 15.1 For illustration, we assume that an optimal **phase II/III program for a new antidepressant** is to be determined. It has to be stressed that the optimal program crucially depends on the chosen parameter values and that therefore the results presented below are not universal but hold true for the considered setting. In depression trials, efficacy is commonly measured using the 17-item version of the Hamilton Depression Rating Scale (HAM-D; Hamilton 1960). We assume the total score of the HAM-D at treatment end to be normally distributed with standard deviation $\sigma = 8$, which is a typical value observed in clinical depression trials. According to Montgomery (1994), it may be argued that a difference of 4 points on the HAM-D total score "*represents an effect that can be accepted as unequivocally clinically relevant and a difference of 3 points as probably clinically relevant.*" For an effect of $\Delta_A = 4$ points and the go/no-go scenario with threshold $\kappa = 2$ points, Table 15.1 shows the expected total sample size in phase III, the probability to go to phase III, and the probability of a successful program in dependence of the total sample size chosen for phase II. Sample size calculation for phase III is performed to assure a power of $1 - \beta = 0.90$ for the phase III trial where a one-sided level of $\alpha/2 = 0.025$ is applied in the analysis.

It can be seen that with increasing phase II sample size, the expected sample size in phase III conditional on a go-decision increases. The reason for this phenomenon is as follows. Due to the applied go/no-go decision rule, the program is only continued if the observed treatment effect lies above the threshold κ. Therefore, the treatment effect tends to be overestimated for those phase II trials for which a go-decision is

Table 15.1 Expected total sample size in phase III $E[n_3|\text{go}]$, probability to proceed to phase III p_{go}, and probability of a successful program POSP for various total sample sizes in phase II n_2 ($\Delta_A = 4, \sigma = 8, \kappa = 2, \alpha/2 = 0.025$, one-sided, $1 - \beta = 0.90, r = n_E/n_C = 1$; see Example 15.1)

n_2	$E[n_3|\text{go}]$	p_{go}	POSP
20	151	0.71	0.47
50	180	0.81	0.62
100	195	0.89	0.73
150	198	0.94	0.79
200	198	0.96	0.82

made. For large phase II trials, this upward bias is smaller due to the higher precision of the estimator. As a consequence, for higher n_2 the treatment effect $\widehat{\Delta}_2$ used for sample size calculation for phase III is less often "too large" leading overall to higher expected sample sizes $E[n_3|\text{go}]$. Furthermore, Table 15.1 shows that both the probability to proceed to phase III and the probability of a successful program are increasing with increasing sample size in phase II. This means that choosing a higher sample size in phase II results in a higher expected phase III sample size and thus in higher expected costs of the program. On the other side, these higher investments involve a higher probability of a successful program and with it higher expected revenues. Maximizing the expected utility which takes into account both costs and gain enables to find an optimal trade-off between these two aspects.

Motivated by the appraisal of Montgomery, we defined the lower boundaries of the regions defining small, medium, and large treatment effect as $s = 0, m = 3$, and $l = 5$ points, respectively. The assumed study costs are based on figures given by Sertkaya et al. (2014) who reported for central nervous system clinical trials total per-study costs of about US\$15 million in phase II and of about US\$20 million in phase III. Assuming a portion of 10% of the total budget for the fixed costs leads to $c_{02} = \text{US\$1.5 million}$ and $c_{03} = \text{US\$2 million}$, respectively. For phase II trials with 200 patients and phase III trials with 250 patients, this results in per-patient costs of $c_2 = \text{US\$67,500}$ and $c_2 = \text{US\$72,000}$. The benefit parameters are specified based on the assumption of a five-year income period and a profit margin of 20%. Moreover, it is supposed that a new antidepressant with an effect classified as at most medium will hardly achieve a higher turnover than drugs that are already on the market but lost patient exclusivity and for which annual sales in the magnitude of US\$300 million are reported (Glöckler 2015). For promising new antidepressants with outstanding product characteristics, annual revenues of US\$1 billion can be expected (Glöckler 2015). Therefore, the following two exemplary benefit scenarios are considered with associated benefit parameter values expressed in units of 10^5 US \$: $(b_s, b_m, b_l) = (500, 1000, 10{,}000)$ and $(b_s, b_m, b_l) = (750, 3000, 10{,}000)$.

Table 15.2 shows for the above-described scenarios and treatment effects $\Delta_A = 3, 4$, and 5 the optimal threshold and the optimal phase II sample size together with the related program characteristics for the optimal design: expected total sample size

Table 15.2 Optimal threshold κ^*, optimal total phase II sample size n_2^*, related expected total phase III sample size $E[n_3|go]$, probability to proceed to phase III p_{go}, probability of a successful program POSP, and maximal expected utility $E[u(n_2^*, \kappa^*)]$ in 10^5 US \$ for various treatment effects Δ_A ($\sigma = 8$, $c_{02} = 15$, $c_{03} = 20$, $c_2 = 0.675$, $c_3 = 0.72$ in 10^5 US \$, $\alpha/2 = 0.025$, one-sided, $1 - \beta = 0.90$, $r = n_E/n_C = 1$; see Example 15.1)

Benefit in 10^5 US\$						
b_s, b_m, b_l	κ^*	n_2^*	$E[n_3|go]$	p_{go}	POSP	$E[u(n_2^*, \kappa^*)]$
$\Delta_A = 3$						
(500, 1000, 10,000)	2.0	36	203	0.65	0.41	69.8
(750, 3000, 10,000)	1.6	110	314	0.82	0.64	236.5
$\Delta_A = 4$						
(500, 1000, 10,000)	1.6	86	221	0.92	0.75	235.0
(750, 3000, 10,000)	0.8	104	293	0.98	0.82	710.5
$\Delta_A = 5$						
(500, 1000, 10,000)	1.3	88	161	0.99	0.84	587.6
(750, 3000, 10,000)	0.8	76	191	0.99	0.83	1306.3

in phase III given a go-decision, probability to proceed to phase III, probability of a successful phase II/III program, and maximal expected utility.

It can be seen from Table 15.2 that for $\Delta_A = 3$ and 4, the more profitable scenario with the larger values for the benefit parameters b_1 and b_2 corresponding to larger revenues leads to larger phase II sample sizes, smaller (i.e. more liberal) thresholds, and larger probabilities for a go-decision. These findings reflect the fact that larger investments (which lead to higher program success probabilities in these scenarios) can be better compensated in case of larger returns. For larger treatment effects, for example $\Delta_A = 5$, p_{go} is already quite high for moderate values of n_2. Increasing the sample size in phase II (and with it the costs) is then no more counterbalanced to the same amount by larger revenues. This explains why n_2^* may then be smaller for the more profitable benefit scenario. For increasing treatment effect Δ_A, the maximal expected utility becomes larger which is due to an increased probability of a successful program and an increased probability of observing larger treatment effects (with the consequence of higher returns).

Figure 15.1a shows for $\Delta_A = 4$ and $(b_s, b_m, b_l) = (500, 1000, 10,000)$ the expected utility depending on the phase II sample size and the threshold.

In the above-presented approach for optimizing phase II/III programs, a fixed value Δ_A for the treatment effect was assumed. Instead, a prior distribution can be used to model the knowledge on the effect thus taking uncertainty into account (see Sect. 3.5). Figure 15.1b shows the consequence of modelling the treatment effect by a $N(\Delta_A, \tau^2)$ distribution with $\Delta_A = 4$ and $\tau^2 = 2.96$. Note that due to the relationship $\tau^2 = [\Delta_A/\Phi^{-1}(\Pr(\Delta < 0))]^2$ the value 2.96 for τ^2 corresponds for $\Delta_A = 4$ to the prior probability $\Pr(\Delta < 0) = 0.01$. Due to the introduced vagueness

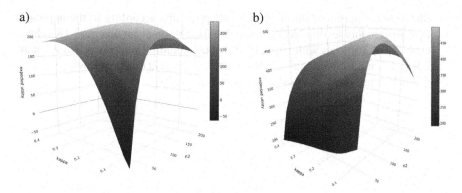

Fig. 15.1 Expected utility depending on the sample size in phase II n_2 and threshold κ used for the go/no-go decision after phase II $((b_s, b_m, b_l) = (500, 1000, 10,000)$ in 10^5 US \$; other parameter values as given in legend of Table 15.2). **a** fixed effect $\Delta_A = 4$; **b** prior distribution for effect: $N(\Delta_A, \tau^2)$ with $\Delta_A = 4$ and $\tau^2 = 2.96$ (see Example 15.1)

of the effect, the probability to go to phase II (0.80 vs. 0.92) as well as the probability of a successful program (0.59 vs. 0.75) are for the optimal design smaller when using a prior as compared to assuming a fixed value $\Delta_A = 4$. However, on the other side, treatment effects in the large-effect category are observed more frequently in phase III for the prior approach leading overall to a larger expected utility for the optimal design (498.5 vs. 235.0).

Remarks

1. Kirchner et al. (2016) applied the approach of using a prior distribution for the effect to optimize phase II/III programs to the setting of time-to-event outcomes. They considered priors that are weighted sums of two normal distributions expressing an optimistic or pessimistic view about the treatment effect.
2. Typically, the budget available for a phase II/III program is restricted which limits the costs at which a program can be performed. Such a restriction can be implemented in the above-described approach by considering only parameter values (n_2, κ) fulfilling the constraint $E[c(n_2, \kappa)] \leq C$ with upper bound C for the expected costs when maximizing the expected utility.
3. Due to the go/no-go decision rule, only promising phase II results lead to initiation of a phase III trial. Therefore, the treatment effect estimates from phase II trials that lead to go-decisions generally overestimate the true effect. If this estimate is used for sample size calculation of the phase III trial (as assumed in the approach described above), the study is underpowered. This leads to the idea of "discounting" phase II results when planning phase III trials (Kirby et al. 2012). Erdmann et al. (2020) integrated various ways of discounting in the framework of phase II/III planning and optimized the program not only over the sample size of the phase II trial and the threshold of the go/no-go decision but also over the parameter defining the extent of discounting.

4. Methods for optimization of phase II/III programs according to the approach described in this section are also available for time-to-event outcomes (Kirchner et al. 2016), for trials with normally distributed multiple endpoints (Kieser et al. 2018), for multi-armed trials (Preussler et al. 2020), and for the case where several phase III trials are performed (Preussler et al. 2019).

Chapter 16
Simulation-Based Sample Size Calculation

As we saw in the preceding chapters, closed-form sample size formulas exist for many clinical trial situations. However, there are complex study designs and analysis methods for which derivation of a closed-form expression for the required sample size is extremely challenging or is not possible at all. For example, if multiple imputation (Rubin 1996) or pattern-mixture models (Little 1993, 1994) are applied in the analysis to deal with missing values, no closed-form expressions are available for sample size calculation. Another example is the application of a stepwise multiple test procedure in the analysis, such as the Bonferroni-Holm procedure (see Sect. 11.1), where the significance levels to be applied to the single hypotheses are data-dependent and are thus not known in the planning phase. In such situations, simulations are a useful and flexible tool for determining the sample size (Arnold et al. 2011; Landau and Stahl 2012). The general simulation-based approach for a sample size calculation can be described by the following steps:

Step 1 Specify the values for the parameters, that define the data-generating model for the assumed alternative, as well as the allocation ratio. Fix the number of replications r and a starting value for the sample size n (see below for guidance for the choice of these values).

Step 2 Generate n random values from the distributions adhering to the allocation ratio defined in step 1.

Step 3 Perform the analysis specified in the trial protocol on the generated dataset and determine whether or not the null-hypothesis can be rejected at the significance level to be applied.

Step 4 Repeat steps 2 and 3 r times and estimate the power by the proportion of the number of rejections from the number of replications r.

Step 5 If the estimated power is larger than the aspired value $1 - \beta$: decrease n and repeat steps 2 through 4.
  If the estimated power is smaller than the aspired value $1 - \beta$: increase n and repeat steps 2 through 4.

© Springer Nature Switzerland AG 2020
M. Kieser, *Methods and Applications of Sample Size Calculation and Recalculation in Clinical Trials*, Springer Series in Pharmaceutical Statistics,
https://doi.org/10.1007/978-3-030-49528-2_16

The loop of steps 2 through 4 stops if the smallest n with estimated power of at least $1 - \beta$ is determined.

A reasonable starting value for n in step 2 can be obtained by using a sample size formula for a simplified model. For example, when applying a complex method for dealing with missing values one may start with the sample size required in case of no missing values which is then forecasted based on the assumed drop-out rate. When the sample size for the Bonferroni-Holm procedure is determined by simulations, one may start with the sample size required in case of Bonferroni adjustment (see Sect. 11.2.2a2), the latter being an upper bound for the former. For step 4, the following considerations can be used to determine the number of replications required to obtain a power estimate of appropriate accuracy. If we denote the true power for sample size n by $1 - \beta$ and the estimated power for sample size n based on r replications by $(\widehat{1 - \beta})_{(r)}$, then $\mathrm{Var}((\widehat{1 - \beta})_{(r)}) = (1 - \beta) \cdot \beta / r$. Therefore, an approximate two-sided $(1 - \alpha)$ confidence interval for $1 - \beta$ is given by $\left(\widehat{1 - \beta}\right)_{(r)} \pm z_{1-\alpha/2} \cdot \sqrt{\frac{(1-\beta) \cdot \beta}{r}}$. For example, if a power $1 - \beta$ is desired and the simulation study shall provide a 95% confidence interval with specified length l, the number of required replications is proportional to $1/l^2$. For $1 - \beta = 0.80$ and $l = 0.05, 0.01$, and 0.005, the up-wards rounded number of replications is in the range of 1000, 25,000, and 100,000, respectively. Finally, step 5 can be performed in a computationally more efficient way by, for example, application of the bisectional method instead of just decreasing or increasing, respectively, the sample size.

Example 16.1 The **SYNCHRONOUS** trial comparing resection of the primary tumor versus no resection prior to chemotherapy in patients with metastatic colon cancer was introduced in Example 6.3, Sect. 6.3. There, the required sample size for the primary outcome variable "overall survival" was calculated under the assumption of proportional hazards. However, it could not be ruled out in the planning stage that resection of the primary tumor for which an advantage was stipulated may show no or even a negative effect during the early follow-up phase, for example due to surgical complications. To assess the robustness of the power for the sample size calculated under the proportional hazards assumption, a simulation study was performed (Rahbari et al. 2012). For this, various scenarios were considered reflecting such a situation. An instantaneous increase of the mortality in the group with resection and a linear increase in the group without resection between 0.05 and 0.10 during the first three months of the follow-up were assumed with crossing of the survival times after three months and exponential survival thereafter such that the same median survival times result as initially assumed. The test based on Cox's model without covariates, which is asymptotically equivalent to the log-rank test, was applied in the simulations which were performed with 100,000 replications for each scenario (resulting in standard errors of 0.0022 for a true type I error rate of 0.05 and of 0.0036 for a rate of 0.85, respectively). The sample size calculated to achieve a power of 0.85 under the assumption of proportional hazards provided estimated power values in the range from 0.845 to 0.903 under the scenarios described above thus demonstrating robustness of the power to the assumptions made. Furthermore, as the original test based on the Cox model (asymptotically) only assures control of the specified level for proportional hazards, Lin and Wei (1989) developed a robust test that allows valid statistical inference even if the Cox proportional hazard model is misspecified. Therefore, this test was defined for the analysis and the above simulations were repeated by implementing it. Additionally, the actual type I error rate was assessed by considering the same scenarios but under the null-hypothesis of identical survival functions in both groups. The estimated power values were in the range from 0.846 to 0.905 and the observed type I error rate for the nominal two-sided significance level $\alpha = 0.05$ lay between 0.0478 and 0.0498; for the test based on the standard Cox model, the range of estimate type I error rates was 0.0495 to 0.0508. In summary, the simulations demonstrated that for the considered scenarios the standard test based on the Cox model and the robust test show similar power values as under exponential survival and that in particular the robust test leads to appropriate type I error rate even under departures from the proportional hazards assumption. As no analytic solution exists for these questions, the simulation-based approach is the only feasible one.

Remarks

1. Maruo et al. (2018) improved the efficiency and accuracy of the above-described simulation algorithm by fitting the probit model to data obtained from simulation results. This approach is based on the observation that many (approximate) sample size formulas can be written as $n = C \cdot \left(z_{1-\alpha/2} + z_{1-\beta}\right)^2$ with some constant $C > 0$. The equivalent equation $1 - \beta = \Phi\left(-z_{1-\alpha/2} + \sqrt{1/C} \cdot \sqrt{n}\right)$ has the structure of so-called probit models. This relationship between sample size and power can be used to decrease the computation time of the simulations and increase the accuracy of the estimated sample size substantially.
2. While the general approach of simulation-based sample size determination is simple, there are a number of points that have to be well-considered when designing, conducting, and reporting simulation studies. Burton et al. (2006) and Morris et al. (2019) give comprehensive overviews and offer guidance on issues, that have to be taken into account when performing simulation studies in the field of medical statistics.

Part III
Sample Size Recalculation

Chapter 17
Basic Concepts of Sample Size Recalculation

The methods for sample size calculation presented in Part II correspond to the so-called fixed sample size design. Here, all quantities required for determination of the sample size have to be specified in the planning phase and the trial is then conducted by including the number of patients resulting from these parameter values. If the assumptions made are correct, everything is fine. However, it is the exception rather than the rule that reliable and precise knowledge on the value of all parameters affecting the sample size is available when planning a study. To express it in an exaggerated way: If all quantities were known already in the planning stage, there would be no need for the trial whatsoever. An obvious approach to address this challenge is by taking data that arises during the course of the trial into account when determining the final sample size. A first step in this direction are sequential designs [see, for example, Whitehead (1997), Jennison and Turnbull (2000), Wassmer and Brannath (2016), and Sect. 27.1 of this book]. Here, interim analyses are performed when the outcome of each newly included patient (strictly sequential designs) or of groups of patients (group sequential designs) are available. If the test statistic falls below or above pre-specified boundaries, the study is stopped for futility or with early demonstration of efficacy, respectively. By this, the sample size becomes a random variable which depends not only on the planning assumptions but also on the data of the current study. However, in these "classical" sequential designs, the stage-wise sample sizes to be accrued between the interim analyses have to be pre-specified in the planning phase and have to be adhered to in order to assure control of the type I error rate. To address the limitations associated with this restriction, more flexible designs were required. Starting during the nineties of the 20th century, two classes of such designs were developed: internal pilot study designs and adaptive designs. In *internal pilot study designs*, nuisance parameters (i.e. parameters that are not involved in the test problems to be assessed but that affect the sample size) are estimated during the course of the study. This is generally done based on data which is pooled over the intervention groups. The estimated values are used to recalculate the initial sample size and to modify it if required. No interim testing of null-hypotheses is performed

© Springer Nature Switzerland AG 2020
M. Kieser, *Methods and Applications of Sample Size Calculation and Recalculation in Clinical Trials*, Springer Series in Pharmaceutical Statistics,
https://doi.org/10.1007/978-3-030-49528-2_17

but just one analysis is conducted when the data of all included patients are available. As the patients' treatment group allocation needs not to be known for sample size recalculation, blindness can be maintained within this approach. *Adaptive designs* are in some kind extensions of the above-mentioned group sequential designs. In addition to early stopping, the sample size (and also other design elements) can be changed after the interim analyses based on the information available so far. As the treatment group allocation is required for interim testing, this approach is *per se* an unblinded one.

From a regulatory perspective, there are two critical issues with sample size recalculation which concern blindness and type I error control. For example, the ICH E9 Guideline *Statistical Principles for Clinical Trials* states in Section 4.4 "Sample size adjustment": *"The steps taken to preserve blindness and consequences, if any, for the Type I error ... should be explained"* (ICH E9 1998, p. 22). The EMA Reflection Paper on *Methodological Issues in Confirmatory Clinical Trials Planned with an Adaptive Design* points out *"Whenever possible, methods for blinded sample size reassessment that properly control the type I error should be used, especially if the sole aim of the interim analysis is the recalculation of sample size"* (EMA 2007, p. 6). The reasons for these requirements are as follows. If the treatment effect becomes aware during the ongoing trial to persons involved in the study, there is an inherent risk that a bias is induced. Therefore, if adequate in the situation at hand, it is preferable to use blinded data for sample size recalculation. Hereby, any discussion can be avoided about whether or not sufficient measures have to be taken to exclude that any bias may arise. Attention to the type I error rate has to be paid when recalculating the sample size during the ongoing trial, as uncritical application of such methods may lead to an α-inflation. For example, Proschan and Hunsberger (1995) showed that the actual type I error rate of the z-test may fall considerably above the nominal level when the sample size is changed mid-course based on the observed treatment effect. For the cases they considered, Proschan and Hunsberger (1995) demonstrated that a more than doubled maximum actual one-sided level of 0.1146 may occur for a nominal one-sided type I error rate of $\alpha = 0.05$.

In the following, methods for and characteristics of blinded sample size calculation within internal pilot study designs (Chaps. 18–26) and unblinded sample size reassessment in adaptive designs (Chaps. 27–31) are presented and their application is illustrated by clinical trial examples.

Part IV
Blinded Sample Size Recalculation in Internal Pilot Study Designs

Chapter 18
Internal Pilot Study Designs

Nuisance parameters denote quantities which are not included in the formulation of the test problem to be assessed in the confirmatory analysis but which have an effect on the required sample size. Examples are the variance or the coefficient of variation when comparing means or the overall rate when comparing proportions. Commonly, there is considerable uncertainty about the value of nuisance parameters in the planning phase of a study. For example, in a review of clinical trials in patients with major depression (Linde et al. 2008), standard deviations between 3.8 and 9.9 were observed for the Hamilton Depression Rating Scale (see Examples 9.1 and 15.1) after therapy. This translates to a factor of almost 7 for the required sample size (see Sects. 3.3 and 8.2). Furthermore, in their reviews on the reporting of sample size calculation in clinical trial publications, Vickers (2003) and Charles et al. (2009) found that there are frequently large discrepancies between the nuisance parameter values used for sample size calculation and those observed later on in the analysis. Based on data on nuisance parameters reported by Charles et al. (2009), Tavernier and Giraudeau (2015) showed by simulations that, solely due to imprecise specification of nuisance parameters, the sample size is for a substantial portion of clinical trials too small or too large. Therefore, alternatives to the fixed sample size design are desirable that enable correcting inaccurate assumptions on nuisance parameters during the ongoing study.

Basically, in trials with an internal pilot study design, accumulated data are used mid-course to reestimate the value of nuisance parameters and to update the initially selected sample size based on the information obtained so far. The principle idea behind this approach dates back to Stein (1945) who proposed a two-stage design that guarantees exact control of the coverage probability of a confidence interval with fixed width for a normal mean by estimating the unknown variance after the first stage. Wittes and Brittain (1990) readopted this approach, showed its usability for clinical trials, and introduced the terminology "internal pilot study".

Formally, the steps of an internal pilot study design can be described as follows. Let np denote the (vector of) nuisance parameter(s) affecting the sample size needed

© Springer Nature Switzerland AG 2020
M. Kieser, *Methods and Applications of Sample Size Calculation and Recalculation in Clinical Trials*, Springer Series in Pharmaceutical Statistics,
https://doi.org/10.1007/978-3-030-49528-2_18

for the test problem at hand and N the required total sample size for the true value of np.

Step 1   Planning phase of the trial

The initial sample size is calculated based on the trial specifications (for example α, $1-\beta$, clinically relevant effect, etc.) and on the current knowledge about the nuisance parameter(s) captured by np_0. This leads to an initial total sample size n_0. Furthermore, the portion of patients π, $0 < \pi < 1$, to be included in the sample size review has to be defined, i.e., the sample size is recalculated when the data of $n_1 = \pi \cdot n_0$ patients are available.

Step 2   Sample size recalculation

Based on the data of n_1 patients which constitute the internal pilot study, the nuisance parameter (vector) np is reestimated and the sample size is recalculated based on the same trial specifications as in the planning phase but now using the estimate $\widehat{np}$ instead of the initially assumed np_0. This results in a recalculated total sample size $\widehat{N}_{\text{recalc}}$ (which is a random variable due to the randomness of $\widehat{np}$). The recalculated sample size may then be adjusted according to a pre-specified rule to obtain the final total sample size. Examples for such rules are as follows:

$$- \ \widehat{N} = \max\left(n_0, \widehat{N}_{\text{recalc}}\right): \text{``restricted design'' (Wittes and Brittain 1990)} \qquad (18.1)$$

$$- \ \widehat{N} = \max\left(n_1, \widehat{N}_{\text{recalc}}\right): \text{``unrestricted design'' (Birkett and Day 1994)} \qquad (18.2)$$

$$- \ \widehat{N} = \min\left(\max\left(n_1, \widehat{N}_{\text{recalc}}\right), n_{\max}\right): \text{``unrestricted design with upper boundary''} \qquad (18.3)$$

where $n_{\max}$ is an upper bound for the sample size resulting, for example, from restrictions of the recruitment resources or of the budget.

Step 3   Inclusion of further $\widehat{N}_2 = \widehat{N} - n_1$ patients and analysis based on the data of all $\widehat{N}$ patients.

To avoid unnecessary notational burden, the notations n and n_2 will in the following sometimes be used instead of $\widehat{N}$ and $\widehat{N}_2$, respectively. Nevertheless, it should be kept in mind that these quantities are random in internal pilot study designs.

Remarks

1. It is a distinct characteristic of the internal pilot study design, that a hypothesis test is performed only once, namely when the data of all patients are available. In particular, no interim analysis is conducted and no early stopping based on information about the treatment effect is possible. These are fundamental differences to the class of adaptive designs presented in Part IIIb.

2. Wittes and Brittain (1990) proposed to apply the restricted design as defined in (18.1) which means that the initial sample size can only be corrected upwards. However, if the assumptions on the nuisance parameters made in the planning phase turn out to be too pessimistic after the internal pilot study, the final sample size will inevitably be too high. Therefore, Birkett and Day (1994) suggested to abandon this constraint leading to the unrestricted design defined by (18.2). Other rules for determining the final sample size than those given by (18.1)–(18.3) can be implemented. For example, in addition to an upper bound, a lower boundary for the sample size n_{min} with $n_1 < n_{min} < n_{max}$ may be implemented, possibly to assure enough exposed patients for studying the safety profile. The final sample size is then determined by the rule $\widehat{N} = \min(\max(n_{min}, \widehat{N}_{recalc}), n_{max})$.

In their seminal paper, Wittes and Brittain (1990) considered the two-sample t-test situation with the population variance as nuisance parameter. They proposed to recalculate the sample size based on the pooled variance estimator obtained from the unblinded data of the internal pilot study. Comprehensive research has been done on sample size recalculation based on the pooled variance estimator [see, for example, Birkett and Day (1994), Coffey and Muller (1999), Denne and Jennison (1999), Wittes et al. (1999), Zucker et al. (1999), Kieser and Friede (2000a), Proschan and Wittes (2000), Miller (2005), and Proschan (2005)]. However, regulatory guidelines clearly prefer that blinding is preserved during the course of the trial, especially when sample size recalculation is performed (see quotations from guidelines given in Chap. 17). Otherwise, adequate measures have to be taken to withhold any information arising from unblinded data from all persons involved in the conduct of the trial. This requires implementation of an independent Data Monitoring Committee (DMC) with defined procedures and communication channels. These are considerable expenditures for just estimating nuisance parameters. However, it could be shown for many test problems addressed in clinical trials that estimation of the nuisance parameters based on the data pooled over the intervention groups results in internal pilot study designs with excellent performance, sometimes even better than when using unblinded data. Therefore, we restrict in the following to methods for blinded sample size recalculation. Besides the importance of keeping the blinding, guidelines point out the requirement of type I error rate control when recalculating the sample size during the course of the trial. This is due to the fact that the simulations performed by Wittes and Brittain (1990) and subsequent work showed that the type I error rate may be inflated in internal pilot study designs. For this reason, we present in the following for the most common clinical trial designs and test problems methods for blinded estimation of the involved nuisance parameters, the type I error rate characteristics of the related sample size recalculation procedure within the internal pilot study design and, if not assured anyway, techniques how to control the nominal level. Furthermore, results on the robustness of the achieved power with respect to the planning assumptions concerning the nuisance parameters are shown, as this feature is the motivation for implementing an internal pilot study design. Moreover, as another important characteristic of sample size recalculation procedures, properties of the sample size distribution are presented.

Chapter 19
A General Approach for Controlling the Type I Error Rate for Blinded Sample Size Recalculation

In fixed sample size designs, permutation tests control the nominal type I error rate under the strong null hypothesis that the effect of the experimental treatment is not different from that of the control. No further distributional assumptions are required to assure validity of this result. In randomized trials, the null hypothesis of no treatment group difference can be tested with a permutation test by building all permutations of group assignments which are possible under the applied randomization scheme (and which are equally plausible under the null hypothesis) and by calculating the values of an appropriate test statistic for all assignments. By this, a reference distribution can be constructed which can be used to define the rejection region for a valid level-α test [see, for example, Good (2013) for an overview and Berger (2000) for a critical appraisal of discussions about the application of permutation tests in clinical trials]. The permutation model conditions on the data and the treatment group assignments are the only random elements. As blinded sample size recalculation does not utilize the group assignments, in the framework of permutation tests the sample size is considered as fixed and therefore independent of the test statistic. Therefore, if a permutation test is applied in the analysis of clinical trials with blinded sample size recalculation, the type I error rate is controlled, irrespective of which specific recalculation rule is used. This result was first mentioned by Zucker et al. (1999) for the t-test and later by Kieser and Friede (2003) for the general situation. The permutation test approach thus provides a framework allowing to conduct blinded sample size recalculation in internal pilot study designs under preservation of the basic regulatory requirement of level control. We will come back to this approach in the subsequent Sects. 20.1.3 and 21.3 when presenting methods for specific scale levels of the outcomes and related test problems.

© Springer Nature Switzerland AG 2020
M. Kieser, *Methods and Applications of Sample Size Calculation and Recalculation in Clinical Trials*, Springer Series in Pharmaceutical Statistics,
https://doi.org/10.1007/978-3-030-49528-2_19

Chapter 20
Comparison of Two Groups for Normally Distributed Outcomes and Test for Difference or Superiority

20.1 t-Test

20.1.1 Background and Notation

We assume the same statistical model as in the fixed sample size design (see Sect. 3.3), i.e. mutually independent random variables X_{Cj} and X_{Ej} denoting the outcomes under the control and the experimental treatment, respectively, with expectation μ_C and μ_E and common unknown variance σ^2. For simplicity of notation, we consider in the following the one-sided test problem for superiority of E versus C:

$$H_0: \mu_E - \mu_C \leq 0$$
$$\text{versus} \qquad (20.1)$$
$$H_1: \mu_E - \mu_C > 0$$

The results with respect to type I error rate (Sect. 20.1.3) and power (Sect. 20.1.4) for the related two-sided test for difference can be obtained by taking in the derivations the two-tailed shape of the rejection region into account. As in the fixed sample size design, the t-test statistic T is applied for the assessment of test problem (20.1):

$$T = \sqrt{\frac{r}{(1+r)^2} \cdot n} \cdot \frac{\bar{X}_E - \bar{X}_C}{S} \qquad (20.2)$$

with means $\bar{X}_E$ and $\bar{X}_C$ in group E and C, respectively, pooled standard deviation S given by (3.8a) and (3.8b) (see Sect. 3.3), and total sample size n allocated to the groups with ratio $r = n_E/n_C$. In the internal pilot study design, the sample size n is random due to the recalculation performed mid-course. Therefore, the distribution of T in this setting is not evident. The null-hypothesis is rejected if $t > t_{1-\alpha/2, n-2}$,

© Springer Nature Switzerland AG 2020
M. Kieser, *Methods and Applications of Sample Size Calculation and Recalculation in Clinical Trials*, Springer Series in Pharmaceutical Statistics,
https://doi.org/10.1007/978-3-030-49528-2_20

where $t_{1-\alpha/2,n-2}$ denotes the $(1 - \alpha/2)$-quantile of the central t-distribution with $n - 2$ degrees of freedom.

20.1.2 Blinded Sample Size Recalculation

Sample size recalculation is based on the approximation formula (3.5), where a blinded estimator S^2 is inserted instead of the population variance σ^2:

$$\hat{N}_{recalc} = \frac{(1+r)^2}{r} \cdot \left(z_{1-\alpha/2} + z_{1-\beta}\right)^2 \cdot \left(\frac{S}{\Delta_A}\right)^2. \tag{20.3}$$

Here, $\alpha/2$ is the one-sided significance level, $1 - \beta$ the desired power, z_γ denotes the γ-quantile of the standard normal distribution, and $\Delta_A = (\mu_E - \mu_C)_A \neq 0$ is the pre-specified alternative. The simplest approach for blinded variance estimation is by just ignoring the treatment group allocation and using the variance estimator obtained from the pooled sample. If n_{E1}, n_{C1}, and n_1 denote the group-wise or total sample size of the internal pilot study, respectively, and $X_{Ej}, j = 1, \ldots, n_{E1}$, and $X_{Cj}, j = 1, \ldots, n_{C1}$, the outcomes, the blinded one-sample variance estimator is given by

$$S^2_{OS} = \frac{1}{n_1 - 1} \cdot \sum_{i=C,E} \sum_{j=1}^{n_{i1}} \left(X_{ij} - \bar{X}_1\right)^2, \tag{20.4}$$

where $\bar{X}_1 = \frac{1}{n_1} \cdot \sum_{i=C,E} \sum_{j=1}^{n_{i1}} X_{ij}$ denotes the overall mean of the internal pilot study. Using this variance estimator for sample size recalculation in internal pilot study designs turned out to provide overall the best performance with respect to achieved power and sample size variability amongst its competitors (see Remarks below). Therefore, we concentrate on this approach in the following.

Remarks

1. In case of an existing treatment group difference, i.e. if $\Delta = \mu_E - \mu_C \neq 0$, the one-sample variance estimator is biased upwards:

$$E\left(S^2_{OS}\right) = \sigma^2 + \frac{r}{(1+r)^2} \cdot \frac{n_1}{(n_1 - 1)} \cdot \Delta^2. \tag{20.5}$$

Gould and Shih (1992) and Zucker et al. (1999) proposed to correct the one-sample variance estimator by inserting the value Δ_A assumed under the alternative thus obtaining an estimator which is unbiased under $H_{1,A}$:

$$S^2_{corr} = S^2_{OS} - \frac{r}{(1+r)^2} \cdot \frac{n_1}{(n_1 - 1)} \cdot \Delta^2_A. \tag{20.6}$$

For appropriate choices of n_1, the expected sample size when using S_{corr}^2 for sample size recalculation is in the unrestricted design equal to the sample size required in the fixed design. Furthermore, under the alternative $H_{1,A}$ the additional expected total sample size that is required when using S_{OS}^2 instead of S_{corr}^2 is given by

$$\hat{N}_{\text{recalc,OS}} - \hat{N}_{\text{recalc,corr}} = \left(z_{1-\alpha/2} + z_{1-\beta}\right)^2 \cdot \frac{n_1}{n_1 - 1} \approx \left(z_{1-\alpha/2} + z_{1-\beta}\right)^2$$

which amounts to 8 or 12 patients for the common values $\alpha/2 = 0.025$ and $1 - \beta = 0.80$ or 0.90, respectively. This approach seems to be appealing at first sight. However, Kieser and Friede (2000a) showed that, in contrast to when applying S_{OS}^2 (see description and results in Sect. 20.1.4), using S_{corr}^2 for sample size recalculation results in an expected power that is lower than the desired one. The slight increase in sample size produced by the one-sample variance thus just compensates the marginally reduced efficiency as compared to the fixed design. This can be seen as the price to be paid for the robustness against planning misspecifications gained by sample size recalculation.

2. Gould and Shih (1992) proposed to use a procedure based on the expectation-maximization (EM) algorithm to estimate the variance without unblinding by treating the group allocation of the patients as missing data. However, this method is not only much more complicated than calculating S_{OS}^2 but, much more serious, it also turned out to have a number of critical flaws (Friede and Kieser 2002; Waksman 2007). Even if rectified according to Waksman (2007), the EM-algorithm-based estimator (and with it the recalculated sample size) has higher variability as compared to S_{OS}^2 (Waksman 2007). It should therefore not be used for blinded sample size recalculation.

3. Xing and Ganju (2005) proposed another estimator for blinded variance estimation which is unbiased under $H_{1,A}$ and which uses the randomization block information. As for the corrected one-sample variance estimator (see Remark 1 above), which is also unbiased under the alternative, applying this method may lead to a lower power than desired. Moreover, the variability may be higher than for the simple one-sample estimator (Friede and Kieser 2013). Therefore, this approach is not recommended for application.

4. Shih (2009) proposed an approach for variance estimation that he denoted as "perturbed unblinding". It requires unblinding of a person who adds values e and c to each observation of group E and C, respectively. Another person, who does not know the values e and c, calculates the pooled variance estimator which is identical to the pooled variance of the original data set and which is thus unbiased. By adding the constants e and c, the person(s) performing the sample size recalculation remain masked with respect to the treatment group differences. However, this is not true for the person changing the values in the database who has to be unblinded. Furthermore, as for the other unbiased variance estimators mentioned in Remarks 1 and 3 above, the desired power may not be achieved when using it for sample size recalculation. As another argument against this

method, the type I error rate of the sample size recalculation procedure is inflated when using the pooled variance estimator [see, for example, Wittes et al. (1999) and Kieser and Friede (2000a)]. Thus, to assure control of the nominal level, the critical value (Denne and Jennison 1999; Kieser and Friede 2000a) or the variance estimate inserted in the t-test statistic (Miller 2005) have to be adjusted. It will be shown below that this is not necessary when using the one-sample variance estimator (see the following Sect. 20.1.3). Therefore, this approach is not recommended to be applied in clinical trials.

20.1.3 Type I Error Rate

Kieser and Friede (2003) presented the following approach for calculating the actual type I error rate of the t-test in internal pilot study designs with blinded sample size recalculation based on the one-sample variance estimator S_{OS}^2. The idea is to formulate the test statistic in terms of components for which the (conditional) distribution is known, to determine the joint density under the null-hypothesis, and to compute the type I error rate by numerical integration. For this purpose, the following abbreviations are introduced:

$$
Z_i = \sqrt{\frac{r}{(1+r)^2} \cdot n_j} \cdot \frac{\bar{X}_{iE} - \bar{X}_{iC} - \Delta}{\sigma}, \quad i = 1, 2
$$

$$
V_i = \frac{(n_i - 2) \cdot S_i^2}{\sigma^2}, \quad i = 1, 2,
$$

where n_i, $\bar{X}_{ij}$, and S_i denote the total sample size, the mean in group j, $j = C, E$, and the pooled variance, respectively, based on the data captured before ($i = 1$) or after ($i = 2$) the data review. Note that while n_1 is fixed, the sample sizes n_2 and $n = n_1 + n_2$ are random. With the above notations, the one-sample variance can be written as $S_{OS}^2 = \frac{\sigma^2}{n_1 - 1} \cdot \left[V_1 + \left(Z_1 + \sqrt{\frac{r}{(1+r)^2} \cdot n_1} \cdot \frac{\Delta}{\sigma} \right)^2 \right]$. Denoting by N the required total sample size for the true value of σ, the recalculation rule (20.3) can be written as $\hat{N}_{\text{recalc}} = N \cdot \frac{S_{OS}^2}{\sigma^2}$ and thus

$$
\hat{N}_{\text{recalc}}(Z_1, V_1) = \frac{N}{(n_1 - 1)} \cdot \left[V_1 + \left(Z_1 + \sqrt{\frac{r}{(1+r)^2} \cdot n_1} \cdot \frac{\Delta}{\sigma} \right)^2 \right]. \tag{20.7}
$$

Furthermore, the following notations are used:

$$
Z_1^* = \sqrt{\frac{r}{1+r} \cdot \frac{n_1 \cdot n_2}{n}} \cdot \frac{\bar{X}_{E1} - \bar{X}_{C1}}{\sigma}
$$

$$Z_2^* = \sqrt{\frac{r}{1+r} \cdot \frac{n_1 \cdot n_2}{n}} \cdot \frac{\bar{X}_{E2} - \bar{X}_{C2}}{\sigma}$$

$$V_2^* = V_2 + \left(Z_1^*\right)^2 + \left(Z_2^*\right)^2.$$

Then, the test statistic of the *t*-test can be written as a function of Z_1, Z_2, V_1, and V_2^*:

$$T = \frac{\sqrt{\frac{n_1}{n}} \cdot Z_1 + \sqrt{\frac{n_2}{n}} \cdot Z_2 + \sqrt{\frac{r}{(1+r)^2}} \cdot n \cdot \frac{\Delta}{\sigma}}{\sqrt{\frac{1}{n-2} \cdot \left(V_1 + V_2^*\right)}}. \tag{20.8}$$

It holds that

$$Z_1 \sim N(0, 1) \tag{20.9a}$$

$$V_1 \sim \chi^2_{n_1-2}, \tag{20.9b}$$

where χ^2_{df} denotes the central chi-square distribution with *df* degrees of freedom. Furthermore, Z_1 and V_1 are independent, and conditional on Z_1 and V_1 it holds true that

$$Z_2|(Z_1, V_1) \sim N(0, 1). \tag{20.9c}$$

Moreover, it was argued (see Remark 2 below) that, conditional on Z_1 and V_1, $V_2^* \sim \chi^2_{n_2}$ and Z_2 and V_2^* are independent. Therefore, setting $\Delta = 0$ the joint density of (Z_1, V_1, Z_2, V_2^*) under H_0 can be written as $f(z_1, v_1, z_2, v_2^*) = f_{N(0,1)}(z_1) \cdot f_{\chi^2_{n_1-2}}(v_1) \cdot f_{N(0,1)}(z_2) \cdot f_{\chi^2_{n_1-2}}(v_2^*)$, where $f_{N(0,1)}$ denotes the density of the standard normal distribution and $f_{\chi^2_{df}}$ the density of the central chi-square distribution with *df* degrees of freedom. The actual type I error rate α_{act} of the *t*-test in an internal pilot study design with blinded sample size recalculation based on the one-sample variance estimator can now be computed by integration of the density $f(z_1, v_1, z_2, v_2^*)$ over the rejection region of the test statistic T (i.e. over $]t_{n-2,1-\alpha/2}, \infty[$ for one-sided tests at nominal level $\alpha/2$ or over $]-\infty, -t_{n-2,1-\alpha/2}[\cup]t_{n-2,1-\alpha/2}, \infty[$ for two-sided tests at nominal level α) [see Friede (2000) for details]. Note that, due to the used parametrization, the following important property can be concluded from the above derivation. When using S_{OS}^2 for sample size recalculation, the actual type I error rate only depends on n_1, α, and the applied strategy to determine the final sample size from the recalculated one (which are known features for a specific clinical trial) as well as on the unknown true required sample size N. As a consequence of representation (20.7) of the sample size recalculation rule, $1 - \beta$ is already captured by N. Therefore, the maximum actual type I error rate for a specific clinical trial setting can be determined by calculating α_{act} for all possible values of the unknown true required sample size N (or equivalently by considering all possible values of σ^2).

Kieser and Friede (2003) computed the actual type I error rate by numerical integration for all combinations of $\alpha = 0.025, 0.05$ (one-sided), $n_1 = 10, 15, \ldots, 50$, and $N = 10, 20, \ldots, 100$, both for the restricted design (18.1) with minimal sample size $n_0 = 2 \cdot n_1$ and for the unrestricted design (18.2). For these scenarios, the maximum observed increase in type I error rate was only 0.0001 which occurred for the restricted design. Similar results have been obtained when applying alternative approaches for type I error rate calculation or simulation studies, respectively (Glimm and Läuter 2013; Lu 2016; see also Remark 2). Therefore, although this is no formal proof, it is concluded that, if at all, blinded sample size recalculation based on the one-sample variance estimator does not affect the type I error rate to a relevant amount when testing for difference or superiority with the t-test.

Remarks

1. Numerical calculations and simulations on the actual type I error rate have been performed for a multitude of design scenarios. None of these investigations revealed any relevant inflation of the type I error rate when applying the t-test for difference or superiority in internal pilot study designs with blinded sample size recalculation based on the one-sample variance. There is another argument supporting the conjecture that this result may hold generally true for the setting considered in this section. In Chap. 19, it was shown that the type I error rate is controlled for any blinded sample size recalculation procedure if a permutation test is applied in the analysis. As permutation distributions are very well approximated by normal distributions [see Proschan et al. (2014) and references given there], it is not unexpected that the type I error rate is also preserved for the t-test if the sample size is recalculated in a blinded fashion.

2. Glimm and Läuter (2013) stated that the conditional distribution of V_2^* given Z_1 and V_1 is not a central chi-square distribution with n_2 degrees of freedom (as supposed in the above-described derivation) but a mixture of the central chi-square distribution with $n_2 - 1$ degrees of freedom and a "rescaled" non-central chi-square distribution. Lu (2016) showed that in fact it holds true that

$$V_2^* | (Z_1, V_1, Z_2) \sim \chi_{n_1-1}^2 + \left(\sqrt{\frac{n_2}{n}} \cdot Z_1 - \sqrt{\frac{n_1}{n}} \cdot Z_2 \right)^2. \tag{20.9d}$$

From (20.9d), the result derived by Glimm and Läuter (2013) follows, namely $V_2^* | (Z_1, V_1) \sim \chi_{n_2-1}^2 + \frac{n_1}{n} \cdot \chi_{1, \frac{n_2}{n_1} \cdot z_1^2}^2$, where $\chi_{df,nc}^2$ denotes the non-central chi-square distribution with df degrees of freedom and non-centrality parameter nc. By (20.9a)–(20.9d), the joint distribution of (Z_1, V_1, Z_2, V_2^*) is defined. This result derived by Lu (2016) allows to characterize the exact distribution of the t-test statistic in internal pilot study designs after blinded sample size recalculation based on the one-sample variance estimator. Notwithstanding the differences between the approaches, the consequences for the calculated type I error rates are extremely small and irrelevant for practice [see Glimm and Läuter (2013) and Lu (2016)]. Thus, any of the above methods can be used to determine the actual type I error rate of the *t*-test following blinded sample size recalculation.

3. The representations given above for the sample size recalculation rule and the *t*-test statistic by means of the random variables Z_1, V_1, Z_2, and V_2^*, whose (conditional) distributions are known, enable not only to determine the actual type I error rate and power by numerical integration but also by an efficient Monte Carlo simulation algorithm. For the latter one, it is not necessary to generate and process the single observations. Instead, it is sufficient to generate the random variables Z_1, V_1, Z_2, and V_2^* according to the (conditional) distributions they follow under the null- or alternative hypothesis, respectively, which requires much less computational effort [see Lu (2016) for a description of the algorithm].

20.1.4 Power and Sample Size

The expected power of the *t*-test in internal pilot study designs with sample size recalculation based on the one-sample variance can be calculated analogously to the type I error rate (see preceding Sect. 20.1.3) by now considering scenarios under the alternative hypothesis.

The expected sample size for the unrestricted design can be determined by distinguishing the cases $\hat{N}_{\text{recalc}} \leq n_1$ and $\hat{N}_{\text{recalc}} > n_1$. Due to the representation (20.7) for $\hat{N}_{\text{recalc}}$, it follows that $\hat{N}_{\text{recalc}} \leq n_1$ occurs with probability $F_{\chi^2;n_1-1,0}\left(\frac{(n_1-1)}{N} \cdot n_1\right)$, where $F_{\chi^2;n_1-1,0}$ denotes the distribution function of the central chi-square distribution with $n_1 - 1$ degrees of freedom. Therefore

$$E\left(\hat{N}_{\text{recalc}}\right) = F_{\chi^2;n_1-1,0}\left(\frac{(n_1-1)}{N} \cdot n_1\right) \cdot n_1$$

$$+ \int_c^\infty \frac{N}{(n_1-1)} \cdot V_1^* \cdot f_{\chi^2;n_1-1,0}\left(v_1^*\right) dv_1^*,$$

where $c = \frac{(n_1-1)}{N} \cdot n_1$, $V_1^* = V_1 + \left(Z_1 + \sqrt{\frac{r}{(1+r)^2} \cdot n_1} \cdot \frac{\Delta_A}{\sigma}\right)^2$ and where $f_{\chi^2;n_1-1,0}$ denotes the density of the central chi-square distribution with $n_1 - 1$ degrees of freedom. By appropriate modifications, the calculations of expected power and sample size can analogously be done for alternative variance estimators as, for example, the corrected one-sample variance estimator S_{corr}^2 given by (20.6), as well as for alternative recalculation rules as, for example, the restricted design (18.1). Kieser and Friede (2003) calculated expected power and sample size for a one-sided level $\alpha/2 = 0.025$, desired power $1 - \beta = 0.80$, required total sample sizes $N = 64$, 126, and 348, and various choices for the pilot sample size n_1. They considered both the restricted (with minimal sample size $n_0 = 2 \cdot n_1$) and the unrestricted design and both the crude one-sample variance estimator S_{OS}^2 and its corrected version S_{corr}^2; for the latter, $\Delta = \Delta_A$ was assumed making S_{corr}^2 under the specified alternative to an unbiased estimator of σ^2. Consequently, for values $2 \cdot n_1$ or n_1 in the restricted or unrestricted design, respectively, which are smaller and not too narrow to the truly required sample size N, the corrected variance estimator results in expected sample sizes close to N. In contrast, the crude one-sample variance estimator leads to an expected excess in total sample size of approximately $\left(z_{1-\alpha/2} + z_{1-\beta}\right)^2$ (see Remark 1 in Sect. 20.1.2) which amounts to 8 in the situation considered here. Using the one-sample variance estimator S_{OS}^2 results in an expected power equal or very close to the desired one. Thus, the (small) additional expected sample size resulting from application of the one-sample variance estimator just compensates the (small) efficiency loss caused by the option for sample size modification. In contrast, the corrected variance estimator provides expected power values below $1 - \beta$. This behavior has already been observed by Wittes et al. (1999) for unblinded sample size recalculation based on the pooled variance estimator which is mimicked by S_{corr}^2.

For small values of $2 \cdot n_1$ or n_1 as compared to N, the final sample size of the restricted or unrestricted design, respectively, is mainly determined by the applied variance estimator and not by constraints imposed by the choice of the internal pilot study sample size n_1. Therefore, unrestricted and restricted designs show similar performance in these cases. If the minimum total sample size $2 \cdot n_1$ or n_1 of the restricted or unrestricted design, respectively, is already larger than N, sample size recalculation cannot prevent the expected power to fall above $1 - \beta$.

Remarks

1. *Choice of sample size of internal pilot study*: As mentioned above, n_1 should of course lie below (and not too narrow to) the required sample size as otherwise the desired power is inevitably exceeded. Furthermore, if n_1 is small, the variability of the variance estimator and with it that of the recalculated sample size are large. Moreover, the investigations by Kieser and Friede (2003) show that the expected power may fall below the desired value if the sample size of the internal pilot study is small. This result is not only observed for the t-test but also for other settings [see, for example, Friede and Kieser (2011a)]. Statistical considerations on the choice of n_1 were, amongst others, presented by Sandvik et al. (1996) and Denne and Jennison (1999). The method proposed by Sandvik et al. (1996) assumes that an estimate of σ^2, which arises from a random sample of the study population, is available when planning the trial with internal pilot study design. This approach aims at making the internal pilot study sample size as large as possible while at the same time ensuring that the probability of including more patients than required for the entire study is smaller that a pre-specified value. Denne and Jennison (1999) proposed to choose n_1 such that the ratio of the expected sample size in the internal pilot study to the sample size required in the fixed design is minimized for the true value of the population variance. They suggested a strategy that provides a value for the ratio close to the minimum even under misspecification of σ^2. Besides statistical aspects, practical issues have to be taken into account. For example, the speed of recruitment and the period between entering the study and measurement of the outcome determine

the time at which the data for sample size recalculation is available for a specific number of patients and how many further patients have meanwhile entered the study. To avoid over-recruitment of patients above the required sample size N, the internal pilot study has to be sized appropriately. Singer (1999) modified the proposal of Sandvik et al. (1996) accordingly.

2. It was mentioned in Sect. 3.3, Remark 1, that, due to the skewness of the distribution of the sample variance, there is a substantial risk of undershooting the desired power if the point estimate of the variance from an *external* pilot study is used for sample size calculation. Kieser and Friede (2000a) showed by numerical computations for a number of scenarios that the same holds true in *internal* pilot study designs and sample size recalculation with the unblinded pooled variance estimator. Furthermore, the calculations indicate that using the upper boundary of the one-sided $(1 - \gamma)$ confidence interval for σ^2 for sample size recalculation leads to a probability of at least $1 - \gamma$ to reach the planned power $1 - \beta$. The probability increases with increasing internal pilot sample size. Therefore, if it is aspired to achieve the desired power with high probability, an inflated value of the variance estimate should be used for recalculation. An appropriate inflation factor can be determined by calculations analogous to those performed in Kieser and Friede (2000a).

3. Frequently, the primary endpoint of a clinical trial is not only measured once at the end of the follow-up phase but also at an intermediate time point. Taking into account this short-term data and including also those patients that have not completed the trial but for which this intermediate information is available may reduce the variability of the variance estimator and with it that of the recalculated sample size. Wüst and Kieser (2003) proposed such a method for normally distributed outcomes, where short- and long-term data are merely assumed to be correlated but where no specific model describing the relationship between them is supposed. It was shown that the variance of the common variance estimator based on long-term data is asymptotically reduced by the factor $\left(1 - \frac{n_{long}}{n_{short}}\right) \cdot \rho^4$, where n_{long} and n_{short} is the sample size of those patients with both long- and short-term data and short-term data only, respectively, and ρ is the correlation between short- and long-term outcome.

Example 20.1 Acute myocardial infarction (AMI) may lead to worsening of cardiac function and terminal heart failure. Based on promising pre-clinical studies, the **SITAGRAMI trial** was conducted to investigate whether combined application of granulocyte colony stimulating factor and Sitagliptin is efficacious with respect to global cardiac function after an AMI that was successfully revascularized (Brenner et al. 2016). The SITAGRAMI trial was performed as a double-blind, randomized, placebo-controlled study. There were two hierarchically ordered primary endpoints (see Sect. 11.2.2b), namely the absolute changes in global left and right ventricular ejection fractions, respectively, between screening and 6 months follow-up. Sample size calculation was performed for analysis with the two-sample *t*-test at one-sided significance level $\alpha = 0.025$ to achieve a power $1 - \beta = 0.80$ for each of the endpoints for a difference between groups of $\Delta_A = 3.5$. A standard deviation of $\sigma_0 = 5.5$ was assumed, and the sample size should be balanced between the two groups. These specifications lead to an initial total sample size of $n_0 = 78$ patients based on the z-test approximation formula (3.5). Due to existing uncertainty with respect to the true standard deviation, a blinded sample size recalculation was planned which was based on the one-sample variance estimator S_{OS}^2 and which was intended to be performed close to the completion of the recruitment of the initially planned number of patients. For illustration, we consider the scenarios $n_1 = 20, 40$, and 60 for the sample size of the internal pilot study and assume that an unrestricted design with a maximum total sample size $n_{max} = 2 \cdot n_0 = 156$ was specified. Figures 20.1, 20.2 and 20.3 were prepared based on simulations with 1,000,000 replications for each parameter constellation resulting in a standard error of 0.00016 for a true type I error rate of 0.025 and of 0.0004 for a rate of 0.80, respectively.

Figure 20.1 shows the simulated type I error rate of the internal pilot study design for the above specifications depending on the population standard deviation σ. The marginal deviations from the nominal level reflect the random variations that are an inherent characteristics of simulation studies. Especially, there is no indication for an inflated type I error rate.

The simulated power of the internal pilot study design and the related fixed sample size design (where $\sigma = 5.5$ is assumed) are shown in Fig. 20.2. It is obvious that, as compared to the fixed-size design, the power characteristics of the design with blinded sample size recalculation are markedly more stable with respect to the value of σ. For small σ, the power is higher than aspired as for these scenarios the internal pilot study sample size n_1 is already higher than the required sample size. Conversely,

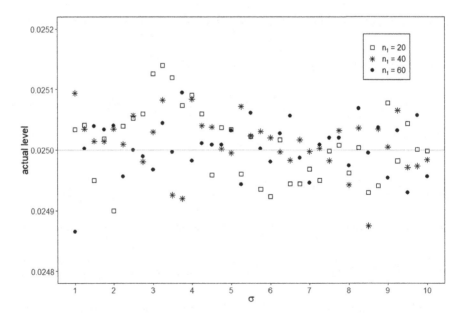

Fig. 20.1 Actual level of the t-test at one-sided nominal level $\alpha/2 = 0.025$ for the internal pilot study design described in Example 20.1 depending on the standard deviation σ

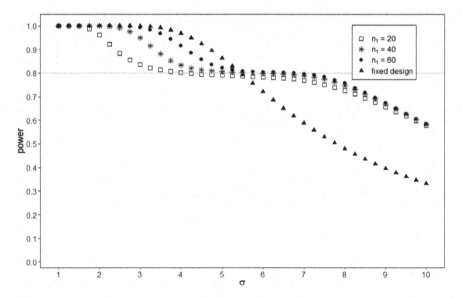

Fig. 20.2 Power of the *t*-test at one-sided nominal level $\alpha/2 = 0.025$ for the internal pilot study design described in Example 20.1 and for the related fixed sample size design (assuming $\sigma = 5.5$) depending on the standard deviation σ

the upper bound for the sample size $n_{\max} = 156$ prevents from reaching a power of 0.80 for large σ. Considerations of this kind are helpful in the planning stage to support an appropriate choice of the values for n_1 and $n_{\max}$.

Another important performance measure of designs with sample size recalculation is the distribution of the sample size under the specified alternative. Besides the mean value, the variability of the sample size plays a crucial role: The smaller the variation, the better the predictability of the final sample size and with it that of the required resources. Figure 20.3a shows box-and-whisker plots of the recalculated total sample size for $\sigma = 5.5$ depending on the internal pilot sample size. The average sample size amounts to about 86 and is thus 8 patients higher than the sample size required in the fixed-size design. This was to be expected based on the considerations presented in Sect. 20.1.2, Remark 1. The variability of the recalculated sample size decreases with increasing n_1, whereby the reduction is especially pronounced when raising the pilot sample size from small to medium. Assuming for illustrative purposes that an internal pilot sample size of $n_1 = 40$ is chosen, Fig. 20.3b shows the associated distribution of the recalculated sample size depending on σ. For small or large σ, respectively, the final sample size is in most cases equal to n_1 or $n_{\max}$. In between, the location of the distribution shifts towards larger values according to the larger sample size required for higher σ. The variability is substantial. For example, for standard deviations between 5 and 7, one has to make provisions for all sample sizes between n_1 and $n_{\max}$ if one wants to be prepared for all scenarios that may actually

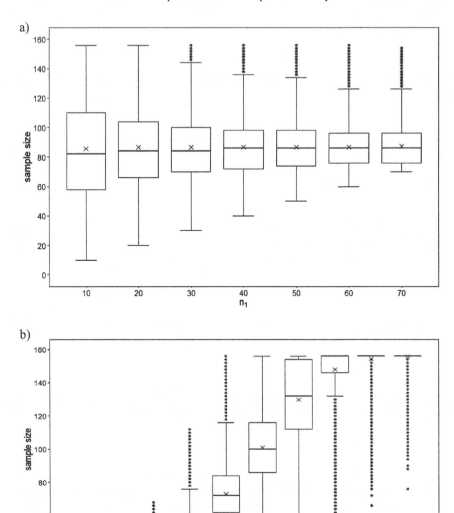

Fig. 20.3 Box-and-whisker plots of the recalculated sample size for the *t*-test at one-sided nominal level $\alpha/2 = 0.025$ in the internal pilot study design described in Example 20.1. **a** $\sigma = 5.5$ fixed, sample size of internal pilot study n_1 varying; **b** $n_1 = 40$ fixed, σ varying

occur. The middle fifty per cent of the distribution lie in a range of about 20 (for $\sigma = 5$) and 40 (for $\sigma = 7$).

20.2 Analysis of Covariance

20.2.1 Background and Notation

We assume the same statistical model as in the fixed sample size design and consider the situation of a single random covariate (see Sect. 3.4). For the case of multiple random covariates see the Remark at the end of Sect. 20.2. We suppose that in each group i, $i = C, E$, the outcome variable X and the covariate W follow a bivariate normal distribution with unknown variances σ^2 and σ_W^2, respectively, and correlation ρ. The statistical model is

$$X_{Cj} = \tau_C + \beta \cdot \left(W_{Cj} - \bar{W}\right) + \varepsilon_{Cj}, \quad j = 1, \ldots, n_C \tag{20.10a}$$

$$X_{Ej} = \tau_E + \beta \cdot \left(W_{Ej} - \bar{W}\right) + \varepsilon_{Ej}, \quad j = 1, \ldots, n_E, \tag{20.10b}$$

where the same notation is used as in (3.12a) and (3.12b). The one-sided test problem is given by

$$\begin{aligned} H_0&: \tau_E - \tau_C \le 0 \\ &\text{versus} \\ H_1&: \tau_E - \tau_C > 0, \end{aligned} \tag{20.11}$$

where $\tau_E - \tau_C$ denotes the adjusted treatment effect. As in the fixed sample size design, the ANCOVA test statistic T_{ANCOVA} is used for assessing (20.11):

$$T_{ANCOVA} = \frac{\hat{\tau}_E - \hat{\tau}_C}{\sqrt{\widehat{Var}\left(\hat{\tau}_E - \hat{\tau}_C\right)}},$$

where $\widehat{Var}\left(\hat{\tau}_E - \hat{\tau}_C\right)$ is defined by (3.14).

20.2.2 Blinded Sample Size Recalculation

Sample size recalculation is based on the Guenther-Schouten formula (3.18), where an appropriate estimator S_{res}^2 of $(1 - \rho) \cdot \sigma^2$ is inserted:

$$\hat{N}_{\text{recalc}} = \frac{(1+r)^2}{r} \cdot \left(z_{1-\alpha/2} + z_{1-\beta}\right)^2 \cdot \frac{S_{\text{res}}^2}{\Delta_A^2} + \frac{\left(z_{1-\alpha/2}\right)^2}{2}, \qquad (20.12)$$

where r denotes the sample size allocation ratio $r = n_E/n_C$, z_γ the γ-quantile of the standard normal distribution, and $\Delta_A = (\tau_E - \tau_C)_A$ the alternative specified for the adjusted treatment group difference. An unbiased estimator of $(1 - \rho^2) \cdot \sigma^2$ based on the data of the internal pilot study with total sample size n_1 is given by $\left(\frac{n_1-2}{n_1-3}\right) \cdot \left(1 - \hat{\rho}^2\right) \cdot S^2$, which is equal to the residual variance. Here, $\hat{\rho}$ denotes the estimator of the correlation between outcome and covariate and S^2 the pooled variance estimator given in Sect. 3.3 by (3.8a) and (3.8b). Calculation of this estimator requires unblinding. However, the following blinded variant of the above estimator can be used for blinded sample size recalculation. Here, the observations X_{ij} and W_{ij} are treated as arising from a single sample. Then, the regression model (20.10a) and (20.10b) with $\tau_E = \tau_C$ is fitted based on the pooled data of group E and C, and the related one-sample residual variance S_{OS}^2 is inserted for S_{res}^2 in (20.12):

$$S_{OS}^2 = \frac{1}{n_1 - 2} \cdot \sum_{i=C,E} \sum_{j=1}^{n_{i1}} \left(X_{ij} - \hat{\tau}_{OS} - \hat{\beta}_{OS} \cdot \left(W_{ij} - \bar{W}\right)\right)^2, \qquad (20.13)$$

where n_{i1} is the sample size in the internal pilot study in group i, $i = C, E$, $\hat{\tau}_{OS}$ and $\hat{\beta}_{OS}$ are the estimators of the model parameters of the regression model (20.10a) and (20.10b) with $\tau_E = \tau_C$ and computed based on the pooled data, and $\bar{W} = \frac{1}{n_1} \sum_{i=C,E} \sum_{j=1}^{n_{i1}} W_{ij}$ is the overall mean.

Remarks

1. For $\tau_E - \tau_C \neq 0$, the residual variance obtained from the pooled sample is an upward-biased estimator of $(1 - \rho^2) \cdot \sigma^2$. However, it was shown for the t-test that using an unbiased estimator of the nuisance parameter may lead to an undershooting of the required sample size (see Sect. 20.1.4). Having this result in mind, no corrected versions of the one-sample residual variance reducing or eliminating the bias were considered in the literature.
2. In their simulation study on the performance characteristics of the above-described sample size recalculation procedure, Friede and Kieser (2011a) included further sample size formulas for recalculation, amongst others Eq. (3.17) proposed by Frison and Pocock (1999). However, these alternative approaches lead to too high or too low sample sizes, respectively, and are therefore not considered here.

20.2.3 Type I Error Rate

At the time when writing this book, no analytical results on the distribution of T_{ANCOVA} when applied in the internal pilot study design were available. Friede and

Kieser (2011a) investigated the characteristics of the blinded sample size recalculation procedure described above (i.e. sample size recalculation based on Eq. (20.12) with use of the one-sample residual variance as blinded estimator of the nuisance parameters) by means of a simulation study. They considered the setting of a one-sided test at level $\alpha/2 = 0.025$, a desired power $1 - \beta = 0.80$, balanced sample size (i.e. $r = 1$), correlations $\rho = 0, 0.1, \ldots, 0.8$, variances $\sigma^2 = \sigma_Y^2 = 1$, and (standardized) treatment group differences $(\tau_E - \tau_C)_A = 0.3, 0.5$, and 0.7 to be used for sample size recalculation. The sample size of the internal pilot study was half the total sample size N obtained from the approximation formula (3.17) by inserting the true value of $(1 - \rho^2) \cdot \sigma^2$, i.e. $n_0 = N$ and $n_1 = 0.5 \cdot n_0$. An unrestricted design with an upper boundary $n_{\max} = 4 \cdot n_0$ was applied [see (18.3)]. In the fixed sample size design, these scenarios cover total sample sizes between 24 and 348. For investigation of the type I error rate, the true treatment group difference was set to $(\tau_E - \tau_C) = 0$. For these parameter constellations, the mean (minimum to maximum) actual level amounts to 0.0250 (0.0248 to 0.0253). Taking into account the number of 1,000,000 simulated trials with a standard error of 0.00016 for a true level of 0.025, there was no indication for an inflated type I error rate. This is consistent with the results reported in Sect. 20.1.3 for blinded sample size recalculation based on the one-sample variance and analysis with the t-test.

20.2.4 Power and Sample Size

The power and sample size characteristics of the blinded sample size recalculation procedure were investigated in the above-described simulation study of Friede and Kieser (2011a) by setting $\Delta = \Delta_A$. For all considered scenarios, the achieved power matches very well with the desired value. Furthermore, it turned out that the expected total sample size required to achieve the same power as in the fixed sample size design is by 6 to 7 patients higher for the internal pilot study design. This is similar to the t-test (here, the increase amounts to 8 for $\alpha/2 = 0.025$ and $1 - \beta = 0.80$) and can again be seen as the (small) price to be paid to achieve robustness of power with respect to assumption about the nuisance parameters.

In a further exploration within their simulation study, Friede and Kieser (2011a) varied the "timing" of the sample size recalculation by considering ratios of the internal pilot study sample size and the required sample size $n_1/N = 0.2, 0.3, \ldots, 1.2$, correlations $\rho = 0, 0.5$, and the same values for $\Delta = \Delta_A$ as above. For all parameter settings, the achieved power increases with increasing ratio n_1/N and falls for all scenarios below the aspired $1 - \beta = 0.80$ for $n_1/N = 0.2$ and above this value for $n_1/N = 1.0$. For fixed n_1/N, the difference between actual and desired power decreases with increasing n and the magnitude of the deviations is small. For example, for $0 \leq n_1/N \leq 1$ the deviation is less than 0.01 for $\Delta = \Delta_A = 0.3$ and $\rho = 0, 0.5$ ($N = 348$ or 262, respectively) and less than 0.025 for $\Delta = \Delta_A = 0.7$, $\rho = 0, 0.5$ ($N = 64$ or 48, respectively). Overall, the achieved power is rather robust with respect to the sample size of the internal pilot study. This is an important feature as in practice, other than in the simulation study, the actually required sample size N is unknown and with it the timing of the sample size recalculation in terms of the ratio n_1/N.

Remark Zimmermann et al. (2020) proposed a blinded sample size recalculation procedure for ANCOVA models with multiple random covariates. Its characteristics were investigated in a comprehensive simulation study considering the cases of two and three covariates. The results correspond to those reported above for a single covariate. In particular, the empirical type I error rates are close to the nominal level, the achieved power is robust against misspecifications in the planning phase, and the additional expected total sample size required as compared to the fixed design is for balanced groups and two covariates on average between 6 and 7.

Chapter 21
Comparison of Two Groups for Binary Outcomes and Test for Difference or Superiority

21.1 Background and Notation

We assume the same statistical model and notations as for the design with fixed sample size considered in Chap. 5. The event rates in the experimental (E) and the control group (C) are denoted as p_E and p_C, respectively. If n_E and n_C are the sample sizes in group E and C, respectively, the sum of events X_E and X_C follow the binomial distribution, i.e. $X_i \sim \text{Bin}(n_i, p_i)$, $i = C, E$. The sample size allocation ratio is again denoted by $r = n_E/n_C$. In the design with internal pilot study, we further need the notations n_{1i} and X_{1i}, $i = C, E$, respectively, for the group-wise sample size and the number of events observed until sample size recalculation occurs. As in the fixed sample size design, we consider the situations that the treatment effect is expressed in terms of the rate difference $\Delta = p_E - p_C$, the risk ratio $RR = \frac{p_E}{p_C}$, and the odds ratio $OR = \frac{p_E \cdot (1-p_C)}{p_C \cdot (1-p_E)}$. In case of testing for difference or superiority, the null-hypotheses formulated in terms of Δ, RR, and OR are equivalent and can thus be assessed by the same statistical test. In the following Sect. 21.2, we describe in detail the methods and characteristics of blinded sample size recalculation for the normal approximation test (which is equivalent to the chi-square test) and the case that the effect is expressed in terms of the difference in event rates. This enables an understanding of the principles underlying this approach for the situation of binary data. Based on this fundament, the cases of using the risk ratio or the odds ratio are sketched subsequently. Section 21.3 presents the methods for the Fisher-Boschloo test.

21.2 Asymptotic Tests

In this section, we consider the situation that the normal approximation test with test statistic

© Springer Nature Switzerland AG 2020
M. Kieser, *Methods and Applications of Sample Size Calculation and Recalculation in Clinical Trials*, Springer Series in Pharmaceutical Statistics,
https://doi.org/10.1007/978-3-030-49528-2_21

$$U = \sqrt{\frac{n_C \cdot n_E}{n_C + n_E}} \cdot \frac{\hat{p}_E - \hat{p}_C}{\sqrt{\hat{p} \cdot (1 - \hat{p})}} \qquad (21.1)$$

is applied for the assessment of difference or superiority, respectively. Here, $\hat{p}_i = \frac{X_i}{n_i}$, $i = C, E$, and $\hat{p} = \frac{X_C + X_E}{n_C + n_E}$. It was already mentioned in Sect. 5.2.1 that the normal approximation test and the chi-square test lead to the same test decision and to the same required sample size.

21.2.1   Difference of Rates as Effect Measure

In the following, we consider the two-sided test problem for assessing a difference between rates

$$H_0: p_E - p_C = 0$$
$$\text{versus} \qquad (21.2\text{a})$$
$$H_1: p_E - p_C \neq 0$$

or the one-sided test problem for demonstrating superiority of E versus C

$$H_0: p_E - p_C < 0$$
$$\text{versus} \qquad (21.2\text{b})$$
$$H_1: p_E - p_C > 0$$

where higher event rates are associated with advantageous results.

21.2.1.1   Blinded Sample Size Recalculation

In the setting considered here, the nuisance parameter determining the variance in case of fixed rate difference $\Delta_A = (p_E - p_C)_A$ is the overall rate $p = \frac{p_C + r \cdot p_E}{1+r}$. Based on the data of the internal pilot study, this quantity can be estimated in a blinded way by

$$\hat{p} = \frac{X_{1C} + X_{1E}}{n_{1C} + n_{1E}}.$$

For the fixed sample size design (see Sect. 5.2.1), the total required sample size is given by

$$n = \frac{1+r}{r}$$
$$\cdot \frac{z_{1-\alpha/2} \cdot \sqrt{(1+r) \cdot p_A \cdot (1-p_A)} + z_{1-\beta} \cdot \sqrt{r \cdot p_{C,A} \cdot (1-p_{C,A}) + p_{E,A} \cdot (1-p_{E,A})}}{\Delta_A^2}$$

$$(21.3)$$

with $p_A = \frac{p_{C,A} + r \cdot p_{E,A}}{1+r}$. Due to

$$p_{C,A} = p_A - \frac{r}{1+r} \cdot \Delta_A \qquad (21.4a)$$

and

$$p_{E,A} = p_A + \frac{\Delta_A}{1+r}, \qquad (21.4b)$$

the estimated overall rate $\hat{p}_A = \frac{X_{1E} + X_{1C}}{n_{1E} + n_{1C}}$ calculated based on the pooled data of the internal pilot study can be used to derive the estimators

$$\hat{p}_{C,A} = \hat{p}_A - \frac{r}{1+r} \cdot \Delta_A$$

$$\hat{p}_{E,A} = \hat{p}_A + \frac{\Delta_A}{1+r}.$$

These can be inserted in Eq. (21.3) to obtain the recalculated sample size:

$$\hat{N}_{\text{recalc}} = \frac{1+r}{r}$$
$$\cdot \frac{z_{1-\alpha/2} \cdot \sqrt{(1+r) \cdot \hat{p}_A \cdot (1-\hat{p}_A)} + z_{1-\beta} \cdot \sqrt{r \cdot \hat{p}_{C,A} \cdot (1-\hat{p}_{C,A}) + \hat{p}_{E,A} \cdot (1-\hat{p}_{E,A})}}{\Delta_A^2}.$$

$$(21.5)$$

The final sample size $\hat{N}$ is determined from $\hat{N}_{\text{recalc}}$ by applying, for example, one of the strategies (18.1)–(18.3) described in Chap. 18.

21.2.1.2 Type I Error Rate

Asymptotic tests such as the normal approximation test do not assure strict control of the type I error rate even in the fixed sample size design. Therefore, this behavious can also be expected when applied in a trial with internal pilot study design. A fair benchmark for the actual level of the test when applied in a recalculation design is thus the actual type I error rate within the fixed sample size design.

It was already shown in Sect. 5.2.1 that in the fixed sample size design with group-wise sample sizes n_C and n_E, respectively, and equal event rates $p_C = p_E = p$, the probability for rejecting the null-hypothesis in (21.2b) with the normal approximation test at one-sided nominal level $\alpha/2$ is given by

$$\alpha_{act}^{fix} = \sum_{x_C=0}^{n_C} \sum_{x_E=0}^{n_E} \binom{n_C}{x_C} \cdot \binom{n_E}{x_E} \cdot p^{x_C+x_E} \cdot (1-p)^{n-x_C-x_E} \cdot I_{\{u>z_{1-\alpha/2}\}}, \qquad (21.6)$$

where $I_{\{\cdot\}}$ denotes the indicator function. To determine the actual type I error rate of the normal approximation test in internal pilot study designs with blinded sample size recalculation as described above, one has to consider the realizations of contingency tables that can occur in the internal pilot study (with fixed sample sizes n_{1C} and n_{1E} and event frequencies X_{1C} and X_{1E}, respectively) and for the second stage of the trial after sample size recalculation (with random sample sizes n_{2C} and n_{2E} (determined by the estimated overall event rate $\hat{p} = \frac{X_{1C}+X_{1E}}{n_{1C}+n_{1E}}$) and event frequencies X_{2C} and X_{2E}, respectively). Then, the actual type I error rate of the recalculation procedure at one-sided nominal level $\alpha/2$ is for $p_C = p_E = p$ given by the following formula (Friede and Kieser 2004):

$$\alpha_{act}^{recalc} = \sum_{x_{1C}=0}^{n_{1C}} \sum_{x_{1E}=0}^{n_{1E}} \sum_{x_{2C}=0}^{n_{2C}} \sum_{x_{2E}=0}^{n_{2E}} \binom{n_{1C}}{x_{1C}} \cdot \binom{n_{1E}}{x_{1E}} \cdot \binom{n_{2C}}{x_{2C}} \cdot \binom{n_{2E}}{x_{2E}}$$
$$\cdot p^x \cdot (1-p)^{n-x} \cdot I_{\{u>z_{1-\alpha/2}\}}, \qquad (21.7)$$

where $x = \sum_{i=C,E} \sum_{j=1,2} x_{ij}$ and $n = \sum_{i=C,E} \sum_{j=1}^{2} n_{ij}$.

Figure 21.1 shows the actual type I error rate of the normal approximation test at one-sided nominal level $\alpha/2 = 0.025$ in the fixed sample size and in the internal pilot study design with unrestricted sample size recalculation, Fig. 21.1a for balanced groups and Fig. 21.1b for the allocation ratio $r = n_E/n_C = 3$. All combinations of the parameter scenarios $\Delta_A = 0.15, 0.20, 0.25, 0.30$ and $p_A = 0.30$ to 0.50 by 0.01 are considered. For the recalculation design, internal pilot study sample sizes equal to 25%, 50%, or 75%, respectively, of the sample size required in the fixed design are investigated leading in total to $3 \cdot 84 = 252$ constellations.

Overall, the actual levels in the fixed sample size design and the design with sample size recalculation are quite similar. For balanced groups, the mean level amounts to 0.0257 for both designs. Furthermore, the minimum values are nearly equal for both designs (0.0233 for fixed versus 0.0235 for recalculation design) while the maximum actual level is slightly larger for the fixed-size design (0.0308 versus 0.0289). Whether sample size recalculation is performed after 25 (left), 50 (middle), or 75% of the sample size required in the fixed design does not affect the actual type I error rate of the recalculation design (minimum 0.0238, 0.0236, 0.0235, maximum 0.0279, 0.0286, 0.0289, mean 0.0256, 0.0257, 0.0257, respectively). For the allocation ratio $r = 3$, the mean actual levels are even smaller than for balanced allocation and fall below the nominal level 0.025 for both designs. Furthermore,

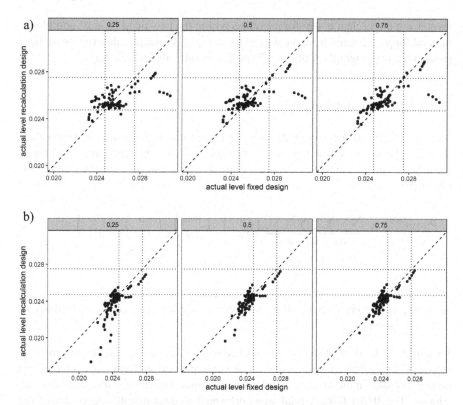

Fig. 21.1 Actual level of the normal approximation test at one-sided nominal level $\alpha/2 = 0.025$ for the fixed sample size design and for the internal pilot study design with blinded sample size recalculation. For the internal pilot study design, sample size recalculation after 25% (left), 50% (middle), and 75% of the sample size required in the fixed design is considered. The vertical and horizontal dashed lines mark deviations of 10 per cent from the nominal level, i.e. actual levels of 0.0225 and 0.0275, respectively. **a** balanced groups ($r = 1$); **b** allocation ratio $r = n_E/n_C = 3$

the actual levels of the internal pilot study design tend to slightly smaller values as compared to the fixed design: The mean (minimum, maximum) actual levels are 0.0243 (0.0214, 0.0279) for the fixed design and 0.0241 (0.0174, 0.0274) for the recalculation design. In summary, it can be concluded that if the extent of the α-inflation of the normal approximation test in the fixed sample size design is deemed to be acceptable, the same holds true for the internal pilot study design with blinded sample size recalculation.

If nevertheless the type I error rate shall be strictly controlled, the algorithms described in detail in Sect. 22.1.3 can be applied. In short, the procedures work as follows. For a specific clinical trial situation, the values for α, $1 - \beta$, r, Δ_A, and n_1 are fixed. Therefore, the actual type I error rate in the internal pilot study design given in (21.7) can be written as a function of p only. The type I error rate can hence be controlled by determining an adjusted level for which the maximum actual level is not larger than α for all $p \in [0, 1]$. An alternative approach is to assure that the maximum level is not larger than $\alpha - \gamma$ for all values within the two-sided $(1 - \gamma)$ confidence interval for p calculated from the data of the completed trial (see description of these procedures in paragraph "Control of type I error rate" in Sect. 22.1.3).

21.2.1.3 Power and Sample Size

The power realized in the internal pilot study design with blinded sample size recalculation can be calculated analogously to the type I error rate by replacing the common value $p_C = p_E = p$ under H_0 by the event rates $p_{C,A}$ and $p_{E,A}$ specified under the alternative:

$$
\mathrm{Power}^{\mathrm{recalc}} = \sum_{x_{1C}=0}^{n_{1C}} \sum_{x_{1E}=0}^{n_{1E}} \sum_{x_{2C}=0}^{n_{2C}} \sum_{x_{2E}=0}^{n_{2E}} \binom{n_{1C}}{x_{1C}} \cdot \binom{n_{1E}}{x_{1E}} \cdot \binom{n_{2C}}{x_{2C}} \cdot \binom{n_{2E}}{x_{2E}}
$$
$$
\cdot\, p_{C,A}^{x_C} \cdot \left(1 - p_{C,A}\right)^{n_C - x_C} \cdot p_{E,A}^{x_E} \cdot \left(1 - p_{E,A}\right)^{n_E - x_E} \cdot I_{\{u > z_{1-\alpha/2}\}}, \quad (21.8)
$$

where $n_i = n_{1i} + n_{2i}$, $i = C, E$. The expected sample size can be obtained by summing up the products of all possible final sample sizes and the probability of their realization.

Example 21.1 Alcohol abuse has numerous serious short- and long-term side effects. When admitted in an intensive care unit, patients suffering from alcoholism commonly sustain the alcohol withdrawal syndrome and show restlessness complications. The **BACLOREA trial** was performed to demonstrate superiority of the muscle-relaxing drug baclofen to placebo in preventing restlessness-related complications in at-risk drinkers for which at time of admission in the intensive care unit the mechanical ventilation duration was expected to be at least 24 h (Vourc'h et al., 2016). The study was conducted as a randomized, double-blind multicentre trial with primary outcome being a composite endpoint (see Remark 4 in Sect. 11.3 for an explanation of this term) of various agitation-related adverse events as, for example, unplanned extubation or falling out of bed. Sample size calculation was performed for analysis with the chi-square test at two-sided level $\alpha = 0.05$. This is equivalent to calculating the sample size for the normal approximation test at one-sided level $\alpha/2 = 0.025$ for which results on the characteristics of the actual type I error rate were reported above. A power of $1 - \beta = 0.80$ was desired for the assumed alternative that baclofen administration reduces the incidence of patients with at least one complication of agitation from $p_{C,A} = 0.42$ under placebo to $p_{E,A} = 0.27$, i.e. $\Delta_A = -0.15$ and $p_A = 0.5 \cdot (p_{C,A} + p_{E,A}) = 0.345$. For balanced sample size allocation, a total sample size of $n_0 = 2 \times 157 = 314$ patients is required in the fixed sample size design. Blinded sample size recalculation was planned after half of the initially planned sample size; $n_1 = 158$ was chosen to enable balanced groups. In the published clinical trial protocol, the sample size recalculation strategy was not specified. For illustration, we assume that an unrestricted design is applied.

Figure 21.2 shows the actual type I error rate for the fixed sample size design and the design with blinded sample size recalculation depending on the true overall rate p for the above-described setting of the BACLOREA trial (Fig. 21.2a) as well as for the same value for p_A but with assumed alternative $\Delta_A = 0.30$ resulting in $n_0 = 2 \times 39 = 78$ and in an internal pilot study sample size $n_1 = 40$ (Fig. 21.2b). The actual levels are symmetrical around $p = 0.5$. For the setting of the BACLOREA trial, the actual type I error rates of the fixed-size design and the design with blinded sample size recalculation are overall very similar with mean (minimum, maximum) values of 0.0248 (0.0236, 0.0256) for the fixed and 0.0250 (0.0235, 0.0267) for the internal pilot study design. If strict control of the type I error rate is deemed

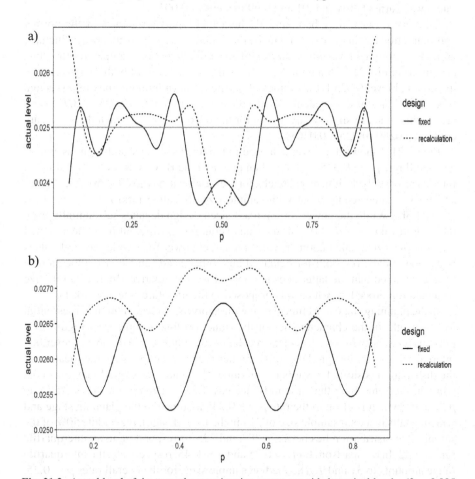

Fig. 21.2 Actual level of the normal approximation test at one-sided nominal level $\alpha/2 = 0.025$ for the fixed sample size design and for the internal pilot study design with blinded sample size recalculation depending on the overall rate p. **a** Fixed design and internal pilot study design as described in Example 21.1; **b** fixed design and internal pilot study design as described in Example 21.1 but with $\Delta_A = 0.30$

to be necessary, the adjusted one-sided nominal level $(\alpha/2)_{adj} = 0.0243$ has to be applied for the fixed sample size design and $(\alpha/2)_{adj} = 0.0247$ for the design with sample size recalculation. For the fixed design, the maximum actual level when applying the nominal level $\alpha/2 = 0.025$ is attained for the rate $p = 0.37$ and when applying the adjusted level $(\alpha/2)_{adj} = 0.0243$ it is achieved for $p = 0.355$. Interestingly, for the recalculation design, the rates where the maxima are attained are quite different, namely $p = 0.075$ for nominal level $\alpha/2 = 0.025$ and $p = 0.40$ for $(\alpha/2)_{adj} = 0.0247$. This is an explanation for the unexpected result that the adjusted level for the recalculation design is larger although at the same time the extent in type I error rate inflation is higher than for the fixed design. For the calculations, the rate p was varied between 0.01 and 0.99 in steps of 0.005.

Contrary to what may be assumed, the actual type I error rates of the normal approximation test in the fixed and in the recalculation design do not always fluctuate around the nominal level when varying the overall rate p. For example, for the same designs as in Fig. 21.2a but with $\Delta_A = 0.30$, the actual levels of both designs exceed the nominal level 0.025 for all values of p. The mean (minimum, maximum) actual type I error rates of the recalculation design amount to 0.0268 (0.0259, 0.0273) and are for this scenario slightly larger than for the fixed design for which the values are equal to 0.0260 (0.0253, 0.0268).

Figure 21.3 shows the power achieved in the fixed sample size design (assuming an overall rate $p_A = 0.345$ as done when planning the BACLOREA trial) and in the internal pilot study design, respectively, for the specifications given above (Fig. 21.3a) and for the same scenario but with sample size allocation ratio $r = n_E/n_C = 3$ (Fig. 21.3b). While the power for the fixed-sized design deviates substantially from the aspired value $1 - \beta = 0.80$ if the true overall rate is different from the assumed one, sample size recalculation provides the target power for a wide range of values for p. This holds true both for balanced and unbalanced sample size allocation. It should be noted that the latter does not apply for an alternative blinded sample size recalculation procedure which was proposed in the literature (see Remark 1).

In the planning stage of a clinical trial with sample size recalculation, it is essential to get insight in the characteristics of the (random) final sample size. Figure 21.4 presents box-and-whisker plots of the recalculated sample size in the above-described internal pilot study design. Due to the higher required sample size, the location of the distribution is shifted towards larger values the nearer the overall rate p is to 0.5 while at the same time the variability decreases. For example, for the overall rate $p = 0.35$ (which is close to the rate $p_A = 0.345$ assumed in the planning stage and which results in a total sample size of 314 in the fixed design), the middle 50% of the sample size distribution lies between 303 and 325 corresponding to an interquartile range of 22. In comparison, for $p = 0.25$ and $p = 0.45$, respectively, the interquartile range amounts to 31 and 7. The median sample sizes for the overall rates $p = 0.25$, $p = 0.35$, and $p = 0.45$ are 258, 315, and 343, respectively.

Remarks

1. Friede and Kieser (2004) considered a simplified version of the approximate sample size formula (21.3) for sample size calculation and recalculation:

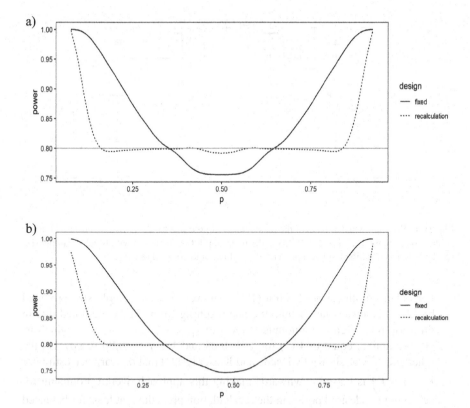

Fig. 21.3 Power of the normal approximation test at one-sided nominal level $\alpha/2 = 0.025$ for the fixed sample size design (assumed overall rate $p_A = 0.345$) and for the internal pilot study design with blinded sample size recalculation depending on the overall rate p. **a** Fixed design and internal pilot study design as described in Example 21.1 ($r = 1$); **b** fixed design and internal pilot study design as described in Example 21.1 but with allocation ratio $r = n_E/n_C = 3$

$$n = \frac{(1+r)^2}{r} \cdot \left(z_{1-\alpha/2} + z_{1-\beta}\right)^2 \cdot \frac{p \cdot (1-p)}{\Delta_A^2}, \qquad (21.9)$$

where $p = \frac{p_C + r \cdot p_E}{1+r}$ denotes the overall rate. As compared to (21.3), the variance under the alternative is in formula (21.9) replaced by that under the null-hypothesis. This is why (21.9) is in the literature denoted as the "homogeneous variance formula". Inserting the estimator $\hat{P}$ instead of p in Eq. (21.9) provides the recalculated total sample size

$$\hat{N}_{\text{recalc}} = \frac{(1+r)^2}{r} \cdot \left(z_{1-\alpha/2} + z_{1-\beta}\right)^2 \cdot \frac{\hat{P} \cdot \left(1 - \hat{P}\right)}{\Delta_A^2}. \qquad (21.10)$$

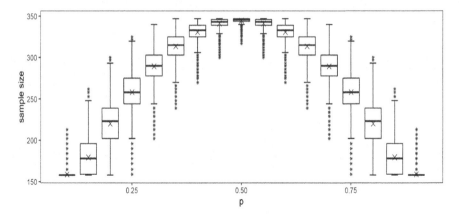

Fig. 21.4 Box-and-whisker plots of the recalculated sample size for the normal approximation test at one-sided nominal level $\alpha/2 = 0.025$ in the internal pilot study design with blinded sample size recalculation as described in Example 21.1 depending on the overall rate p

For balanced allocation, Lachin (1981) showed that the sample size provided by (21.9) is always larger than the one resulting from (21.3). He noted that in this case the difference amounts at most to $z_{1-\beta}^2 + 2 \cdot z_{1-\alpha/2} \cdot z_{1-\beta}$, which is, after rounding, equal to 4 (7) for $\alpha = 0.05$ and $1 - \beta = 0.80$ (0.90). On the other side, it was shown by Friede and Kieser (2004) that the variance estimator $\hat{P} \cdot \left(1 - \hat{P}\right)$ is biased downwards making this approach overall promising for achieving the desired power in the recalculation procedure, at least for balanced allocation. Friede and Kieser (2004) investigated the actual type I error rate and power of this approach over a wide range of parameter values. The actual levels for the fixed sample size design and for the recalculation design turned out to be very similar with mean values close to the nominal level for both designs and with comparable maximum and minimum values. Furthermore, for balanced sample size the achieved power is close to the desired value irrespective of the assumption on the overall rate. However, in case of unbalanced allocation, the "homogeneous variances sample size formula" (21.9) is not accurate. This translates to an unfavorable behavior of the sample size recalculation procedure based on it [see, for example, Fig. 21.3 in Friede and Kieser (2004)]. Therefore, application of the sample size recalculation method based on sample size Eq. (21.3) that is presented in the preceding section is generally recommended, at least for the case of unbalanced sample size.

2. Gould (1992) investigated the performance characteristics of blinded sample size recalculation procedures based on the estimated overall rate for the cases that the treatment effect is measured by the rate difference, the risk ratio, or the odds ratio. In contrast to the methodology presented above, simulation studies (instead of numerical computations) were used. Furthermore, the actual type I error rates were compared to the nominal level (instead of the actual level in the fixed sample size design).

3. Shih and Zhao (1997) proposed an alternative method for sample size recalculation in internal pilot study designs with binary data. Here, artificial "dummy" strata are built within which patients are allocated to the groups in an unbalanced way such that overall the desired allocation ratio is achieved. Using the known strata-wise allocation ratios and the estimated overall rates within the strata, the group-wise event rates can be estimated and inserted in the sample size formula for recalculation. On the patient level, no unblinding is required for this approach. However, an estimate of the treatment group difference becomes available during the course of the trial. To prevent that any (conscious or unconscious) bias is introduced in the study, processes and institutions have to be implemented assuring a strict firewall between the group of people performing sample size recalculation and the personnel involved in the trial. Therefore, although the treatment code is not broken for individual patients, the same precautions have to be taken as in case of unblinded sample size recalculation.

4. Wüst and Kieser (2005) proposed a method for the inclusion of short-term information in internal pilot study designs with binary data which is analogous to the one described in Sect. 20.1.4, Remark 3, for the case of normally distributed endpoints. Numerical computations and simulations showed that the actual type I error rate of the procedure is generally not larger than for the fixed sample size design and that the inclusion of short-term data can reduce the variance of the recalculated sample size considerably. As a consequence of the latter, the risk of obtaining an unacceptably low or a too high power is reduced as compared to utilizing the long-term information only.

21.2.2 Risk Ratio or Odds Ratio as Effect Measure

Methods for sample size calculation for the risk ratio or odds ratio as effect measure, respectively, and various asymptotic tests were presented in Sects. 5.2.2 and 5.2.3 for the fixed sample size design. Analogously to the case of the rate difference described in the preceding Sect. 21.2.1, blinded sample recalculation can be performed for these effect measures by using the estimated overall rate. For this, the relationship between overall rate, specified alternative, and rates under the alternative is used which is for the risk ratio given by

$$p_{C,A} = \frac{1+r}{1+r \cdot RR_A} \cdot p_A \tag{21.11a}$$

$$p_{E,A} = \frac{(1+r) \cdot RR_A}{1+r \cdot RR_A} \cdot p_A \tag{21.11b}$$

and for the odds ratio by

$$p_{C,A} = \frac{1+r}{1+r \cdot OR_A} \cdot p_A \tag{21.12a}$$

$$p_{E,A} = \frac{(1+r) \cdot OR_A}{1 + r \cdot OR_A} \cdot p_A. \tag{21.12b}$$

Inserting the estimated overall rate $\hat{p}$ instead of p_A in (21.11a) and (21.11b), or (21.12a) and (21.12b), respectively, results in estimates $\hat{p}_{C,A}$ and $\hat{p}_{E,A}$ which can be employed in the respective sample size formula to obtain the recalculated sample size. The characteristics of the resulting internal pilot study design with blinded sample size reassessment (type I error rate, power, and sample size) can then be investigated numerically just as shown for the rate difference in Sect. 21.2.1 or alternatively by simulations.

For the risk ratio and analysis with the chi-square test, Pobiruchin and Kieser (2013) investigated in a comprehensive simulation study the above-described approach based on sample size formula (5.12). Both balanced and unbalanced allocation ($r = 2$) were evaluated by considering 263 and 289 parameter scenarios, respectively. It turned out that the internal pilot study design shows actual levels close to the nominal ones and maximum type I error rates even slightly smaller as compared to the fixed sample size design. Furthermore, the target power of $1 - \beta = 0.80$ was achieved for a wide range of parameter scenarios, and the observed power was higher than 0.78 if the internal pilot study sample size was at least $n_1 = 35$.

There are currently no systematic investigations available on the performance characteristics of blinded sample size recalculation procedures based on the other tests and sample size formulas presented in the Sects. 5.2.2 and 5.2.3. However, due to the similar performance of the asymptotic tests and the exact Fisher-Boschloo test in the fixed design and due to the results presented in the following Sect. 21.3 for blinded sample size recalculation and subsequent analysis with the Fisher-Boschloo test, it can be expected that the results with respect to the actual type I error rate are very similar to those presented above for the chi-square test with risk difference and risk ratio as effect measure. Furthermore, the results with respect to power should also reflect those reported above for alternatives Δ_A and RR_A. In a specific clinical trial situation, it is recommended to perform the above-described calculations for the specified values of α, $1 - \beta$, allocation ratio, and alternative for various pilot sample sizes n_1 to get an impression of the performance and to support an appropriate choice of the internal pilot sample size.

21.3   Fisher-Boschloo Test

The Fisher-Boschloo test and the related method for sample size calculation in the fixed sample size design were described in Sect. 5.2.3. The Fisher-Boschloo test (as Fisher's exact test on which its construction is based) conditions on the total number of observed events. Therefore, it follows from the permutation principle presented in Chap. 19 that the type I error rate is controlled in internal pilot studies for any blinded sample size recalculation rule if the Fisher-Boschloo test is applied in the analysis. Regardless of which measure for the treatment effect is used, the two rates $p_{C,A}$ and $p_{E,A}$ under the alternative have to be specified. Alternatively, the overall rate $p_A = \frac{p_{C,A}+r \cdot p_{E,A}}{1+r}$ and the value for the alternative Δ_A, RR_A, and OR_A, respectively, can be defined from which the rates under the alternative can be determined according to (21.4a) and (21.4b), (21.11a) and (21.11b), and (21.12a) and (21.12b). Blinded sample size recalculation can thus be performed by determining the estimates $\hat{p}_{C,A}$ and $\hat{p}_{E,A}$ from the overall rate obtained from the pooled data and by recalculating the sample size based on these updated values. As determination of the sample size for the Fisher-Boschloo test is computationally intensive, there are currently no investigations available evaluating the power characteristics of this procedure. However, due to the similarity of the characteristics of the Fisher-Boschloo test to those of the asymptotic tests in the fixed design, it can be expected that the performance of these two tests is also similar for the design with blinded sample size recalculation.

Chapter 22
Comparison of Two Groups for Normally Distributed Outcomes and Test for Non-Inferiority

22.1 t-Test

22.1.1 Background and Notation

We assume the same statistical model as for the assessment of non-inferiority with the t-test in the fixed sample size design (Sect. 8.2), i.e. mutually independent random variables X_{Cj} and X_{Ej} with expectation μ_C and μ_E, respectively, and common unknown variance σ^2. The sample size is allocated with a ratio of $r : 1$ between E and C. The test problem to demonstrate non-inferiority of E versus C (assuming that higher values of the outcome refer to more beneficial results) is given by

$$H_0: \mu_E - \mu_C \leq -\delta$$
$$\text{versus} \tag{22.1}$$
$$H_1: \mu_E - \mu_C > -\delta$$

with pre-specified non-inferiority margin $\delta > 0$. The null-hypothesis can be tested by applying the shifted t-test with test statistic

$$T = \sqrt{\frac{r}{(1+r)^2} \cdot n} \cdot \frac{\bar{X}_E - \bar{X}_C + \delta}{S} \tag{22.2}$$

with group means $\bar{X}_E$ and $\bar{X}_C$, pooled standard deviation S given in Sect. 3.3 by (3.8a) and (3.8b), and final total sample size n (see Sect. 8.2.1).

22.1.2   Blinded Sample Size Recalculation

Sample size recalculation is performed based on the approximation formula for the shifted z-test:

$$\hat{N}_{\text{recalc}} = \frac{(1+r)^2}{r} \cdot \left(z_{1-\alpha/2} + z_{1-\beta}\right)^2 \cdot \left(\frac{S}{\Delta_A - \delta}\right)^2, \qquad (22.3)$$

where $\alpha/2$ is the one-sided significance level, $1 - \beta$ the desired power, $\Delta_A = (\mu_E - \mu_C)_A$ the specified alternative, and S^2 an appropriate blinded estimator of the population variance σ^2. As for the assessment of difference or superiority, respectively, the one-sample variance estimator obtained from the pooled data of groups C and E is used for blinded sample size recalculation:

$$S_{\text{OS}}^2 = \frac{1}{n_1 - 1} \cdot \sum_{i=C,E} \sum_{j=1}^{n_{i1}} \left(X_{ij} - \bar{X}_1\right)^2, \qquad (22.4)$$

where $\bar{X}_1 = \frac{1}{n_1} \cdot \sum_{i=C,E} \sum_{j=1}^{n_{i1}} X_{ij}$ denotes the overall mean of the internal pilot study based on the data of n_1 patients and n_{i1}, $i = C, E$, are the group-wise internal pilot study sample sizes.

Remarks

1. In non-inferiority assessment, the alternative of equal efficacy of C and E is commonly used for sample size calculation, i.e. $\Delta_A = 0$. Note that due to (20.5), S_{OS}^2 is an unbiased estimator for σ^2 under this alternative.
2. The non-inferiority test problem (22.1) can be formulated as a superiority test problem for transformed data by defining $\mu_E^* = \mu_E + \delta$ and $\mu_C^* = \mu_C$ which results in

$$\begin{aligned} H_0 &: \mu_E^* - \mu_C^* \le 0 \\ &\text{versus} \\ H_1 &: \mu_E^* - \mu_C^* > 0 \end{aligned} \qquad (22.5)$$

Test problem (22.5) can be assessed by using the common t-test statistic based on the observations $X_{Ej}^* = X_{Ej} + \delta$ and $X_{Cj}^* = X_{Cj}$. The following relationship between the one-sample variance estimators for the untransformed and the transformed data, S_{OS}^2 and $\left(S_{\text{OS}}^*\right)^2$, can be derived:

$$\begin{aligned} S_{\text{OS}}^2 = \left(S_{\text{OS}}^*\right)^2 &- \frac{r}{(1+r)^2} \cdot \frac{2 \cdot n_1}{(n_1 - 1)} \cdot \delta \cdot \left(\bar{X}_{1E} - \bar{X}_{1C}\right) \\ &- \frac{r}{(1+r)^2} \cdot \frac{n_1}{(n_1 - 1)} \cdot \delta^2, \end{aligned} \qquad (22.6)$$

where $\bar{X}_{1i}, i = C, E$ denotes the mean in group i in the internal pilot study. Equation (22.6) links the variance estimators to be used for blinded sample size recalculation when assessing non-inferiority based on the untransformed data and superiority based on the transformed data (see paragraph "Reasons for type I error rate inflation" in the following Sect. 22.1.3).

22.1.3 Type I Error Rate

22.1.3.1 Actual Type I Error Rate

The actual type I error rate for the shifted *t*-test statistic (22.2) after sample size recalculation based on the one-sample variance estimator can be determined analogously to the approach described in Sect. 20.1.3 for the assessment of difference or superiority. The sample size recalculation rule as well as the test statistic can be written in terms of random variables whose joint distribution is known. The actual type I error rate can then be calculated by numerical integration or estimated by simulations (Friede and Kieser 2003; Lu 2016). The sample size recalculation rule when written in terms of the true required sample size N is equal to the one for testing difference or superiority, respectively, and is thus given by (20.7). Furthermore, the following decomposition of the test statistic holds true:

$$T = \frac{\sqrt{\frac{n_1}{n}} \cdot Z_1 + \sqrt{\frac{n_2}{n}} \cdot Z_2 + \sqrt{\frac{r}{(1+r)^2}} \cdot n \cdot \left(\frac{\Delta - \delta}{\sigma}\right)}{\sqrt{\frac{1}{n-2} \cdot (V_1 + V_2^*)}}, \qquad (22.7)$$

where n_i denotes the total sample size before ($i = 1$) or after ($i = 2$) sample size reassessment and where Z_i, $i = 1, 2$, V_1, and V_2^* are defined as in Sect. 20.1.3. Furthermore, the (conditional) distributions of Z_i, $i = 1, 2$, V_1, and V_2^* are as given in (20.9a) to (20.9d) (Lu 2016) which allows computation or simulation of the actual type I error rate. Note that, due to the representation of the sample size recalculation rule (20.7) and the test statistic (22.7), the actual type I error rate can be written as a function of α, n_1, N, and δ/σ. Comprehensive investigations by means of numerical integration (Friede and Kieser 2003) and simulation studies (Lu 2016) showed that the type I error rate may be considerably inflated for non-inferiority testing following blinded sample size recalculation (see also Example 22.1 below). This is in contrast to the assessment of superiority and difference and gives cause to two questions: Where does this α-inflation come from, and how can control of the nominal level be achieved?

22.1.3.2 Reasons for Type I Error Rate Inflation

Friede and Kieser (2003) showed that the component $\sqrt{\frac{n_1}{n}} \cdot Z_1 + \sqrt{\frac{n_2}{n}} \cdot Z_2$ of the numerator of the t-test statistic T in (22.7) is positively biased under the null-hypothesis which leads to an excess of the rejection probability. However, the bias of $\sqrt{\frac{n_1}{n}} \cdot Z_1 + \sqrt{\frac{n_2}{n}} \cdot Z_2$ approaches zero when δ/σ increases while at the same time the actual type I error rate remains above α. Therefore, Friede and Kieser (2003) stated that this bias is not the only reason for the level-inflation. In fact, based on Eq. (22.6), Glimm and Läuter (2013) pointed out another aspect and gave a further intuitive explanation that the nominal type I error rate may be exceeded. Assessing non-inferiority by testing for superiority based on the transformed data (see Remark 2 in Sect. 22.1.2) and recalculating the sample size based on the related one-sample variance estimator $\left(S_{OS}^*\right)^2$ keeps the nominal level according to the results presented in Sect. 20.1.3. Note, however, that unblinding is required to calculate $\left(S_{OS}^*\right)^2$ as the observations of one of the groups have to be transformed (subtraction of δ from all outcomes in C or addition of δ to all outcomes in group E, respectively). Rearranging Eq. (22.6), the following relationship between S_{OS}^2 and $\left(S_{OS}^*\right)^2$ follows:

$$S_{OS}^2 = \left(S_{OS}^*\right)^2 - \frac{r}{(1+r)^2} \cdot \frac{2 \cdot n_1}{(n_1 - 1)} \cdot \delta \cdot \left(\bar{X}_{1E} - \bar{X}_{1C} + \frac{\delta}{2}\right).$$

Therefore, for sample means of the internal pilot study with $\bar{X}_{1E} - \bar{X}_{1C} > -\frac{\delta}{2}$, which are in favor of the alternative hypothesis, sample size recalculation leads to fewer additional patients to be included as compared to blinded sample size recalculation for superiority testing. The reverse is true for interim effects in favor of the null-hypothesis. Overall, this leads to an increased rejection probability under H_0 and thus to an inflated type I error rate.

22.1.3.3 Control of Type I Error Rate

It was already mentioned above that for a defined sample size recalculation rule the actual type I error rate can be represented as a function of α, n_1, N, and δ/σ. In a specific clinical trial situation, the sample size recalculation rule as well as α, n_1, and δ are known while σ (that determines N and *vice versa*) is the only unknown quantity. To control the actual type I error rate at the nominal level, an adjusted level α_{adj} is to be determined such that the related maximum actual type I error rate $\alpha_{act}^{max}\left(\alpha_{adj}\right)$ does not exceed α over the range of possible values of σ. As the maximum actual level α_{act}^{max} is a monotonous function of the nominal type I error rate, the adjusted level can be found by solving the equation $\alpha_{act}^{max}\left(\alpha_{adj}\right) = \alpha$. The solution can be determined by applying the following iterative algorithm (Kieser and Friede 2000a):

Start: $\alpha_{adj}^{(0)} = \alpha$ (initial value for α_{adj})

Step k, $k \geq 1$: $\alpha_{adj}^{(k)} = \alpha_{adj}^{(k-1)} \cdot \frac{\alpha}{\alpha_{act}^{max}\left(\alpha_{adj}^{(k-1)}\right)}$

Stop if $\alpha_{\text{act}}^{\max}\left(\alpha_{\text{adj}}^{(s)}\right) - \alpha \leq \varepsilon$ for a specified precision $\varepsilon > 0$; define $\alpha_{\text{adj}} = \alpha_{\text{adj}}^{(s)}$.

Note that when conducting this algorithm, the adjusted significance level is not only to be applied in hypothesis testing but has also to be inserted in the formula used for sample size recalculation.

Remarks

1. Lu (2016) determined via simulation $\alpha_{\text{act}}^{\max}$ and α_{adj} for a range of scenarios which, however, are restricted to the alternative $\Delta_A = 0$ to be used for sample size (re)calculation.

2. In the situation considered in the previous remark, the maximum type I error rate may be achieved for a value of σ which is far away from the estimated pooled standard deviation obtained from the data of the completed trial. Then, little evidence exists that this worst case actually occurs. Therefore, the following approach may be seen as an attractive alternative for level control: A value $0 < \gamma < \alpha$ is pre-specified (for example $\gamma = 0.0001$) and an adjusted level is determined such that the maximum actual level does not exceed $\alpha - \gamma$ for all values of σ which are included in the $(1 - \gamma)$ confidence interval for σ based on the data of the completed study. Then, the type I error rate is controlled by α when this adjusted level is applied for hypothesis testing (Friede and Kieser 2011b). Note that the strategy to determine α_{adj} has to be completely pre-specified, in particular the value of γ in case of using the confidence interval based approach. Furthermore, although the maximum actual level within the $(1 - \gamma)$ confidence interval is, of course, at most as large as (and potentially smaller than) the global maximum, the related adjusted level is not necessarily larger than the global counterpart due to the stricter requirement of keeping the maximum actual level below $\alpha - \gamma$ instead of α. Example 23.1 illustrates practical application of this approach.

22.1.4 Power and Sample Size

The expected power can be calculated analogously to the actual type I error rate by considering scenarios under the alternative. The expected sample size can be determined along the same lines as described for superiority and difference testing in Sect. 20.1.4.

It was already mentioned in Sect. 22.1.2, Remark 1, that the estimator S_{OS}^2 used for sample size recalculation is unbiased for the alternative $\Delta_A = 0$ which is commonly applied in non-inferiority trials. At first sight, this may be seen as an appealing feature. However, it was pointed out in Sect. 20.1 that the slight upward-bias of S_{OS}^2 in case of a difference between groups was a welcome effect as it compensates the small loss in efficiency due to sample size recalculation. In superiority trials, this upward-bias leads to an additional total sample size of 8 or 12 patients, respectively, for the common values of $\alpha/2 = 0.025$ and $1 - \beta = 0.80$ or 0.90 (see Remark 1 in Sect. 20.1.2). The effect of this sample size overshooting on power is noticeable only for small sample sizes. As non-inferiority trials commonly require a large number of patients to be included, the smaller bias of S_{OS}^2 as compared to superiority trials has usually no relevant impact on power. In case of a small sample size, an inflation factor as the one presented in Sect. 22.2 may be used to counterbalance the fact that the bias of S_{OS}^2 in non-inferiority trials is reduced as compared to superiority trials.

Example 22.1 The **SITAGRAMI trial** (see Example 20.1) compared combined application of granulocyte colony stimulating factor and Sitagliptin (E) to placebo in patients with acute myocardial infarction. Let us assume for illustration that the study is performed with an active control instead of placebo and that non-inferiority of E versus this control is to be shown for the endpoint "change in ventricular ejection fraction between screening and 6 month follow-up". We suppose a pre-specified margin $\delta = 1.5$ and assume equal efficacy of the comparators for sample size (re)calculation, i.e. $\Delta_A = 0$. All other specifications are set equal to those of the original trial, i.e. $\alpha = 0.025$, one-sided, $1 - \beta = 0.80$, and $\sigma_0 = 5.5$. Then, the approximation formula based on the shifted z-test leads to total required sample size of $n_0 = 424$ patients. Figure 22.1 shows the simulated type I error rate for the case that a blinded sample size recalculation based on S_{OS}^2 is performed after the data of $n_1 = 20, 200$ or 300 patients are available, respectively, and that the maximum total sample size is fixed at $n_{\max} = 2 \cdot n_0 = 848$ (1,000,000 replications for each parameter constellation with a standard error of 0.00016 for a true type I error rate of 0.025). While the type I error rate is hardly affected for $n_1 = 200$ and 300, there is a marked inflation for $n_1 = 20$ with a maximum actual level of 0.0307 occurring for $\sigma = 1.5$. To assure control of the type I error rate at 0.025 for the internal pilot sample size $n_1 = 20$, an adjusted level $\alpha_{\mathrm{adj}} = 0.0199$ has to be applied. The maximum actual level occurs for a value of σ that is far from the anticipated one. Therefore, one may instead pre-specify the confidence interval based approach for level control that may

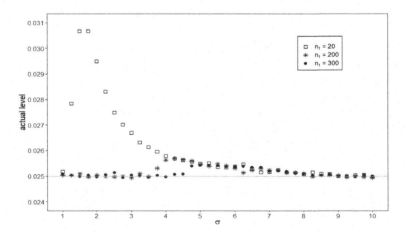

Fig. 22.1 Actual level of the shifted t-test at one-sided nominal level $\alpha/2 = 0.025$ for the internal pilot study design described in Example 20.1 depending on the standard deviation σ

be less conservative if the observed standard deviation will be within the expectations (see Example 23.1 for an illustration of this method).

22.2 Analysis of Covariance

We assume the same statistical model as considered in Sect. 20.2 for the assessment of difference or superiority with ANCOVA, i.e. (20.6a) and (20.6b). The test problem for demonstrating non-inferiority of the experimental treatment E to the control C is given by

$$H_0: \tau_E - \tau_C \leq -\delta$$
$$\text{versus} \tag{22.8}$$
$$H_1: \tau_E - \tau_C > -\delta$$

where $\delta > 0$ denotes the pre-specified non-inferiority margin. The shifted ANCOVA test statistic

$$T_{ANCOVA} = \frac{\hat{\tau}_E - \hat{\tau}_C + \delta}{\sqrt{\widehat{Var}(\hat{\tau}_E - \hat{\tau}_C)}}$$

with $\widehat{Var}(\hat{\tau}_E - \hat{\tau}_C)$ given by (3.14) can be applied for testing H_0. Blinded sample size recalculation can be performed by inserting in the approximate sample size formula (8.7) an appropriate estimator of $(1 - \rho^2) \cdot \sigma^2$. Under the alternative $\Delta_A = 0$, which is commonly assumed in non-inferiority trials, the one-sample residual variance estimator S_{OS}^2 given in Eq. (20.13) is unbiased. However, we know from Chap. 20 that a (slightly) upwards-biased variance estimator is advantageous for sample size recalculation to compensate the loss in power which may occur due to the option for sample size modification. Motivated by a proposal made by Zucker et al. (1999) for sample size recalculation based on the (unbiased) pooled variance estimator in the *t*-test situation, Friede and Kieser (2011b) suggested to multiply the sample size obtained from (8.7) with the "inflation factor"

$$IF = \left(\frac{t_{1-\alpha/2, n_1-2} + t_{1-\beta, n_1-2}}{t_{1-\alpha/2, \hat{N}_{recalc}-2} + t_{1-\beta, \hat{N}_{recalc}-2}} \right)^2. \tag{22.9}$$

This leads to the following formula for the recalculated total sample size:

$$\hat{N}_{recalc} = IF \cdot \frac{(1+r)^2}{r} \cdot \left(z_{1-\alpha/2} + z_{1-\beta}\right)^2 \cdot \frac{S_{OS}^2}{(\Delta_A + \delta)^2} + \frac{\left(z_{1-\alpha/2}\right)^2}{2} \tag{22.10}$$

Friede and Kieser (2011b) performed an extensive simulation study to investigate the characteristics of this sample size recalculation procedure. Overall, more than 200 scenarios with $\alpha/2 = 0.025$, $1 - \beta = 0.80$, and balanced groups were considered covering required total sample sizes N between 20 and 600. It turned out that the type I error rate may be considerably inflated, especially for small N, with actual levels up to 0.045. For total sample sizes N of at least 200, the inflation, if any, is small with actual type I error rates below 0.026 (1,000,000 simulated trials with a standard error of the estimated level of 0.00016 for a true type I error rate of $\alpha/2 = 0.025$). For small N, the inflation is larger for smaller internal pilot study sample size n_1, while there is no apparent effect of n_1 for large N. Note that in order to investigate the actual type I error rate it is sufficient to consider its dependence on the residual variance $(1 - \rho^2) \cdot \sigma^2$ instead of looking at all three variance-covariance parameters σ^2, σ_Y^2, and ρ^2 (Friede and Kieser 2011b). This enables to control the actual type I error rate at the nominal level by applying the methods described in Sect. 22.1.3 for the shifted t-test. The maximum actual type I error rate has now to be searched over the range of $(1 - \rho^2) \cdot \sigma^2$ (instead of σ^2 for the t-test) and adjustment of the level has to be done accordingly.

Chapter 23
Comparison of Two Groups for Binary Outcomes and Test for Non-Inferiority

23.1 Background and Notation

The same statistical model is assumed and the same notations are used as for the assessment of non-inferiority with binary endpoints in the fixed sample size design (see Sect. 8.4). Denoting with p_i the event rate in group i, $i = C, E$, and by n_i the group-wise sample size, the sum of events in group i is binomially distributed as $X_i \sim \text{Bin}(n_i, p_i)$, $i = C, E$. Furthermore, n_{1i} and X_{1i}, $i = C, E$, are the sample sizes and the number of events, respectively, in the internal pilot study and $r = n_E/n_C = n_{1E}/n_{1C}$. We restrict in the following sections to the case that asymptotic tests are applied in the analysis. The case of using the exact unconditional Fisher-Boschloo test is addressed in a Remark at the end of this chapter.

23.2 Difference of Rates as Effect Measure

For the difference of rates, the test problem for assessing non-inferiority of E as compared to C is given by

$$H_0: p_E - p_C \le -\delta$$
$$\text{versus} \tag{23.1}$$
$$H_1: p_E - p_C > -\delta$$

where $\delta > 0$ is the pre-specified non-inferiority margin and where larger rates relate to more beneficial effects. The test statistic proposed by Farrington and Manning (1990), which is given in (8.15) and (8.16), is used for the assessment of (23.1).

© Springer Nature Switzerland AG 2020
M. Kieser, *Methods and Applications of Sample Size Calculation and Recalculation in Clinical Trials*, Springer Series in Pharmaceutical Statistics,
https://doi.org/10.1007/978-3-030-49528-2_23

23.2.1 Blinded Sample Size Recalculation

Sample size calculation and recalculation is based on sample size formula (8.18):

$$n = \frac{1+r}{r}$$
$$\cdot \frac{\left(z_{1-\alpha/2} \cdot \sqrt{r \cdot \tilde{p}_C \cdot (1 - \tilde{p}_C) + \tilde{p}_E \cdot (1 - \tilde{p}_E)} + z_{1-\beta} \cdot \sqrt{r \cdot p_{C,A} \cdot (1 - p_{C,A}) + p_{E,A} \cdot (1 - p_{E,A})} \right)^2}{(\Delta_A + \delta)^2}$$

$$(23.2)$$

with $\Delta_A = p_{E,A} - p_{C,A}$ and $\tilde{p}_C$, $\tilde{p}_E$ denoting the large sample approximations of the restricted maximum likelihood estimators (see Sect. 8.4.1.1). As for the test for difference or superiority, respectively, the overall event rate is estimated from the pooled internal pilot study data by $\hat{p}_A = \frac{x_{1E} + x_{1C}}{n_{1E} + n_{1C}}$. Then, again as in the case of difference or superiority testing, the event rates are estimated based on Eqs. (21.4a) and (21.4b): $\hat{p}_{C,A} = \hat{p}_A - \frac{r}{1+r} \cdot \Delta_A$ and $\hat{p}_{E,A} = \hat{p}_A + \frac{\Delta_A}{1+r}$. Inserting these values instead of $p_{C,A}$ and $p_{E,A}$ when calculating the restricted rate estimates results in the quantities $\hat{\tilde{p}}_C$ and $\hat{\tilde{p}}_E$. Blinded sample size recalculation can now be performed by employing the values $\hat{p}_{C,A}$ and $\hat{p}_{E,A}$ (instead of $p_{C,A}$ and $p_{E,A}$) and $\hat{\tilde{p}}_C$ and $\hat{\tilde{p}}_E$ (instead of $\tilde{p}_C$ and $\tilde{p}_E$) in sample size formula (23.2). This procedure was proposed by Friede et al. (2007) who also investigated its characteristics.

23.2.2 Type I Error Rate

As the Farrington-Manning test is an asymptotic test and thus does not guarantee strict control of the type I error rate in the fixed sample size design, the actual level of the recalculation procedure is not only to be compared to the nominal level but also to the actual type I error rate of the Farrington-Manning test for fixed sample size. The computations are very similar to those presented in Chap. 21 for the normal approximation test taking into account that the rejection probabilities have now to be determined for $p_E - p_C = -\delta$. The actual type I error rate for nominal one-sided type I error rate $\alpha/2$ is for the fixed sample size design therefore given by

$$\alpha_{act}^{fix} = \sum_{x_C=0}^{n_C} \sum_{x_E=0}^{n_E} \binom{n_C}{x_C} \cdot \binom{n_E}{x_E} \cdot p_C^{x_C} \cdot (1 - p_C)^{n_C - x_C}$$
$$\cdot p_E^{x_E} \cdot (1 - p_E)^{n_E - x_E} \cdot I_{\{u > z_{1-\alpha/2}\}}$$

$$(23.3)$$

and in the internal pilot study design with blinded sample size recalculation by

$$\alpha_{act}^{recalc} = \sum_{x_{1C}=0}^{n_{1C}} \sum_{x_{1E}=0}^{n_{1E}} \sum_{x_{2C}=0}^{n_{2C}} \sum_{x_{2E}=0}^{n_{2E}} \binom{n_{1C}}{x_{1C}} \cdot \binom{n_{1E}}{x_{1E}} \cdot \binom{n_{2C}}{x_{2C}} \cdot \binom{n_{2E}}{x_{2E}}$$

$$\cdot p_C^{x_C} \cdot (1 - p_C)^{n_C - x_C} \cdot p_E^{x_E} \cdot (1 - p_E)^{n_E - x_E} \cdot I_{\{u > z_{1-\alpha/2}\}}, \tag{23.4}$$

where $p_E = p_C - \delta$, u denotes the Farrington-Manning test statistic, and $I_{\{.\}}$ is the indicator function.

Friede et al. (2007) evaluated the Eqs. (23.3) and (23.4) for the unrestricted design (18.2) covering a wide range of parameter settings ($\alpha/2 = 0.025$, $1 - \beta = 0.80$, $r = 1, 2, 3$, $\delta = 0.1$, $\Delta_A = 0$ for sample size recalculation, overall event rate $p = 0.3, 0.31, \ldots, 0.7$, internal pilot sample size equal to half of the truly required fixed sample size). They found that the actual type I error rate of the Farrington-Manning test in the fixed-size and the internal pilot study design are very close both being equal or near to the nominal level. Therefore, in the considered scenarios, blinded sample size recalculation did not lead to an inflation of the type I error rate beyond that occurring in the fixed sample size design. Especially, the actual one-sided type I error rate was in all considered scenarios for both designs smaller than 0.0265.

23.2.3 Power and Sample Size

The actual power of the Farrington-Manning test in the fixed sample size and in the blinded sample size recalculation design, respectively, can be calculated according to the Eqs. (23.3) and (23.4) just by inserting the specified alternative defined by $p_{E,A}$ and $p_{C,A}$, where $\Delta_A = p_{E,A} - p_{C,A} > -\delta$. Characteristics of the distribution function of the recalculated sample size as, for example, the expected sample size or quantiles can be obtained by considering all possible final sample sizes together with their probability of realization. Analogously, the expected sample size can be calculated by summing up the products of all possible final sample sizes and their occurrence probabilities.

For the same parameter scenarios as for the evaluation of the actual type I error rate (restricting p to the values 0.3, 0.4, ..., 0.7), Friede et al. (2007) showed by computations based on (23.3) and (23.4) that the blinded sample size recalculation procedure works well and, in contrast to the fixed-size design, achieves the desired power $1 - \beta$ irrespective of whether or not the assumed overall rate matches to the true value. If the sample size of the internal pilot study design is smaller than half of the sample size required for the fixed design, the achieved power may fall slightly below the target value. This effect was already mentioned in Chap. 21 for difference or superiority testing and is due to the downward-bias of the blinded variance estimator (Friede and Kieser 2004).

When planning a clinical trial with blinded sample size recalculation, one should always evaluate the distribution of the sample size in the preparation stage to get an impression of its characteristics (see, for example, Figs. 20.3 and 21.4). For the parameter scenarios given above, Friede et al. (2007) showed that the expected sample size of the sample size recalculation design is very similar to the sample size required in the fixed design when the planning assumptions are correct. The variability of the recalculated sample size depends on the parameter setting. For example, the difference between the 0.95- and the 0.05-quantile of the sample size distribution varies substantially between 6 (for $p = 0.5$ and $r = 1$, i.e. $N = 780$) and 145 (for $p = 0.7$ and $r = 3$, i.e. $N = 815$) for the setting $\alpha/2 = 0.025$, $1 - \beta = 0.80$, $\delta = 0.1$, and $\Delta_A = 0$. Such considerations are important to support the logistical planning of clinical trials with sample size recalculation.

Example 23.1 Atrial fibrillation (AF) is a common form of cardiac arrhythmia which is associated with increased mortality, a greater risk of stroke, and a substantial burden of symptoms. Pulmonary vein isolation using radiofrequency energy is an effective treatment for paroxysmal AF which, however, is technically challenging and may cause severe complications. As a more recent alternative, a cryoballoon catheter may be used for pulmonary vein isolation. At the time when planning the **FreezeAF trial** (Luik et al. 2010, 2015), no randomized comparison between these two treatment options was available. It was the aim of this study to demonstrate that isolating the pulmonary veins using cryoballoon ablation (E) is non-inferior to radiofrequency ablation (C) in patients with paroxysmal AF. Absence of atrial arrhythmias in combination with absence of persistant complications during the 6- and 12-month follow-up period were defined as co-primary endpoints (see Sect. 11.3.2a for an explanation of this term). For illustration purposes, we assume that the 12-month outcome was the single primary endpoint. The non-inferiority margin was pre-specified as $\delta = 0.15$, and the Farrington-Manning test at one-sided level $\alpha/2 = 0.025$ was to be applied for the assessment of non-inferiority. On the basis of available data on the two treatments, equal rates $p_{C,A} = p_{E,A} = 0.78$ were assumed for sample size calculation. According to Eq. (23.2), an initial total sample size of $n_0 = 2 \times 122 = 244$ patients was calculated for an aspired power of 0.80. Due to the uncertainty with respect to the overall rate, a blinded sample size recalculation was pre-specified in the protocol that was performed after the endpoint data of 100 patients were available. No explicit upper bound for the final sample size was fixed. However, it should be noted that an implicit upper bound is defined by the total sample size calculated for the "worst case" overall rate of 0.5, which is equal to 344 patients. Figure 23.1 shows the actual type I error rate for this internal pilot study design as well as for the associated fixed sample size design depending on the (unknown) true overall rate p. The maximum actual type I error rate inflation is larger for the fixed design (maximum level 0.0281) as compared to the design with sample size recalculation (maximum level 0.0264).

If strict control of the type I error rate at the nominal level $\alpha/2 = 0.025$ over all possible values of p shall be guaranteed, an adjusted level $(\alpha/2)_{\text{adj}} = 0.0221$ has to be applied for the fixed sample size design and $(\alpha/2)_{\text{adj}} = 0.0241$ for the internal pilot study design. Calculations considered all values the rate p can reach under the null-hypothesis, i.e. the range from 0.075 to 0.925, in steps of 0.05. The maximum actual type I error rate for the internal pilot study design occurs for the rate $p = 0.865$ that lies above the anticipated overall rate $p_A = 0.78$. Therefore, to

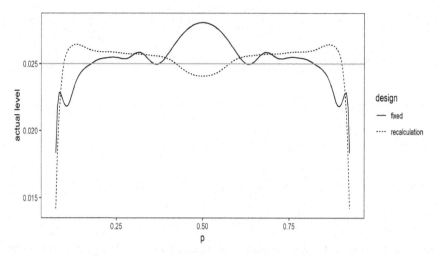

Fig. 23.1 Actual level of the Farrington-Manning test at one-sided nominal level $\alpha/2 = 0.025$ for the fixed sample size design and for the internal pilot study design with blinded sample size recalculation as described in Example 23.1 depending on the overall rate p

achieve strict type I error rate control, one may alternatively consider application of the confidence interval based approach (see Remark 2 in Sect. 22.1.3). For illustration, we assume that the coverage probability $1 - \gamma = 0.9999$ was pre-specified in the protocol. Then, it has to be assured that the maximum actual level within the $(1 - \gamma)$ confidence interval for p is not larger than $\alpha/2 - \gamma = 0.0249$. The exact Pearson-Clopper confidence interval guarantees that the nominal coverage is not undershot. Notwithstanding its conservativeness, application of Pearson-Clopper confidence interval may thus be preferred to an approximate confidence interval as, for example, the Wilson score confidence interval (see Sect. 12.3), if strict control of the type I error rate is aspired. The overall rate for the primary outcome estimated in the analysis of the FreezeAF trial was $\hat{p} = 210/291 = 0.722$ and the related two-sided 95 per cent Pearson-Clopper confidence interval for p is given by [0.611; 0.816]. Note that the confidence interval does not include the value for p where the global maximum of the actual level is attained. Within this confidence interval, the maximum actual type I error rate amounts to 0.0260, which is only marginally smaller than the global maximum of 0.0264. To assure that the maximum actual level is at most 0.0249 within the confidence interval, an adjusted level of $(\alpha/2)_{\text{adj}} = 0.0241$ is to be applied. Thus, for this example, both approaches lead to the same adjusted nominal levels. The calculations for the confidence interval based approach were performed for all values of p between 0.610 and 0.815 in steps of 0.05.

Figure 23.2 shows the power reached in the fixed sample size when assuming an overall rate $p_A = 0.78$, as done when planning the FreezeAF trial, and in the internal pilot study design described above. For the design with fixed sample size, the power deviates considerably from the aspired value 0.80 if the true overall rate

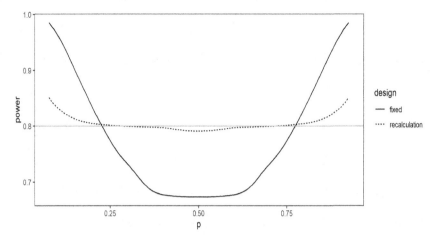

Fig. 23.2 Power of the Farrington-Manning test at one-sided nominal level $\alpha/2 = 0.025$ for the fixed sample size design (assumed overall rate $p_A = 0.78$) and for the internal pilot study design with blinded sample size recalculation as described in Example 23.1 depending on the overall rate p

differs from the assumed one. In contrast, the design with sample size recalculation provides power values close to the target power over a wide range of values for p.

Figure 23.3 shows the distribution of the recalculated sample size in the internal pilot study design by means of box-and-whisker plots. As the required sample size increases when p approaches the value 0.5, the location of the distribution moves to larger values. Note that the shape of the distribution is, in particular near the

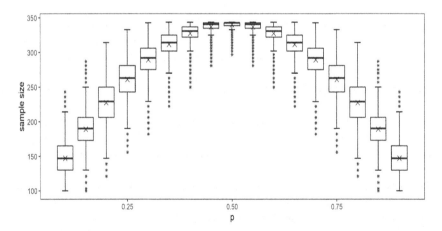

Fig. 23.3 Box-and-whisker plots of the recalculated sample size for the Farrington-Manning test at one-sided nominal level $\alpha/2 = 0.025$ in the internal pilot study design with blinded sample size recalculation as described in Example 21.1 depending on the overall rate p

boundaries of the range for p and for values around 0.5, affected by the lower boundary imposed by the internal pilot sample size ($n_1 = 100$) or the maximum sample size that is possible for the specifications made ($n = 344$), respectively.

23.3 Risk Ratio or Odds Ratio as Effect Measure

For the risk ratio, the test problem for the assessment of non-inferiority is given by (8.20). The null-hypothesis can be tested by using the test statistic (8.21) proposed by Farrington and Manning (1990). In the fixed sample size design, formula (8.24) can be applied to determine the required sample size. As for the case of the difference of rates, the sample size formula includes the assumed event rates under the alternative $p_{E,A}$ and $p_{C,A}$ as well as the large sample approximations of the restricted maximum likelihood estimators $\tilde{p}_E$ and $\tilde{p}_C$. For blinded sample size recalculation, the same procedure can be applied as the one described in the preceding section for non-inferiority hypotheses formulated in terms of the rate difference: Inserting the estimated overall rate $\hat{p}$ in the Eqs. (21.11a) and (21.11b) results in estimated group-wise event rates $\hat{p}_{E,A}$ and $\hat{p}_{C,A}$, from which the related constrained estimates $\hat{\tilde{p}}_E$ and $\hat{\tilde{p}}_C$ are obtained. These values are then inserted in the sample size formula (8.24) to get the recalculated sample size. Evaluation of actual type I error rate, power, and sample size can be done analogously to the rate difference case. Currently, no investigations on the performance characteristics of this blinded sample size recalculation strategy are available in the literature. However, as the ingredients are very similar to the case of the test problem formulated in terms of the rate difference, the performance can expected to be similar as well.

The non-inferiority test problem for the odds ratio as effect measure is given by (8.25) and can be assessed by using the test statistic (8.26) proposed by Wang et al. (2002). Approximate sample size calculation for the fixed-size design can be performed with formula (8.27) on which blinded sample size recalculation can be based. For this purpose, the event rates in the groups E and C can be estimated by inserting the overall rate $\hat{p}$ estimated from the pooled data of the internal pilot study in the Eqs. (21.12a) and (21.12b), respectively. Inserting these estimates in (8.27) results in the recalculated sample size. Again, no investigations on the performance of this strategy in the internal pilot study design have been published up to now. However, if this approach is planned to be implemented in a clinical trial, the calculations can be done analogously to those described above for the case that the rate difference is used as effect measure.

Remark In Sect. 8.4.2, exact unconditional tests for non-inferiority and related sample size calculation methods were described. In contrast to the asymptotic tests considered above, these tests assure strict control of the type I error rate in the fixed sample size design. Therefore, they are also an attractive option for sample size recalculation designs. Depending on the effect measure used, the group-wise event rates can be estimated from the estimate of the overall rate by using Eqs. (21.4a) and

(21.4b), (21.11a) and (21.11b), or (21.12a) and (21.12b) for the rate difference, the risk ratio, or the odds ratio, respectively. The sample size can then be recalculated by employing these values instead of $p_{E,A}$ and $p_{C,A}$. There is currently no published work on the performance of exact unconditional tests in non-inferiority trials with blinded sample size recalculation. In view of the discrepancy between the results observed for the internal pilot study design in the non-inferiority setting for normally distributed outcomes (inflation of the type I error rate) and for binary data analyzed with asymptotic tests (no inflation of the type I error rate beyond that observed for the fixed design), it would be interesting to evaluate the actual type I error rate of exact unconditional tests under blinded sample size recalculation. However, due to the considerable computational burden, such investigations are challenging.

Chapter 24
Comparison of Two Groups for Normally Distributed Outcomes and Test for Equivalence

In this chapter, we consider the most common situation in equivalence assessment for normally distributed outcomes, namely that the treatment group difference is expressed in terms of the difference of means. For the fixed sample size design, this setting was described in Sect. 10.2. For the most common case of a symmetrical equivalence margin, the test problem is given by

$$H_0: \mu_E - \mu_C \leq -\delta \text{ or } \mu_E - \mu_C \geq \delta$$
$$\text{versus}$$
$$H_1: -\delta < \mu_E - \mu_C < \delta,$$

where μ_i, $i = C, E$, denotes the expectation of the outcome in treatment group i and $\delta > 0$ the pre-specified half-width of the equivalence margin. The null-hypothesis can be tested by the two one-sided tests (TOST) procedure applying the test statistics of the shifted t-tests given in (10.2a) and (10.2b). Approximate sample size (re)calculation can be conducted using the Eqs. (10.5) and (10.6), with or without utilization of the Guenther-Schouten correction term.

As for superiority and non-inferiority assessment, blinded sample size recalculation can be based on the one-sample variance estimator S_{OS}^2 (see Eqs. (20.4) and (22.4), respectively). Calculation of the actual type I error rate can be done analogously to the non-inferiority case (see Sect. 22.1.3). This is due to the facts that both test statistics involved in the equivalence assessment can be decomposed according to (22.7) and that the sample size recalculation rule can be expressed accordingly (Friede and Kieser 2003). As for non-inferiority testing, the type I error rate may be seriously inflated in equivalence assessment following blinded sample size recalculation with actual levels larger than 0.07 for nominal $\alpha = 0.05$ (Golkowski et al. 2014). The actual type I error rate can again be represented as a function of α, n_1, N, and δ/σ. For a concrete clinical trial, the sample size recalculation rule, α, n_1, δ as well as the power, the alternative, and the allocation ratio are fixed. Therefore,

© Springer Nature Switzerland AG 2020

M. Kieser, *Methods and Applications of Sample Size Calculation and Recalculation in Clinical Trials*, Springer Series in Pharmaceutical Statistics, https://doi.org/10.1007/978-3-030-49528-2_24

due to the one-to-one relationship between N and σ for specified α, $1 - \beta$, δ, r, and alternative, the actual type I error rate only depends on the unknown σ. Hence, the worst case scenario for the maximum actual type I error rate can be determined. The other way round, an adjusted level can be identified such that applying this level assures control of the type I error rate at α (see paragraph "Control of type I error rate" in Sect. 22.1.3). Note that the TOST procedure requires rejection of both one-sided null-hypotheses to reject the null-hypothesis of non-equivalence. As a consequence, the actual type I error rate of the TOST procedure is bounded by the minimum of those of the related non-inferiority tests. Furthermore, it should be mentioned that, as in case of non-inferiority assessment, the upward-bias of S_{OS}^2 is expected to be smaller than in superiority trials. Therefore, it is worth to consider using the inflation factor (22.9) for recalculating the sample size in order to bring the power of the procedure closer to the target value, especially for small sample sizes. The computations or simulations, respectively, for investigating the performance characteristics can, after appropriate modifications, be performed using the same tools as for the non-inferiority case.

Remarks

1. Glimm et al. (2019) performed a comprehensive simulation study to investigate how and why factors like the sample size recalculation rule, the equivalence margin, the sample size of the internal pilot study, or the treatment group difference influence the actual type I error rate when conducting blinded sample size adjustment in equivalence trials. They showed, among other things, that the variability of the recalculated sample size has a major impact on the type I error rate. As a consequence, small internal pilot study sample size and liberal or no restrictions on the final sample size encourage an inflation of the type I error rate. The other way round, by choosing a sufficiently large sample size for the internal pilot study and by specifying appropriate lower and upper boundaries for the recalculated sample size, the excess in type I error rate can be kept small. Note that the narrower range of minimum and maximum sample size is, the more similar is the internal pilot study design to the design with fixed sample size and the less is the extent of the level inflation. On the other side, however, this reduces the flexibility for the choice of the final sample size and thus counteracts to some extent the motivation for implementing a design with sample size recalculation.
2. A common area of application of equivalence trials is the field of bioequivalence testing (see the Remark in Sect. 10.2). These studies are typically performed in the CE/EC crossover design. Here, the subjects are allocated to the sequence $C - E$ or $E - C$ and receive both treatments sequentially separated by an appropriate wash-out phase. Golkowski et al. (2014) presented methods for blinded sample size recalculation tailored to this setting and investigated their characteristics.

Chapter 25
Regulatory and Operational Aspects

From a regulatory viewpoint, the issue of a potential level-inflation due to blinded sample size recalculation is a major concern. This topic was already addressed in the ICH E9 Guideline *Statistical Principles in Clinical Trials* (ICH 1998) and was readopted in the EMA Reflection Paper on *Methodological Issues in Confirmatory Clinical Trials Planned with an Adaptive Designs*. There, it is pointed out that "*Analysis methods that control the type I error must be pre-specified.*" (EMA 2007, p. 6). To date, a general proof that blinded sample size recalculation does not inflate the type I error rate is only available for the case that a permutation test is applied in the analysis (see Chap. 19). Therefore, for all other settings, the actual type I error rate of the implemented procedure has to be investigated. In order to provide firm arguments with regard to type I error rate control (possibly after appropriate adjustment of the nominal level), the actual level has to be studied for the setting specified for the trial (including the sample size of the internal pilot study and the sample size recalculation rule) over all plausible values of the nuisance parameter(s). If possible, this can be done by numerical computations or otherwise by simulations. In any case, it is important that the precision of the results can be quantified in order to reliably demonstrate type I error rate control. Mayer et al. (2019) and the FDA Guidance for Industry *Adaptive Designs for Clinical Trials of Drugs and Biologics* (FDA 2019, p. 18ff) provide helpful recommendations when performing simulation studies exploring the operating characteristics of adaptive designs. These principles apply also for internal pilot study designs.

It has already been mentioned that the sample size recalculation rule has to be pre-fixed as otherwise the characteristics of the design cannot be explored. The design properties include actual type I error but of course also the achieved power—this is the reason why sample size recalculation is performed—and the characteristics of the distribution of the recalculated sample size. Figures 20.2 and 21.3 may serve as examples how such investigations may strengthen the confidence that the implemented design in fact provides the desired robustness against misspecifications of nuisance parameters. Furthermore, together with evaluations such as those shown in Figs. 20.3

© Springer Nature Switzerland AG 2020
M. Kieser, *Methods and Applications of Sample Size Calculation and Recalculation in Clinical Trials*, Springer Series in Pharmaceutical Statistics,
https://doi.org/10.1007/978-3-030-49528-2_25

and 21.4 for the sample size distribution, they may support the choice of the final design, for example with respect to the internal pilot sample size or the upper bound imposed on the sample size. By the way, pre-specification of the sample size recalculation strategy is also necessary due to logistical reasons, for example in order to assure sufficient drug supply or to guarantee the availability of further patients in case the sample size is increased. The explorations of the statistical properties of the design as well as their rationale and their interpretation should be documented together with further informations in the study protocol or, in most cases preferable, in a separate document which is finalized before start of the trial. To clarify the process and in particular to demonstrate that blindness is preserved, the following specifications should be made and fixed in a document: Which persons/institutions are involved in the process of sample size recalculation and what are their tasks? Who gets access at which time to which data? How is the process validated (including, for example, the steps of data extraction, data transfer, and sample size recalculation)? How are the communication channels between the involved parties defined? How are the results of the sample size recalculation documented and reported? Either, standard operating procedures (SOPs) exist defining these issues, or a study-specific document has to be prepared. This formal framework is required to demonstrate statistical properness as well as to assure clarity and transparency of the processes thus preventing any doubts on the trial integrity that may be caused due to implementation of blinded sample size recalculation.

Chapter 26
Concluding Remarks

The *Reflection Paper on Methodological Issues in Confirmatory Clinical Trials Planned with an Adaptive Design* of the European Medicines Agency nicely captures in a nutshell the motivation and basic conditions of sample size recalculation: "*In cases where good justification can be provided that uncertainty about the required sample size is not an indicator for insufficient research in earlier stages, sample size reassessment based on results of an ongoing trial is an option. ... Whenever possible, methods for blinded sample size reassessment that properly control the type I error should be used, especially if the sole aim of the interim analysis is the re-calculation of sample size*" (EMA 2007, p. 6). In this part of the book, internal pilot study designs were presented where sample size recalculation is performed based on the data pooled over the intervention groups, i.e. without unmasking the patients' treatment group membership. Besides the situations considered above, related methods exist for a wide range of alternative scenarios. Just to mention a few of them, such procedures were proposed for the comparison of more than two groups by application of ANOVA (Kieser and Friede 2000b), time-to-event outcomes (Todd et al. 2012; Friede et al. 2019), recurrent events (Ingel and Jahn-Eimermacher 2014), count data (Friede and Schmidli 2010), longitudinal data (Wachtlin and Kieser 2013), and composite endpoints (Sander et al. 2017). All these approaches have in common that they are easy to implement and provide substantial robustness of the achieved power against mis-specifications of nuisance parameters in the planning stage. At the same time, they show only marginal efficiency loss as compared to the fixed design with perfect pre-specification. Furthermore, if appropriate precautions are taken with respect to operational and logistical aspects, these designs can straightforwardly be put into practice without inducing any doubts on the trial integrity.

Before applying an internal pilot study design, the performance characteristics have to be comprehensively investigated. In this regard, the aspect of type I error rate control is essential from a regulatory viewpoint. As shown in Chaps. 22 and 24 for the cases of normally distributed outcomes and assessment of non-inferiority and equivalence, respectively, performing sample size recalculation in a blinded manner

© Springer Nature Switzerland AG 2020

M. Kieser, *Methods and Applications of Sample Size Calculation and Recalculation in Clinical Trials*, Springer Series in Pharmaceutical Statistics, https://doi.org/10.1007/978-3-030-49528-2_26

does not guarantee that the level is controlled; in contrast, it may be seriously inflated. In such cases, appropriate adjustment of the nominal level has to be made to ensure type I error rate control. The other way round, if the uncorrected level is used in the analysis, it has to be justified why an adjustment is not deemed to be required. From the position of the study team, the characteristics of the design with respect to the achieved power and the recalculated sample size are of major importance. Such evaluations are also necessary to guide the choice of design elements such as the internal pilot study sample size or the maximum overall sample size. For these reasons, the required resources for planning a trial with internal pilot study will generally be higher as compared to a fixed sample size design. However, these investments will pay off by a reduced risk of ending up with an inappropriate sample size.

Part V
Unblinded Sample Size Recalculation in Adaptive Designs

Chapter 27
Group Sequential and Adaptive Designs

In Part IIIA of the book, it is shown how to deal with uncertainty with respect to nuisance parameters by blinded sample size recalculation in internal pilot study designs. However, there is commonly also considerable uncertainty with respect to the parameter for which the test problem is formulated, namely the treatment group difference. Expressed pointedly: If the treatment effect would be known, the study has not to be conducted anymore. It is inherent to the blinded approach that internal pilot study designs cannot correct for false assumptions with respect to the treatment group difference. As in fixed sample size designs, the sample size will be too high or too low, respectively, if the treatment effect was understated or overstated in the planning stage.

It is an obvious idea to estimate the treatment effect mid-course and to use it for sample size recalculation. Amongst others, Shun et al. (2001) and Chen et al. (2004) showed for the z-test without and with early stopping, respectively, that such an approach may lead to a substantial inflation of the type I error rate if the usual fixed sample size test statistic and the unadjusted level are applied in the final analysis. The other way round, they demonstrated that the type I error rate can be controlled by restricting to specific sample size recalculation rules. However, due to the imposed constraints, these sample size adjustment strategies do not enable full flexibility and in addition they turned out to be inefficient (Jennison and Turnbull 2015). Therefore, this class of methods will not be considered further but is only sketched by shortly discussing in Remark 3 of Chap. 28 a design that is based on the work of Chen et al. (2004).

Another intuitive approach for sample size recalculations is to integrate them in pre-planned unblinded interim analyses which enable early stopping for futility (in case of a too small or even detrimental effect) or with early rejection of the null-hypothesis (in case of an overwhelming efficacy result). Although in principle such interim analyses may be performed after each patient, this is in most cases not practicable and too costly in clinical trials. Therefore, group sequential designs, where interim analyses are performed when the outcome is available for a defined number

© Springer Nature Switzerland AG 2020

M. Kieser, *Methods and Applications of Sample Size Calculation and Recalculation in Clinical Trials*, Springer Series in Pharmaceutical Statistics, https://doi.org/10.1007/978-3-030-49528-2_27

of patients, have become the common approach. These designs also constituted the starting point for the development of adaptive designs.

Below, the principle of group sequential designs—and subsequently of adaptive designs—will be sketched for the most frequent case of two-stage designs with one interim analysis. This matches the fact that regulatory guidelines express a preference for a single sample size reassessment in adaptive designs [see, for example, EMA (2007), p. 6]. Furthermore, to avoid technical difficulties in the presentation, the methods are not described in their full generality. Instead, we restrict to the case of normally distributed outcomes with known variance to elucidate the principle of the approaches. Therefore, we consider in the following normally distributed outcomes X_{ijk}, $i = C, E, j = 1, 2, k = 1, \ldots, n_{ij}$, with common and known variance σ^2. Here, $j = 1$ denotes the data included in the first stage of the study before the interim analysis and $j = 2$ the disjoint data set of stage two. The superiority test problem to be assessed is given by

$$
\begin{aligned}
H_0 &= \mu_E - \mu_C \leq 0 \\
&\text{versus} \\
H_1 &= \mu_E - \mu_C > 0
\end{aligned}
\tag{27.1}
$$

and the involved z-test statistics based on the data of the first and the second stage as well as on the pooled data of both stages, respectively, are given by

$$
Z_1 = \sqrt{\frac{n_{E1} \cdot n_{C1}}{n_{E1} + n_{C1}}} \cdot \frac{\overline{X}_{E1} - \overline{X}_{C1}}{\sigma}
\tag{27.2a}
$$

$$
Z_2 = \sqrt{\frac{n_{E2} \cdot n_{C2}}{n_{E2} + n_{C2}}} \cdot \frac{\overline{X}_{E2} - \overline{X}_{C2}}{\sigma}
\tag{27.2b}
$$

$$
Z = \sqrt{\frac{n_E \cdot n_C}{n_E + n_C}} \cdot \frac{\overline{X}_E - \overline{X}_C}{\sigma}.
\tag{27.2c}
$$

Here,

$$
\overline{X}_{ij} = \frac{1}{n_{ij}} \cdot \sum_{k=1}^{n_{ij}} X_{ijk}, \ i = C, E, j = 1, 2,
$$

and

$$
\overline{X}_i = \frac{1}{n_E + n_C} \cdot \sum_{j=1}^{2} \sum_{k=1}^{n_{ij}} X_{ijk}, \ i = C, E, \text{ with } n_i = \sum_{j=1}^{2} n_{ij}.
$$

The comprehensive methodology of group sequential and adaptive designs is only sketched here. For more in-depth presentations, the interested reader is referred to the monographs by Jennison and Turnbull (2000) and Wassmer and Brannath (2016) as well as the comprehensive review of Bauer et al. (2016) on confirmatory adaptive designs.

27.1 Group Sequential Designs

The basic idea of the group sequential approach is to perform interim analyses when specified numbers of observations are available. The test statistics used for decision making include the data accumulated so far and are thus stochastically dependent. Exploiting the joint distribution of the test statistics, decision boundaries can be determined such that the type I error rate is controlled.

For the situation of two normal distributions with known variance, the test statistic Z_1 based on the first-stage data is used for the interim analysis and the test statistic Z is applied in the final analysis (see (27.2a) and (27.2b)). For balanced sample sizes in both stages, i.e. $n_{Ej} = n_j/2, j = 1, 2$, it holds true that

$$Z = \sqrt{\frac{n_1}{n_1 + n_2}} \cdot Z_1 + \sqrt{\frac{n_2}{n_1 + n_2}} \cdot Z_2. \tag{27.3}$$

Furthermore, (Z_1, Z) is bivariate normally distributed with expectation vector $\left(\frac{\sqrt{n_1}}{2} \cdot \frac{\Delta}{\sigma}, \frac{\sqrt{n_1 + n_2}}{2} \right)$, where $\Delta = \mu_E - \mu_C$ and correlation $\rho = \sqrt{n_1/(n_1 + n_2)}$ [see, for example, Wassmer and Brannath 2016, p. 9].

The following decisions are made within the group sequential trial:

- early stop for futility with acceptance of H_0, if the one-sided p-value based on the stage-one data $p_1 \geq \alpha_0$ (where $\alpha_0 > \alpha$), i.e. if $z_1 \leq z_{1-\alpha_0}$
- early stop for efficacy with rejection of H_0, if the one-sided p-value based on the stage-one data $p_1 < \alpha_1$ (where $\alpha_1 < \alpha$), i.e. if $z_1 > z_{1-\alpha_1}$
- if the study is continued after the first stage, i.e. if $\alpha_1 \leq p_1 < \alpha_0$ (equivalent to $z_{1-\alpha_0} < z_1 \leq z_{1-\alpha_1}$):
  - acceptance of H_0, if the one-sided p-value based on the pooled data of stage one and stage two $p \geq \alpha_2$, i.e. if $z \leq z_{1-\alpha_2}$
  - rejection of H_0, if the one-sided p-value based on the pooled data of stage one and stage two $p < \alpha_2$, i.e. if $z > z_{1-\alpha_2}$.

Here, $z_{1-\gamma}$ denotes the $(1 - \gamma)$-quantile of the standard normal distribution.

The level-condition assuring type I error rate control at α is given by

$$\Pr_{H_0}\left(Z_1 > z_{1-\alpha_1}\right) + \Pr_{H_0}\left(z_{1-\alpha_0} < Z_1 \leq z_{1-\alpha_1}, Z > z_{1-\alpha_2}\right) = \alpha. \tag{27.4}$$

For specified α, Eq. (27.4) can be solved for any of the critical boundaries if the other two are fixed. Various classes of boundaries have been proposed in the literature. For example, Pocock (1977) suggested applying the same critical value in all (interim) analyses for the assessment of H_0 meaning in our case to set $\alpha_1 = \alpha_2$. Then, the resulting equal critical boundaries for the p-values at the interim and final analysis are for equal stagewise sample size $n_1 = n_2$ (i.e. $\rho = 1/\sqrt{2}$), $\alpha = 0.025$, one-sided, and $\alpha_0 = 0.3, 0.5$, and 1.0, respectively, given by $\alpha_1 = \alpha_2 = 0.0151, 0.0148$, and 0.0147.

The choices $\alpha_0 = 1.0$ or $\alpha_1 = 0$, respectively, implicate that no early stopping for futility or efficacy is desired. It should be mentioned that early stopping with accepting H_0 can be conducted without inflating the type I error rate even if the boundary α_0 was set equal to 1 in the protocol. However, the other way round, if a value $\alpha_0 < 1$ was specified in the protocol and if this value was taken into account when calculating the critical values (so-called "binding futility boundary"), the trial has to be stopped in any case if $p_1 \geq \alpha_0$ as otherwise the type I error rate is inflated. This is due to the fact that the critical value for the p-value at the final analysis is smaller for larger α_0.

In a way, the application of early stopping rules allows to deal with unexpectedly small or large treatment effects. However, if the study is continued after the interim analysis, group sequential designs require sticking with the initially specified sample size for the subsequent stage(s). In fact, Cui et al. (1999) showed by simulation studies that the type I error rate of group sequential trials may be substantially inflated if the sample size is modified based on an interim estimate of the treatment group difference. A main motivation for the development of adaptive designs was to preserve the options offered by group sequential designs while at the same time enabling data-driven size reassessment based on unblinded interim data without compromising the significance level.

27.2 Adaptive Designs

The adaptive designs presented below do not only allow sample size recalculation according to arbitrary rules under control of the type I error rate but enable also other design modifications as, for example, data-dependent selection of treatment groups, endpoints, or target populations. To construct designs enabling such flexibility, two interrelated principles can be distinguished which are presented hereafter.

27.2.1 Combination Function Approach

The basic idea behind the combination function approach is to calculate the stage-wise p-values p_1 and p_2 separately based on the data of stage one and stage two, respectively, and to aggregate them to a test statistic $C(p_1, p_2)$. The combination function $C(p_1, p_2)$ is continuous in p_2 and non-decreasing in both arguments. For the combination test, the early stopping boundaries for p_1 are denoted as for group sequential designs by α_0 (stopping for futility) and α_1 (stopping for efficacy). At the final analysis, H_0 is rejected if $C(p_1, p_2) < c$. If under the null-hypothesis the p-values p_1 and p_2 are independently and uniformly distributed on the interval $[0, 1]$, then the type I error rate is controlled by α if

$$\alpha_1 + \int_{\alpha_1}^{\alpha_0} \int_0^1 I_{\{C(p_1,p_2)<c\}} \, dp_2 dp_1 = \alpha. \tag{27.5}$$

The combination test based on C is a level-α test even if the design of the second stage is modified based on the results of the interim analysis (or external data). Actually, the assumption of independently and uniformly distributed p-values can be attenuated: To assure type I error rate control, it is sufficient that under the null-hypothesis the distribution of p_1 is stochastically larger than or equal to the uniform distribution and that the conditional distribution of p_2 given p_1 is also stochastically larger than or equal to the uniform distribution (so-called "p clud property"; Brannath et al. 2002).

In their seminal papers introducing the adaptive design methodology, Bauer (1989) and Bauer and Köhne (1994) proposed applying Fisher's product test with combination function

$$C(p_1, p_2) = p_1 \cdot p_2. \tag{27.6}$$

For this combination test, the level condition (27.5) reads

$$\alpha_1 + c \cdot (\ln(\alpha_0) - \ln(\alpha_1)) = \alpha. \tag{27.7}$$

Any constellation of values α_0, α_1, and c fulfilling (27.7) constitutes an adaptive two-stage design that controls the type I error rate at α. Specific choices of the boundaries are discussed in Bauer and Köhne (1994), Bauer and Röhmel (1995), and Wassmer and Brannath (2016, p. 138ff).

Another popular combination test is based on the weighted inverse normal combination function (Cui et al. 1999; Lehmacher and Wassmer 1999). If the stagewise p-values p_1 and p_2 are independently and uniformly distributed, then the random variables $Z_1 = \Phi^{-1}(1 - p_1)$ and $Z_2 = \Phi^{-1}(1 - p_2)$ are independently and standard normally distributed. Here $\Phi(\cdot)$ denotes the cumulative distribution function of the standard normal distribution and $\Phi^{-1}(\cdot)$ its inverse. As a consequence, for any weights w_1, w_2 with $0 < w_i < 1$, $i = 1, 2$, and $w_1^2 + w_2^2 = 1$ and any rule for modifying the design of the second stage, the test statistic $Z = w_1 \cdot Z_1 + w_2 \cdot Z_2$ follows the standard normal distribution. Therefore,

$$C(p_1, p_2) = 1 - \Phi(Z) = 1 - \Phi\big(w_1 \cdot \Phi^{-1}(1 - p_1) + w_2 \cdot \Phi^{-1}(1 - p_2)\big) \tag{27.8}$$

is uniformly distributed on $[0,1]$.

Note that for the choice $w_1 = \sqrt{\frac{n_1}{n_1+n_2}}$, $w_2 = \sqrt{\frac{n_2}{n_1+n_2}}$, the test statistics Z_1 and Z follow the same joint distribution as for the group sequential design for normally distributed outcomes (see Sect. 27.1). Therefore, the decision boundaries α_0, α_1, and α_2 which are calculated for group sequential designs according to (27.4) are for this weighting schemes identical to those for the weighted inverse normal combination test. For other choices of the weights, the boundaries correspond to those of a related group sequential design and can be calculated by evaluating (27.4) for a bivariate standard normally distributed random vector (Z_1, Z) with correlation w_1. For the case of normally distributed outcomes with known variance and balanced samples, the test statistics of the group sequential and the adaptive design are for the above-mentioned weighting algebraically equal if the initially planned sample size for stage two is not changed [see (27.3)]. However, if the stage-two sample size is modified, the specified weighting scheme, which has to be kept fixed, no longer reflects the actual stage-wise sample sizes in the adaptive design.

27.2.2 Conditional Error Function Approach

The conditional error function defines the probability to reject the null-hypothesis if it is in fact true, given the data collected until the interim analysis. For the example of normally distributed outcomes and known variance, any monotonically non-decreasing function $CE(\cdot)$ with

$$\int_{-\infty}^{\infty} CE(z)\phi(z)dz = \alpha \tag{27.9}$$

defines a level-α test if the test after stage two is performed at level $CE(z_1)$. Here, Φ denotes the density of the standard normal distribution. Control of the nominal level α is guaranteed irrespective of whether or not the second study stage has been modified mid-course. Clearly, in case of early acceptance, i.e. $z_1 \leq z_{1-\alpha_0}$, $CE(z_1) = 0$, and in case of early rejection, i.e. $z_1 > z_{1-\alpha_1}$, $CE(z_1) = 1$.

For Fisher's product test and normally distributed data with known variance, $p_1 = 1 - \Phi(z_1)$. Therefore, for $z_{1-\alpha_0} < z_1 \leq z_{1-\alpha_1}$ (i.e. for $\alpha_1 \leq p_1 < \alpha_0$)

$$CE(z_1) = \Pr_{H_0}(p_1 \cdot p_2 < c|z_1)$$
$$= \Pr_{H_0}\left(p_2 < \frac{c}{1 - \Phi(z_1)}|z_1\right)$$
$$= \frac{c}{1 - \Phi(z_1)}.$$

Hence, the conditional error function is given by

$$CE(z_1) = \begin{cases} 0 & \text{if } z_1 \leq z_{1-\alpha_0} \\ \frac{c}{1-\Phi(z_1)} & \text{if } z_{1-\alpha_0} < z_1 \leq z_{1-\alpha_1} \\ 1 & \text{if } z_1 > z_{1-\alpha_1}. \end{cases} \tag{27.10}$$

For the weighted inverse normal combination test, the following holds true for $z_{1-\alpha_0} < z_1 \leq z_{1-\alpha_1}$:

$$CE(z_1) = \Pr_{H_0}\left(Z > z_{1-\alpha_2}|z_1\right)$$

$$= \Pr_{H_0}\left(Z_2 > \sqrt{\frac{n_1+n_2}{n_2}} \cdot z_{1-\alpha_2} - \sqrt{\frac{n_1}{n_2}} \cdot z_1|z_1\right)$$

$$= 1 - \Phi\left(\sqrt{\frac{n_1+n_2}{n_2}} \cdot z_{1-\alpha_2} - \sqrt{\frac{n_1}{n_2}} \cdot z_1\right). \tag{27.11}$$

Again, $CE(z_1) = 0$ for $z_1 \leq z_{1-\alpha_0}$ and $CE(z_1) = 1$ for $z_1 > z_{1-\alpha_1}$.

These examples show that there is a close relationship between the combination test and the conditional error function approach which has been demonstrated to hold true also for other adaptive designs (Posch and Bauer 1999) as well as in a more general sense (Wassmer and Brannath 2016, p. 153ff, and references given there). Notwithstanding their interrelation, it should be noted that there are also differences between the two approaches. For example, the simple p-value combination methods based on Fisher's product test (27.6) or on the weighted inverse normal combination function (27.8) are exact tests even if the variance of normally distributed outcomes is unknown. However, determination of the conditional error function is complex in this case (Posch et al. 2004; Timmesfeld et al. 2007) and in other settings where nuisance parameters are involved (Gutjahr et al. 2011). Therefore, there is a tradeoff between the straightforward application of the combination test approach with the drawback of generally using uncommon test statistics aggregated from stagewise results and the potentially more difficult determination of the conditional error function which, however, may allow using the conventional test statistic at the final analysis.

Chapter 28
Sample Size Recalculation Based on Conditional Power

28.1 Background and Notation

The conditional power is the probability of rejecting the null-hypothesis given the observed interim results. In two-stage trials, the overall power can be regarded as the expectation of the conditional power over all possible outcomes in the interim analysis. However, having observed interim results, it may be seen inadequate to consider the quantity "overall power": By averaging over all interim outcomes, this measure takes into account scenarios that have not been realized in the current study. It may, therefore, be considered as more appropriate to base the power (and the other way round: the sample size recalculation) conditional on what has been observed in the interim analysis. Based on this argumentation, sample size recalculation by applying conditional power arguments has become the most popular approach for sample size reassessment in adaptive designs [see, for example, Proschan and Hunsberger (1995), Cui et al. (1999), Lehmacher and Wassmer (1999), Denne (2001), Friede and Kieser (2001), Shun et al. (2001), Posch et al. (2003), and Mehta and Pocock (2011)].

For illustration, we consider in the following the case of applying the weighted inverse normal method with weights $w_1 = \sqrt{\frac{n_1}{n_1+n_2}}$, $w_2 = \sqrt{\frac{n_2}{n_1+n_2}}$ and the setting of normally distributed outcomes with known variance σ and sample size allocation ratio $r:1$ between groups E and C in both stages. The involved test statistics are then given by

$$Z_1 = \sqrt{\frac{r}{(1+r)^2} \cdot n_1} \cdot \frac{\overline{X}_{E1} - \overline{X}_{C1}}{\sigma} \tag{28.1a}$$

$$Z_2 = \sqrt{\frac{r}{(1+r)^2} \cdot \tilde{n}_2} \cdot \frac{\overline{X}_{E2} - \overline{X}_{C2}}{\sigma} \tag{28.1b}$$

© Springer Nature Switzerland AG 2020
M. Kieser, *Methods and Applications of Sample Size Calculation and Recalculation in Clinical Trials*, Springer Series in Pharmaceutical Statistics,
https://doi.org/10.1007/978-3-030-49528-2_28

$$Z = \sqrt{\frac{n_1}{n_1 + n_2}} \cdot Z_1 + \sqrt{\frac{n_2}{n_1 + n_2}} \cdot Z_2. \tag{28.1c}$$

Here, n_1 and n_2 denote the initially planned sample sizes for stage one and two, respectively, and $\tilde{n}_2$ the actual stage-two sample size that may be a result of recalculation. The null-hypothesis is rejected at the final analysis if $z > z_{1-\alpha_2}$. Conditional on z_1, this is equivalent to $z_2 > \sqrt{\frac{n_1 + n_2}{n_2}} \cdot z_{1-\alpha_2} - \sqrt{\frac{n_1}{n_2}} \cdot z_1$. Furthermore, conditional on z_1, $Z_2 \sim N\left(\sqrt{\frac{r}{(1+r)^2}} \cdot \tilde{n}_2 \cdot \frac{\Delta}{\sigma}, 1\right)$, where $\Delta = \mu_E - \mu_C$. Therefore, the conditional power $CP(z_1)$ is given by

$$CP(z_1) = 1 - \Phi\left(\sqrt{\frac{n_1 + n_2}{n_2}} \cdot z_{1-\alpha_2} - \sqrt{\frac{n_1}{n_2}} \cdot z_1 - \sqrt{\frac{r}{(1+r)^2}} \cdot \tilde{n}_2 \cdot \frac{\Delta}{\sigma}\right). \tag{28.2}$$

Equation (28.2) can be used to calculate the chance of rejecting the null-hypothesis at the final analysis for the stage-two sample size $\tilde{n}_2$. The other way round, setting $CP(z_1)$ equal to the desired conditional power $1 - \beta_{CP}$, the required sample size can be determined by resolving the corresponding equation.

In an analogous manner, the conditional error function approach can be used for sample size recalculation. According to (3.5), if the final test is performed at level $CE(z_1)$, the required stage-two sample size to achieve a conditional power $1 - \beta_{CP}$ is given by

$$\tilde{n}_2 = \frac{(1+r)^2}{r} \cdot \left(z_{1-CE(z_1)} + z_{1-\beta_{CP}}\right)^2 \cdot \left(\frac{\sigma}{\Delta}\right)^2. \tag{28.3}$$

Inserting in (28.3) the expression for $CE(z_1)$ given by (27.11) shows the equivalence of (28.2) and (28.3).

As the true treatment effect Δ is not known, the same holds true for the conditional power. Therefore, the question arises which value to insert for Δ in (28.2) or (28.3), respectively, when calculating the conditional power or the required sample size. In the following, three approaches are presented together with their rationale and the characteristics of the resulting sample size recalculation procedure.

28.2 Using the Interim Effect Estimate

Intuitively, it seems to be appealing to use the estimate of the treatment effect obtained from the interim analysis for sample size recalculation as it represents the up-to-date knowledge on the true effect supported by the trial data. One may argue that this approach may lead to a "hunting for significance" in case of effects that are not worth to be detected. This issue may be at least partly addressed by choosing an appropriate futility boundary using the fact that for fixed n_1 there is a one-to-one relationship between z_1 and the observed interim effect. By this, continuation of the trial with adapted sample size occurs only for a sufficiently large interim effect. Furthermore, as Posch et al. (2003) pointed out, when comparing active treatments any improvement may be of interest if the new experimental treatment has an advantage with respect to another aspect as, for example, safety. Moreover, the judgement of what constitutes a clinically relevant treatment group difference depends on the benefit, risk, and costs of a treatment (Hung 2006) and thus also on the safety profile of the interventions under investigation. Therefore, the opinion may change when new safety data become available during the current trial and may also be guided by the observed effect. However, serious arguments against the use of interim estimates of the treatment effect in sample size recalculation based on conditional power are the poor statistical properties of the resulting procedure. They are due to the fact that the observed interim effect enters twofold in sample size recalculation, namely firstly by conditioning on the interim test statistic, and secondly by inserting the estimate in the conditional power calculation (cf. Eq. (28.3) when replacing there Δ by the interim effect estimate $\widehat{\Delta}$). For this reason, random variations of the interim effect estimate around the true Δ may switch the calculated conditional power to the extreme values 0 or 1. As a consequence, the related sample size recalculation strategy leads in general to large expected sample sizes (Jennison and Turnbull 2003; Posch et al. 2003; Bauer and König 2006) that show at the same time a high variability, especially when the sample size recalculation is performed early during the trial (Bauer and König 2006). The inefficiency of sample size recalculation early in the study was also demonstrated by the evaluations of Gaffney and Ware (2017). Note that large stage-two sample sizes lead to a self-enhancing effect: If the recalculated sample size is much larger than the initially assumed n_2, the sample size of stage two is weighted inefficiently thus requiring a large sample size for the second stage to reach the desired conditional power (Bauer and König 2006). Overall, Bauer and König concluded that *"using the estimated effect size for sample size reassessment seems not to be a recommendable option"* (Bauer and König 2006, p. 35).

28.3 Using the Initially Specified Effect

An alternative to using the interim estimate of the effect is to stick to the initially specified treatment group difference Δ_A. Then, the information about the treatment effect arising from the first-stage data is only included by conditioning on z_1 but is not used for defining the value to be inserted for Δ in Eq. (28.2). In a comprehensive investigation, Bauer and König (2006) studied the characteristics of the density of the conditional power and the corresponding sample size recalculation procedure for this strategy. Although less pronounced, this approach shows the same weaknesses as observed when plugging in the interim estimate. The reason for this is that the conditional power shows an erratic behavior. It may, for example, be uniformly distributed in situations where the true effect is smaller than anticipated in the planning phase and where thus guidance by the interim data is hoped for. Conversely, the conditional power may erroneously indicate a high chance for rejecting the null-hypothesis when it is in fact true. These findings match to the results of Denne (2001). He considered power and expected sample size when recalculating the sample size based on the pre-specified treatment effect (but with the observed variance) in adaptive designs and compared them with those of group sequential designs. It turned out that for smaller effects than the anticipated one, sample size recalculation in adaptive designs results in a somewhat higher power which, however, is associated with a substantial increase of the expected sample size.

28.4 Using Prior Information as Well as the Interim Effect Estimate

From a Bayesian perspective, the knowledge on the treatment effect in the planning stage can be described by a prior distribution which can then be updated by the data of the interim analysis. An appropriate quantity of the resulting posterior distribution, for example the mean or the median, can then be plugged in for Δ when calculating the conditional power with Eq. (28.2). For normally distributed outcomes with known variance σ^2 and sample size allocation ratio E versus C equal to $r{:}1$, the interim effect estimator is distributed as $\widehat{\Delta} \sim N\left(\Delta, \frac{(1+r)^2}{r \cdot n_1} \cdot \sigma^2\right)$. With a normal prior $N\left(\Delta_A, \tau^2\right)$, the posterior distribution of Δ given the interim effect estimate $\widehat{\Delta}$ is again a normal distribution with mean $\left(\frac{\Delta_A}{\tau^2} + \frac{\widehat{\Delta}}{\frac{(1+r)^2}{r \cdot n_1} \cdot \sigma^2}\right) \Big/ \left(\frac{1}{\tau^2} + \frac{1}{\frac{(1+r)^2}{r \cdot n_1} \cdot \sigma^2}\right)$ and variance $\left(\frac{1}{\tau^2} + \frac{1}{\frac{(1+r)^2}{r \cdot n_1} \cdot \sigma^2}\right)^{-1}$ (see, for example, Spiegelhalter et al. 2004, p. 62ff). The value of τ^2 can, for example, be determined by the prior belief of a negative effect: If $\Pr(\Delta < 0) = \varepsilon$, then $\tau^2 = \left(\frac{\Delta_A}{\Phi^{-1}(\varepsilon)}\right)^2$, where $\Phi^{-1}(\cdot)$ denotes the inverse of the standard normal distribution function (see Sect. 3.5.2).

The basic difficulties with conditional power calculations and corresponding sample size recalculation procedures described in the two preceding Sects. 28.2 and 28.3 cannot be eliminated by just inserting another value for Δ in the computations. However, taking into account both prior knowledge and current data leading to a weighted mean of the initially stipulated effect and the effect observed in the interim analysis is in the spirit of adaptive designs, namely updating the knowledge by learning from accumulating data.

Example 28.1 A considerable portion of patients with major depressive disorder fails to respond to available antidepressants. The **TRANSFORM-1 trial** (Fedgchin et al. 2019) included patients who showed nonresponse to at least two antidepressants in the current depression episode. The study investigated whether nasal spray treatment with esketamine, an approved drug for treating patients with treatment-resistant depression, adjunctive to a newly initiated oral antidepressant is superior to treatment with the new oral antidepressant alone. The trial was conducted as a randomized, double-blind, active-controlled study with three groups (new oral antidepressant plus two different doses of esketamine or placebo, respectively). For simplicity of presentation, we assume two groups comparing the combination of nasal spray treatment and oral antidepressant (E) with the oral antidepressant plus placebo nasal spray (C). The primary endpoint was the change in the Montgomery-Asberg Depression Rating Scale (Williams and Kobak 2008) between start and end of the four-week double-blind treatment phase. An adaptive two-stage design was implemented based on the inverse normal method with equal weights for both stages which corresponds to assuming initially equal stage-wise sample sizes $n_1 = n_2$, i.e. $w_1 = w_2 = 1/\sqrt{2}$. At the interim analysis, the study is either stopped for futility or the sample size is recalculated. We assume that a futility boundary of $\alpha_0 = 0.5$ (i.e. $z_{1-\alpha_0} = 0$) for the one-sided p-value of the first stage is applied, i.e., the study is stopped in case of a lacking or detrimental treatment effect. Then, choosing the boundary for early efficacy stopping as $\alpha_1 = 0.00107$ (i.e. $z_{1-\alpha_1} = 3.07$) allows that the level $\alpha_2 = 0.025$ can be applied at the final analysis while controlling the one-sided type I error rate at $\alpha = 0.025$ [calculations based on Eq. (27.4)]. It is assumed that the standard deviation amounts to $\sigma = 12$ and that in the planning stage a treatment group difference of $\Delta_A = 6.5$ is anticipated. To achieve a power of $1 - \beta = 0.90$ for Δ_A when the z-test is applied, a total balanced sample size of $n = 144$ is required for the fixed design according to Eq. (3.5).

For the two-stage design, we assume that the stage-one sample size is chosen as half of the fixed design sample size, i.e. $n_1 = 72$ (and therefore, due to the equal weighting of the stages, also $n_2 = 72$) and that the desired conditional power is set to $1 - \beta_{CP} = 0.90$. If the test statistic in the interim analysis amounts to $z_1 = 1.0$, this corresponds to an observed treatment group difference $\overline{X}_{E1} - \overline{X}_{C1} = 2.8$ and results in a conditional error rate $CE(z_1) = 0.0382$. To achieve a conditional power of 0.90 for this interim effect estimate, a total stage-two sample size of $\tilde{n}_2 = 672$ is required. If instead the interim test statistic is $z_1 = 2.0$, the conditional error rate is 0.2201, the observed treatment group difference amounts to 5.7 and the recalculated sample size for the second stage is $\tilde{n}_2 = 76$. This example nicely illustrates the mechanism of sample size recalculation based on the estimated treatment effect: Small treatment group differences entail at the same time small conditional error rates, and inserting both quantities in sample size recalculation leads to large or even huge stage-two sample sizes. When recalculating the sample size based on conditional power computations for the initially assumed treatment group difference $\Delta_A = 6.5$, the stage-two sample size amounts to $\tilde{n}_2 = 128$ for an interim test statistic $z_1 = 1.0$ and to $\tilde{n}_2 = 58$ for $z_1 = 2.0$. Finally, we assume that both prior information and the observed interim effect are to be used for sample size recalculation. If the belief of a negative effect is in the planning stage quantified as $\Pr(\Delta < 0) = 0.10$, this corresponds in case of a normal prior to a standard deviation $\tau = 5.1$. For an interim test statistic $z_1 = 1.0$ $(z_1 = 2.0)$, the mean of the posterior distribution of the treatment effect amounts to 3.7 (5.9), and the recalculated sample size is $\tilde{n}_2 = 392$

Table 28.1 Total stage-two sample size for the adaptive design described in Example 28.1 when applying different sample size recalculation methods based on conditional power. z_1 denotes the value of the test statistic at the interim analysis

Sample size recalculation method	Interim test statistic z_1	
	1.0	2.0
Using interim effect estimate	672	76
Using initially specified effect	128	58
Using prior information and interim estimate of effect:		
Prior belief $\Pr(\Delta < 0) = 0.10$	392	70
Prior belief $\Pr(\Delta < 0) = 0.20$	504	74

($\tilde{n}_2 = 70$). If there is a higher uncertainty about the treatment group difference which is quantified by $\Pr(\Delta < 0) = 0.20$, this corresponds to $\tau = 7.7$. The resulting values for the posterior mean and the recalculated sample size for the second stage are 3.3 (5.8) and $\tilde{n}_2 = 504$ ($\tilde{n}_2 = 74$), respectively, for an interim test statistic $z_1 = 1.0$ ($z_1 = 2.0$). The results for this example are summarized in Table 28.1.

Remarks

1. When specifying the first-stage sample size in clinical trials with interim analysis, both logistical and statistical aspects have to be taken into account. On the one hand, the interim analysis has to be performed early enough to have a relevant impact on the trial. On the other hand, the interim analysis should be performed late enough to avoid false decisions guided by insufficient data. For example, the probability of falsely stopping for futility should be small for efficacious treatments and corresponding sample size calculations may support the choice of n_1. For a discussion of issues related to the timing of interim analyses see, for example, Anderson (2014). With regard to sample size recalculation based on conditional power, the evaluations by Bauer and König (2006) and Gaffney and Ware (2017) discourage from performing the interim analysis too early.

2. Based on extensive comparisons, Jennison and Turnbull (2003) proposed to apply overpowered group sequential designs instead of sample size recalculation based on conditional power computations. "Overpowered" means that the group sequential design has sufficient power for a pessimistic (but still relevant) treatment effect, say $\Delta_A/2$ or $\Delta_A/3$, resulting in power values considerably above the common values for Δ_A. Posch et al. (2003) showed that similar performance characteristics can be achieved by applying suitable sample size reassessment rules in adaptive designs. The advantage of the latter approach is that it includes the option to react to unforeseen events (as, for example, new information coming from outside of the trial) by design modifications while still controlling the type I error rate.

3. Mehta and Pocock (2011) proposed the so-called "promising zone design" which is based on conditional power calculations for the estimated interim effect. Applying results derived by Chen et al. (2004) and Gao et al. (2008), they constructed three zones for the test statistic Z_1 with corresponding decision rules: If z_1 falls in the "unfavorable zone" (defined by "small" values of the conditional power) or in the "favorable zone" (defined by "large" values of the conditional

power), the original stage-two sample size is kept; if z_1 falls in the "promising zone" in between, the sample size is adjusted to reach the desired conditional power. Then, the conventional fixed sample size design test statistic can be applied in the final analysis at the nominal level α while controlling the type I error rate. Glimm (2012) noted that the promising zone design results as a special case from the conditional error function approach proposed by Müller and Schäfer (2001). The price to be paid for enabling application of the test statistic and the significance level used in the fixed sample size design is a reduced flexibility in sample size recalculation. Furthermore, there was an extensive and controversial debate on the efficiency of the promising zone approach (Emerson et al. 2011; Mehta 2013; Levin et al. 2013). Jennison and Turnbull (2015) improved the sample size reassessment rule used in the promising zone design. However, to control the type I error rate, the inverse normal combination test was applied instead of the conventional test statistic used in the fixed sample size design (see also Mehta and Liu 2016).

4. Herrmann et al. (2020) proposed a conditional performance score that enables measuring the operating characteristics of conditional sample size recalculation strategies. This score consists of four components, namely location and variation of the conditional power and the conditional sample size, respectively. It can be used to compare various sample size recalculation rules and to develop strategies with improved performance.

Chapter 29
Sample Size Recalculation Based on Optimization

Metaphorically speaking, performing sample size recalculation based on conditional power arguments resembles to set sail, to await at which isle the ship ends up, and then—based on considerations as, for example, how far the interstation is from the travel destination and how many time and budget is still available—to decide where to go next. Wouldn't it be more clever to fix already before starting the trip how to continue the journey for each potential intermediate stop? This could then be done in a way that the overall benefit (composed of aspects like pleasure, travel time, required resources, etc.) is maximized if this strategy is applied repeatedly. If unforeseen incidents occur, one can then still change to another strategy. In the above analogy, the first strategy reflects the conditional power approach while the second one mirrors the idea of optimal planning of adaptive designs. Here, considering all potential realizations of the first-stage test statistic, the stage-one sample size, the early stopping boundaries as well as the stage-two sample size and the decision boundary for the test at the final analysis are chosen up-front in such a way that a specified optimality criterion is fulfilled. This unconditional approach adheres to the conventional frequentistic concept of considering performance measures that mirror the long-run performance characteristics.

In its general formulation, the optimization approach was proposed by Pilz et al. (2019a). Following their exposition, a two-stage design is defined by the five quantities n_1, c_0, c_1, $n_2(\cdot)$, and $c_2(\cdot)$, where n_1 denotes the stage-one sample size, c_0 and c_1 the boundaries for early stopping with acceptance or rejection of H_0, respectively, for the first-stage test statistic Z_1 given in (27.2a), $n_2(\cdot)$ is the sample size of the second stage, and $c_2(\cdot)$ the decision boundary for the stage-two test statistic Z_2 given in (27.2b). Note that $n_2(\cdot)$ and $c_2(\cdot)$ are functions of z_1. In order to choose them together with n_1, c_0, and c_1 in an optimal way, a suitable objective criterion has to be selected. In the following, we illustrate the approach for the frequently applied criterion of minimizing the expected sample size under the specified alternative. Readopting the superiority test problem (27.1) for normally distributed data with known variance, the optimal adaptive design $D = (n_1, c_0, c_1, n_2(\cdot), c_2(\cdot))$ can be

© Springer Nature Switzerland AG 2020
M. Kieser, *Methods and Applications of Sample Size Calculation and Recalculation in Clinical Trials*, Springer Series in Pharmaceutical Statistics,
https://doi.org/10.1007/978-3-030-49528-2_29

identified by resolving the following optimization problem:

$$\min_{n_1, c_0, c_1, n_2(\cdot), c_2(\cdot)} E_{\Delta = \Delta_A}(N) \tag{29.1a}$$

under the constraints

$$\Pr_{\Delta = 0}(\text{reject } H_0) \leq \alpha \tag{29.1b}$$

$$\Pr_{\Delta = \Delta_A}(\text{reject } H_0) \geq 1 - \beta. \tag{29.1c}$$

Here, $N = n_1 + n_2(Z_1)$ denotes the total recalculated sample size. The solutions of the optimization problem (29.1a)–(29.1c) can be found by applying a two-step approach:

Step 1   Determine for given n_1, c_0, and c_1 the optimal functions $n_2^*(\cdot)$ and $c_2^*(\cdot)$ (which can be seen as a function of n_1, c_0, and c_1). Note that this is a variational problem as the solutions are functions.

Step 2   Determine the optimal parameter values n_1, c_0, and c_1. This can be done by standard methods for calculus and numerical optimization.

For details of the derivation and computational issues see Pilz et al. (2019a).

It is important to note that planning the adaptive design in an optimal way enables nevertheless to discard the optimal sample size recalculation rule and to apply any alternative sample size recalculation strategy instead. This may, for example, be necessary due to new external information overruling the planning assumptions or due to practical requirements. Described in the picture of the tale at the beginning of this chapter: It is possible to choose another track if, for example, one becomes aware of a new island that was not recorded on the map available at the start of the sailing tour. To control the type I error rate even if the pre-specified optimal sample size recalculation rule $n_2(z_1)$ is changed, just the level $c_2(z_1)$ given by the conditional error function of the optimal design has to be applied in the final analysis. However, it is important to note that due to its definition, any deviation from the optimal design fulfilling (29.1a)–(29.1c) inevitably leads to a sub-optimal performance with respect to the selected objective criterion.

The example presented below illustrates the characteristics of the group sequential design and the adaptive design with sample size recalculation based on conditional power and optimization, respectively, together with their differences.

Example 29.1 We again consider the **TRANSFORM-1 trial** introduced in Example 28.1. For the implemented adaptive two-stage design, boundaries for the one-sided p-value of the first stage were defined as $\alpha_0 = 0.5$ (i.e. $z_{1-\alpha_0} = 0$ for the test statistic) for early futility stopping and as $\alpha_1 = 0.00107$ (i.e. $z_{1-\alpha_1} = 3.07$) for early efficacy stopping. Then, the level $\alpha_2 = 0.025$ can be applied after the second stage. For sample size (re)calculation, it is assumed that a power of 0.90 is desired for a treatment group difference of $\Delta_A = 6.5$ and a standard deviation of $\sigma = 12$ when a one-sided test at level 0.025 is applied.

In the following, we consider the characteristics of three designs with one-sided type I error rate $\alpha = 0.025$ and power $1 - \beta = 0.90$ at $\Delta_A = 6.5$ for $\sigma = 12$: (i) the group sequential design with equal sample sizes for the two stages and the above-mentioned early stopping boundaries; (ii) the adaptive design with inverse normal combination test (equal weights $w_1 = w_2 = 1/\sqrt{2}$) and sample size recalculation based on (28.3) with a conditional power of $1 - \beta_{CP} = 0.90$ at the pre-specified alternative $\Delta_A = 6.5$ and with the above-mentioned early stopping boundaries; (iii) the optimal adaptive design resolving the optimization problem (29.1a)–(29.1c). It is assumed that no upper boundary for the sample size is defined (Example 29.2 below considers the scenario of limiting the maximum sample size).

Figure 29.1 shows the critical values for the second-stage test statistic Z_2 depending on the value of the test statistic at the interim analysis. For the group sequential design and the adaptive design based on conditional power, the same critical values are applied which depend linearly on z_1 in the continuation region $\left[z_{1-\alpha_0}, z_{1-\alpha_1} \right] = [0, 3.07]$. In contrast, for the optimal adaptive design the function $c_2(z_1)$ shows a concave shape with a narrower continuation region $\left[z_{1-\alpha_0}, z_{1-\alpha_1} \right] = [0.28, 2.27]$.

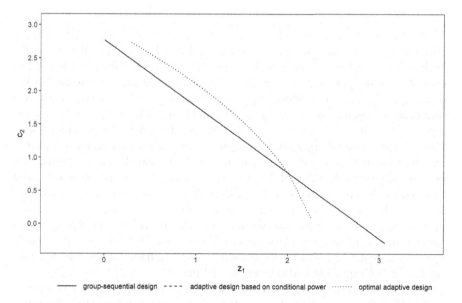

Fig. 29.1 Rejection boundary for the stage-two test statistic in the final analysis depending on the value of the test statistic at the interim analysis for the scenario considered in Example 29.1. Shown are the rejection boundaries for values of the interim test statistic within the continuation region. Below and above the continuation region, the rejection boundaries are $c_2 = \infty$ and $c_2 = -\infty$, respectively. The boundary curves for the group sequential design and for adaptive design based on conditional power are overlapping

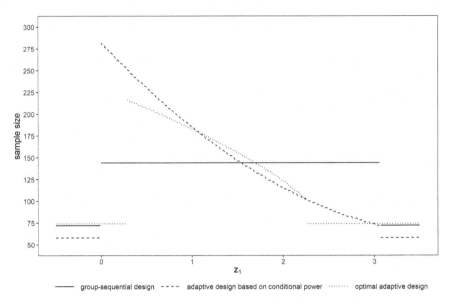

Fig. 29.2 Total sample size depending on the value of the test statistic at the interim analysis for the scenario considered in Example 29.1

The sample size functions $n(z_1) = n_1 + n_2(z_1)$ are shown in Fig. 29.2. For the group sequential design, the equal stagewise total sample size required to assure the desired power of 0.90 at $\Delta_A = 6.5$ is $n_1 = n_2 = 72$. Therefore, $n(z_1) = 72$ in case of early stopping and $n(z_1) = 144$ when the second stage is entered. For the design specifications given above, the power of the adaptive design with sample size recalculation depends only on the stage-one sample size. Choosing $n_1 = 60$ leads to a power of $1 - \beta = 0.90$ for $\Delta_A = 6.5$ thus making the various designs comparable. Inside of the continuation region, the sample size function depending on z_1 is convex and strictly monotonously decreasing with a maximum sample size of 284 which is considerably larger than that of the other two designs. In contrast, the sample size function for the optimal adaptive design is concave in the continuation region (see Remark 2 below). This design starts with a slightly larger stage-one sample size of $n_1 = 74$ and shows a maximum sample size of 216 which is markedly smaller than for the conditional power based design. The excessively large maximum sample size for recalculation based on conditional power is in line with the investigations referenced in Chap. 28 (see also Example 28.1 there).

The expected sample size depending on the treatment group difference is depicted in Fig. 29.3. By construction, the expected sample size for the specified alternative $\Delta_A = 6.5$ is the smallest for the optimal adaptive design ($E_{\Delta=6.5}(N) = 108.2$). The corresponding values for the adaptive design based on conditional power and for the group sequential design are 120.6 and 127.6, respectively.

The optimization approach described above can be modified and extended in various ways (Pilz et al. 2019b). As a major addition, further constraints beyond those

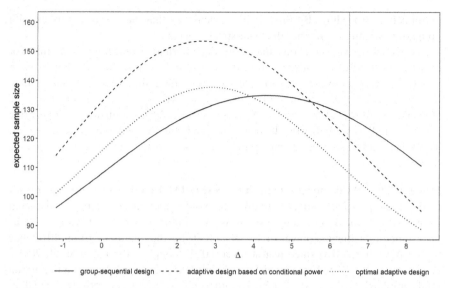

Fig. 29.3 Expected total sample size depending on the treatment group difference for the scenario considered in Example 29.1. The grey vertical line marks the alternative $\Delta_A = 6.5$ for which the expected sample size is minimized by the optimal adaptive design

for type I error rate and power can be imposed reflecting study-specific requirements. For example, a maximum total sample size may have to be fixed due to limitations in time and resources (see Example 29.2 below). If such a constraint is additionally implemented, the optimization problem reads as follows:

$$\min_{n_1, c_0, c_1, n_2(\cdot), c_2(\cdot)} E_{\Delta=\Delta_A}(N) \tag{29.2a}$$

under the constraints

$$\Pr_{\Delta=0}(\text{reject } H_0) \leq \alpha \tag{29.2b}$$

$$\Pr_{\Delta=\Delta_A}(\text{reject } H_0) \geq 1 - \beta \tag{29.2c}$$

$$\max_{z_1}\{n_1 + n_2(z_1)\} \leq n_{\max}. \tag{29.2d}$$

Moreover, the desired power may not only be fixed for the alternative Δ_A but one may in addition require a power $1 - \beta^*$ for a treatment effect $\Delta_A^* < \Delta_A$ which is still worth to be detected. Other constraints may include minimum values for the conditional power or maximum values for the probability of early stopping for futility defined for specified alternatives. Furthermore, design parameters as, for example, the stage-one sample size, may be fixed upfront instead of including them in the optimization procedure. By this, it can be achieved that the design complies with obligations arising from other than statistical aspects.

As a further extension, one may consider alternative objective functions. For example, instead of considering the expected sample size under the alternative, a weighted mean of the expected sample size under the null- and the alternative may be minimized, i.e. $\lambda \cdot E_{\Delta=0} + (1 - \lambda) \cdot E_{\Delta=\Delta_A}$ with $0 \leq \lambda \leq 1$. Furthermore, the variation of the sample size is of great relevance for the logistics of an adaptive trial as a smaller variability results in a better predictability of the required resources. This aspect can be incorporated in optimization by choosing an appropriate objective function.

Example 29.2 We come back to the **TRANSFORM-1 trial** that we already considered in Examples 28.1 and 29.1. In order to avoid inappropriately large recalculated sample sizes, a maximum sample size was fixed in the protocol of this study. It was determined to achieve a power of $1 - \beta = 0.90$ for the treatment group difference $\Delta_A = 6.5$ when a two-sided test at level 0.025 is applied (Fedgchin et al. 2019). According to Eq. (3.5) for the z-test, this results for balanced groups in a maximum total sample size of $2 \times 85 = 170$. To mimic this scenario, we optimize the adaptive design by additionally imposing this maximum total sample size as a further constraint. The related optimal design is obtained by resolving the optimization problem (29.2a)–(29.2d) for $\alpha = 0.025$, $1 - \beta = 0.90$, $\Delta_A = 6.5$, and $n_{max} = 170$. Another aspect of interest is that the anticipated treatment group difference of 6.5 points may be judged as ambitious since even effects of 2 points on the Montgomery-Asberg Depression Rating Scale are seen as clinically relevant (Duru and Fantino 2008). For the group sequential and the two adaptive designs, the power for a treatment group difference of 4 points, which is still clinically relevant, amounts to only 0.514 (group sequential design), 0.550 (conditional power based design), and 0.510 (optimal design), respectively. Since for these designs, the probability to detect this clinically relevant difference is only slightly higher than when using a coin flip for decision making, one may be interested to increase the power for this alternative. Hence, we also consider the case of optimizing the adaptive design under the additional constraint of requiring a power of $1 - \beta^* = 0.60$ at $\Delta_A^* = 4$ points. This leads to the optimization problem

$$\min_{n_1, c_0, c_1, n_2(\cdot), c_2(\cdot)} E_{\Delta=6.5}(N) \tag{29.3a}$$

under the constraints

$$\Pr_{\Delta=0}(\text{reject } H_0) \leq 0.025 \tag{29.3b}$$

$$\Pr_{\Delta=6.5}(\text{reject } H_0) \geq 0.90 \tag{29.3c}$$

$$\Pr_{\Delta=4}(\text{reject } H_0) \geq 0.60. \tag{29.3d}$$

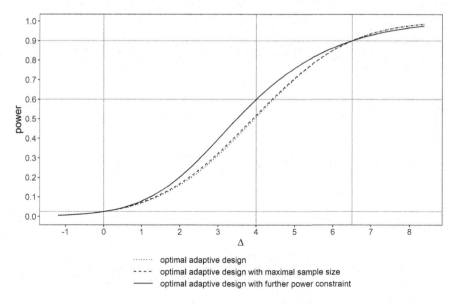

Fig. 29.4 Power depending on the treatment group difference for the scenario considered in Example 29.2. The grey vertical lines mark the alternative $\Delta_A = 6.5$ for which the expected sample size is minimized by the optimal adaptive design, the alternative $\Delta_A^* = 4$ for which the adaptive design with further power constraint achieves a power of $1 - \beta^* = 0.60$, and the null-hypothesis $\Delta = 0$, respectively

Figure 29.4 shows the power curves for the three designs. It can be seen that restricting the maximum sample size to 170 has no marked consequences for the power which is virtually identical for the optimal adaptive designs with and without this constraint. However, requiring a power of $1 - \beta^* = 0.60$ for $\Delta_A^* = 4$ results in a much steeper power curve with considerable power gain for effects smaller than the anticipated value $\Delta_A = 6.5$. This feature has to be paid with a massive increase of the required sample size with a maximum of 486 patients (Fig. 29.5). The optimal adaptive design with the maximum sample size restricted to 170 patients starts with the stage-one sample size of $n_1 = 88$, while the first-stage sample size of the power-constrained design equals $n_1 = 82$. In the continuation region, the sample size is for a wide range constant at this maximum. From Fig. 29.5, it can also be seen that the continuation region is broadest for the optimal design with power constraint ($[z_{1-\alpha_0}, z_{1-\alpha_1}] = [0.10, 2.39]$) and smallest for the design with restricted maximum sample size ($[z_{1-\alpha_0}, z_{1-\alpha_1}] = [0.74, 2.57]$). The expected sample size under the alternative $\Delta_A = 6.5$, which is minimized, is similar for the optimal adaptive design and the design with further constraint on the maximum sample size ($E_{\Delta=6.5}(N) = 108.2$ vs. 112.8) and is moderately increased for the power-constrained design ($E_{\Delta=6.5}(N) = 118.2$) (Fig. 29.6). However, forcing the power to be 0.60 at $\Delta_A^* = 4$ entails a tremendous increase in expected total sample size for treatment effects smaller than Δ_A with a maximum expected sample size of 203.9

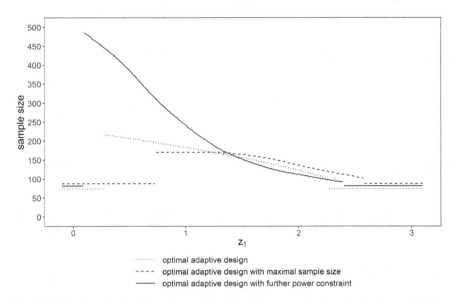

Fig. 29.5 Total sample size depending on the value of the test statistic at the interim analysis for the scenario considered in Example 29.2

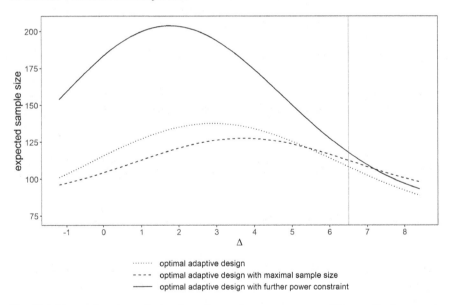

Fig. 29.6 Expected total sample size depending on the treatment group difference for the scenario considered in Example 29.2. The grey vertical line marks the alternative $\Delta_A = 6.5$ for which the expected sample size is minimized by the optimal adaptive design

as compared to 137.7 for the optimal design and 127.6 for the optimal design with maximum sample size 170.

Remarks

1. Optimal group sequential designs can be obtained via the constrained optimization technique by just restricting the solutions of the optimization technique to constant second-stage sample size functions, i.e. by requiring $n_2(z_1) \equiv n_2$ for all z_1. For these designs, the second-stage sample size is always larger than the sample size of the first stage when the objective function is a weighted mean of the expected sample size under the null- and the alternative hypothesis (Pilz et al. 2019b). For considerations on the choice of the first-stage sample size in clinical trials with interim analyses see also Remark 1 in Chap. 28.

2. Another application of the constrained optimization approach is to optimize the timing of the interim analysis for group sequential designs. This can be achieved by fixing the decision boundaries of the group sequential design, restricting the second-stage sample size functions to constants, and searching for the value of n_1 minimizing (for fixed α and $1 - \beta$) the objective function.

3. Jennison and Turnbull (2015) optimized the sample size function $n_2(\cdot)$ for the weighted inverse normal combination function. By this, they restricted to the critical value function $c_2(z_1) = (c - w_1 \cdot z_1)/w_2$, i.e. to linear functions of the first-stage test statistic. By omitting this restriction, the designs obtained by Jennison and Turnbull can be improved. It turns out that the shape of the optimal critical value function $c_2(z_1)$ is not too far away from a linear one (see, for example, Fig. 1 in Pilz et al. 2019a, and Fig. 29.2). Furthermore, it was shown that the optimal sample size function for the optimization problem (29.1a)–(29.1c) is always concave (Pilz et al. 2020). This is different to recalculation based on conditional power, where the stage-two sample size function is generally convex. For illustrative examples of this issue see Figs. 29.2 and 29.5.

Chapter 30
Regulatory and Operational Aspects

In principle, for designs with unblinded sample size recalculation, the same major issues are relevant for regulatory authorities as for the internal pilot study design with blinded sample size reassessment: control of the type I error rate and prevention of introducing a bias into the trial. For the methods presented in Chaps. 28 and 29 above, unblinded sample size recalculation can be performed under strict control of the level at the specified α. In more complex situations, analytical derivations for the actual type I error rate and, if required, analytical adjustment procedures assuring level-control may exist as well. Otherwise, simulations have to be performed which are also a common way to investigate the performance characteristics of the design regarding, for example, power and distribution of sample size [see, for example, the FDA Guidance for Industry *Adaptive Designs for Clinical Trials of Drugs and Biologics* (FDA 2019, p. 18ff)]. Posch et al. (2011) pointed out the challenges of simulation-based procedures tailored to control the type I error rate in adaptive designs, and Mayer et al. (2019) provided a general framework for simulation studies performed to explore the operating characteristics of adaptive designs.

In clinical trials with unblinded interim analyses, treatment effect estimates become available during the course of the study. If trial personnel comes to know information on these interim results, this may introduce biases into the trial, for example, by consciously or unconsciously modifying the patient population to be included or the assessment of the outcome. Therefore, comprehensive documentation of the processes and especially of the implemented firewalls for communicating about observed treatment effects is essential. For an example of steps taken to maintain and to document strict confidentiality of interim results in clinical trials with adaptive design see Mehta et al. (2019). Furthermore, in classical group sequential trials it has become a standard to implement a Data Monitoring Committee (DMC) which is independent of the personnel involved in the study conduct. The DMC ensures that the benefit-risk ratio for the study patients is acceptable by regularly inspecting interim results. For rationale and procedural aspects of DMCs see, for example, the guidelines of the European Medicines Agency (EMA 2005) and the

© Springer Nature Switzerland AG 2020
M. Kieser, *Methods and Applications of Sample Size Calculation and Recalculation in Clinical Trials*, Springer Series in Pharmaceutical Statistics,
https://doi.org/10.1007/978-3-030-49528-2_30

US Food and Drug Administration (FDA 2006) as well as the book on DMCs by Ellenberg et al. (2019). As an assessment of the benefit-risk profile is required for any clinical trial with interim analysis, a DMC has to be established also in adaptive designs. However, as compared to group sequential designs, in adaptive clinical trials there is an additional important task to be handled, namely conducting the analyses required to decide about design modifications as, for example, changing the sample size. There is a debate on whether the DMC should also accomplish this task or whether another board should be implemented [see, for example, Gallo et al. (2014) for arguments for and against a single committee]. Another discussion concerns whether or not the trial sponsor should be involved in the adaptation decision. Such an involvement would make the issue of keeping the interim results confidential even more critical. Sometimes, the decisions to be made are complex and have far-reaching implications, for example with respect to the required budget or the time until market access. Delegating the responsibility completely to a board of independent experts may be seen as not fully appropriate. However, the more detailed the adaptation rules are specified, the less room for ambiguity is and the better the consequences of design modifications can be anticipated. For example, for the optimization approach presented in Chap. 29, the recalculated sample size is defined up-front for each possible interim result. By careful planning, communication, and reconcilement between all stakeholders during the development phase of the protocol, a sponsor involvement may be reduced or even made dispensable.

As a last point, I would like to address the potential of back-calculating the treatment effect from the recalculated sample size and the algorithm on which the recalculation is based on. There is concern that by this mechanism knowledge of the observed treatment effect becomes available via the backdoor although communication firewalls are established. In fact, the more precise the recalculation rule is specified, the better the observed treatment effect can be predicted. However, this danger can be limited by restricting the description of the sample size recalculation procedure in the protocol to specifying just the basic approach. Details of the algorithm are then described exclusively in a separate document that is solely accessible to a defined group of persons. As a further precaution measure, the recalculated sample size may be forwarded only to those persons that absolutely require it, for example the personnel preparing the study medication. All other parties just receive the information they need for a proper conduct of the trial. For example, the additional sample size required in their center may be forwarded to the investigators but not the overall sample size. In the TRANSFORM-1 trial (see Examples 29.1 and 29.2) *the sponsor study team and study site staff were not informed of the adjusted sample size until it had been met to ensure no impact on study conduct*" (Fedgchin et al. 2019, p. 619). Taking such precaution measures, the information that becomes available to trial personnel in adaptive trials with sample size recalculation is very similar to that available in classical group sequential designs. In both cases, from the decision to enter the second study stage it can be concluded that the observed treatment effect lies within a corridor but its exact magnitude cannot be derived. Various aspects concerning DMCs and logistics which have to be kept in mind when performing adaptive designs are discussed by Gallo (2006), Chow et al. (2012), Antonijevic et al. (2013), Sanchez-Kam et al. (2014), and Gallo et al. (2014).

Chapter 31
Concluding Remarks

There is an ongoing debate on the efficiency of unblinded sample size recalculation and on whether or not a potential loss of efficiency may be compensated by advantages with respect to other aspects. Without doubt, situations may arise in clinical trials that were unforeseeable in the planning stage and that require to change the sample size mid-course. Information emerging from external sources have repeatedly been mentioned as a case where classical designs surrender (see, for example, Posch et al. 2003). Classical designs such as the fixed sample size design and group sequential designs cannot handle such scenarios without compromising the type I error rate. Therefore, adaptive designs with their feature of enabling to recalculate the sample size under control of the level are an attractive option. However, as shown above and in the referenced articles, an uncritical application of sample size reassessment rules has its pitfalls. All in all, the following strategy seems to be recommendable. If the study team would like to leave open the option for a change of the sample size during the ongoing trial, an adaptive design has to be implemented. The best choice is an optimal adaptive design according to Chap. 29 as it is most efficient. For this, an objective function and constraints are to be chosen carefully such that they appropriately reflect the study-specific circumstances and requirements. If in fact an unexpected situation occurs during the conduct of the trial, the sample size recalculation rule induced by optimization can be changed and any other arbitrary strategy can be applied instead without compromising the type I error rate. Applying this approach, both the aspects of efficiency and flexibility are properly addressed.

© Springer Nature Switzerland AG 2020   293
M. Kieser, *Methods and Applications of Sample Size Calculation and Recalculation
in Clinical Trials*, Springer Series in Pharmaceutical Statistics,
https://doi.org/10.1007/978-3-030-49528-2_31

Appendix
R Software Code

In this Appendix, R code is provided implementing all methods for sample size calculation presented in this book. In general, for each scenario one sample size function (and if required auxiliary functions) is implemented as well as an exemplary call according to the corresponding book example. The sample size functions are structured consistently and, where feasible, the same output parameters are returned for each scenario. These output parameters are described in Appendix A.1. The applied rounding strategy in order to generate integer sample size values is described in Appendix A.2. The Appendices A.3–A.15 provide the R codes for the methods presented in Part II and Appendices A.16–A.22 those for Part III of the book.

The Appendix was kindly prepared by Laura Benner and Svenja Schüler (Institute of Medical Biometry and Informatics, University of Heidelberg). The R code was implemented by Lukas Baumann (Chapters 21 and 23), Rouven Behnisch (Section 8.4.1), Laura Benner (Chapters 9–13 and Sections 5.2.1, 5.2.2, 5.2.3, 8.2, 8.3, 8.4.1, 8.4.2.2, 8.4.2.3), Samuel Kilian (Sections 5.3.2 and 8.4.2.4), Johannes Krisam (Section 5.2.4), Marietta Kirchner (Chapter 14), Jan Meis (Sections 7.2, 7.3, 7.5), Maximilian Pilz (Chapters 20, 22, 27, 28, 29), Stella Erdmann (Chapter 15), and Svenja Schüler (Chapters 3, 4, 6, 10 and Sections 7.4, 8.2, 8.4.2.2, 8.5), all from the Institute of Medical Biometry and Informatics, University of Heidelberg.

A.1 Output of Sample Size Functions

In general, most implemented sample size functions return the following parameters:

```
n_total = total sample size n
n_E = sample size in experimental group
n_C = sample size in control group
actual_r = allocation ratio of calculated n_E / n_C
actual_power = Power for calculated n_total and actual_r
(determined by converting the sample size formula)
```

© Springer Nature Switzerland AG 2020
M. Kieser, *Methods and Applications of Sample Size Calculation and Recalculation in Clinical Trials*, Springer Series in Pharmaceutical Statistics,
https://doi.org/10.1007/978-3-030-49528-2

For exact tests with binary outcome, the following parameter is returned:

```
Exact_power = Exact power for calculated n_total and actual_r
```

For survival outcomes, the following parameters are returned additionally:

```
p_D = overall probability of observing an event
d_total = number of total events
d_E = number of events in experimental group
d_C = number of events in control group
```

If other output parameters are returned, values are described in the respective R code chapter.

A.2   Remarks on the Rounding Strategy

Different strategies exist for rounding the value obtained from a closed sample size formula to an integer value as described in Sect. 3.3. In the provided R code, the unrounded n is used to calculate $n_C = \left(\frac{1}{1+r}\right) \cdot n$ and $n_E = \left(\frac{r}{1+r}\right) \cdot n$, and both values rounded up to the next integer are returned (n_E and n_C). As the total sample size n_total, the sum of these rounded values is returned. These rounded sample sizes can lead to a slightly different allocation ratio and power value, and thus the actual allocation ratio (actual_r) and the actual power (actual_power) are returned additionally.

It should be noted that in all numerical examples given in the book, r is an integer and the group-wise sample sizes are determined such that their ratio is exactly equal to the specified allocation ratio. Therefore, some sample sizes calculated by the provided R code are not exactly the same as these in the numerical examples given in the book.

All sample size calculations for time-to-event data are based on the calculation of the number of events. In the provided R code, firstly, the number of events are calculated and rounded up to the next integer. Secondly, the sample size is calculated based on the rounded number of events and is afterwards rounded up to the next integer again.

A.3 R Code for Chapter 3

```
# _____ SECTION 3.2 _____

##############################################################################
# Function to calculate the sample size for the z-test according             #
# to formula (3.5)                                                           #
##############################################################################

##############################################################################
# Parameters to specify:
# delta_A  = difference between groups
# sigma = true standard deviation
# alpha = significance level
# beta  = type II error rate
# r     = allocation ratio of sample sizes (n_E/n_C)
##############################################################################

samplesize_zTest <- function(delta_A, sigma, alpha, beta, r){
  n <- (1+r)^2/r * (qnorm(1-alpha/2) + qnorm(1-beta))^2 *(sigma/delta_A)^2
  n_E <- ceiling(r/(1+r) * n)
  n_C <- ceiling(1/(1+r) * n)
  actual_r <- n_E/n_C
  n_total <- n_E + n_C

  actual_power <- pnorm(sqrt((n_E+n_C) * actual_r * delta_A^2/
                        ((1+actual_r)^2 * sigma^2)) - qnorm(1-alpha/2))
  return(data.frame(n_total = n_total, n_E = n_E, n_C = n_C,
                    actual_r = actual_r, actual_power = actual_power))
}

#****************************************************************************#
# Exemplary call according to Example 3.1                                   #
#****************************************************************************#

samplesize_zTest(10, 20, 0.05, 0.1, 1)

##   n_total n_E n_C actual_r actual_power
## 1     170  85  85        1    0.9031373

#
#
# _____ SECTION 3.3 _____

##############################################################################
# Function to calculate the sample size for a t-test                         #
##############################################################################

##############################################################################
# Parameters to specify:
# delta_A  = difference between groups
# sigma = standard deviation
# alpha = significance level
# beta  = type II error rate
# r     = allocation ratio of sample sizes (n_E/n_C)
##############################################################################

#--------------------------------------------------#
# Exact sample size formula according to            #
```

```
# iteratively resolving formula (3.10)                 #
#-------------------------------------------------#
samplesize_tTest <- function(delta_A, sigma, alpha, beta, r){
  power <- 0
  n <- 2
  while(power < 1 - beta){
    n <- n + 1
    n_E <- ceiling(r/(1+r) * n)
    n_C <- ceiling(1/(1+r) * n)
    actual_r <- n_E/n_C
    n_total <- n_E + n_C
    nc <- sqrt(actual_r/(1+actual_r)^2 * n_total) * delta_A/sigma
    power <- 1 - pt(qt(1-alpha/2, n_total-2), n_total-2, nc)
  }

  return(data.frame(n_total = n_total, n_E = n_E, n_C = n_C,
                 actual_r = actual_r, actual_power = power))
}


#-------------------------------------------------#
# Approximate sample size formula according        #
# to Guenther/Schouten (3.11a/b)                   #
#-------------------------------------------------#
samplesize_tTest_GS <- function(delta_A, sigma, alpha, beta, r){

n_C <- ceiling((1+r)/r * (qnorm(1-alpha/2) + qnorm(1-beta))^2 *
        (sigma/delta_A)^2 + (qnorm(1-alpha/2)^2) / (2*(1+r)))
n_E <- ceiling((1+r) * (qnorm(1-alpha/2) + qnorm(1-beta))^2 *
        (sigma/delta_A)^2 + r * (qnorm(1-alpha/2)^2) / (2*(1+r)))
actual_r <- n_E/n_C
n_total <- n_E + n_C
actual_power <- pnorm(1/((1+actual_r) * abs(sigma/delta_A)) *
        sqrt(actual_r*(n_total-(qnorm(1-alpha/2)^2)/
                      (2*(1+actual_r)))) - qnorm(1-alpha/2))
return(data.frame(n_total = n_total, n_E = n_E, n_C = n_C,
           actual_r = actual_r, actual_power = actual_power))
}

#*********************************************************************#
# Exemplary call according to the Example 3.2                        #
#*********************************************************************#

samplesize_tTest(10, 20, 0.05, 0.1, 2)

##   n_total n_E n_C actual_r actual_power
## 1     191 127  64 1.984375    0.9006259

samplesize_tTest_GS(10, 20, 0.05, 0.1, 2)

##   n_total n_E n_C actual_r actual_power
## 1     192 128  64        2    0.9032976

#
#
# _____ SECTION 3.4 _____

#####################################################################
# Functions to calculate the sample size for an ANCOVA              #
#####################################################################
```

```
################################################################
# Parameters to specify:
# delta_A  = difference between groups
# sigma    = standard deviation
# rho = correlation between outcome and covariate
# alpha = significance level
# beta  = type II error rate
# r       = allocation ratio of sample sizes (n_E/n_C)
################################################################

#-------------------------------------------------#
# Exact sample size according to                  #
# iteratively resolving formula (3.15)            #
#-------------------------------------------------#
samplesize_ANCOVA_exact <- function(delta_A, sigma, rho, alpha, beta, r){
  power <- 0
  n <- 2
  while(power < 1 - beta){
    n <- n + 1
    n_E <- ceiling(r/(1+r) * n)
    n_C <- ceiling(1/(1+r) * n)
    actual_r <- n_E/n_C
    n_total <- n_E + n_C
    nc <- sqrt(actual_r/(1+actual_r)^2 * n_total) * delta_A /
      sigma / sqrt(1-rho^2)
    power <- 1 - pt(qt(1-alpha/2, n_total-3), n_total-3, nc)
  }

  return(data.frame(n_total = n_total, n_E = n_E, n_C = n_C,
                actual_r = actual_r, actual_power = power))
}

#-------------------------------------------------#
# Approximate sample size according               #
# to formula (3.18)                               #
#-------------------------------------------------#
samplesize_ANCOVA_approx <- function(delta_A, sigma, rho, alpha, beta, r){

  n_FP <- (1+r)^2/r * (qnorm(1-alpha/2) + qnorm(1-beta))^2 *
        (1-rho^2) * sigma^2 / delta_A^2
  n <- n_FP + qnorm(1-alpha/2)^2/2
  n_E <- ceiling(r/(1+r) * n)
  n_C <- ceiling(1/(1+r) * n)
  n_total <- n_E + n_C
  actual_r <- n_E/n_C
  power <- pnorm(sqrt(actual_r*n_total/(1+actual_r)^2) *
                (delta_A/(sqrt(1-rho^2)*sigma)) - qnorm(1-alpha/2))
  return(data.frame(n_total = n_total, n_E = n_E, n_C = n_C,
                actual_r = actual_r, actual_power = power))
}

#**************************************************************#
# Exemplary call according to the Example 3.3                 #
#**************************************************************#

samplesize_ANCOVA_exact(10, 20, 0.4, 0.05, 0.1, 1)
```

```
##    n_total n_E n_C actual_r actual_power
## 1     144  72  72        1    0.9016359

samplesize_ANCOVA_approx(10, 20, 0.4, 0.05, 0.1, 1)

##    n_total n_E n_C actual_r actual_power
## 1     144  72  72        1    0.9054598

#
#
#_____ SECTION 3.5.2 _____

##############################################################################
# Function to calculate the traditional and expected power for a            #
# two-sided z-test and normal prior according to formulas (3.19) and (3.21) #
##############################################################################

##############################################################################
# Parameters to specify:
# n       = sample size
# delta_A = difference under the alternative or prior expectation
# sigma   = standard deviation
# pr_neg  = probability for delta < 0
# alpha   = significance level
# r       = allocation ratio of sample sizes (n_E/n_C)
##############################################################################

power_expected <- function(n, delta_A, sigma, pr_neg, alpha, r){

  tau <- -delta_A / qnorm(pr_neg)
  # expected Power (according to formula 3.21)
  EP <- 1 - pnorm(sqrt((1+r)^2/(r*n) * sigma^2 / ((1+r)^2/(r*n) * sigma^2 +
       tau^2)) * (qnorm(1-alpha/2) - sqrt(r*n / (1+r)^2) * delta_A/sigma))
  # traditional Power (according to formula 3.19)
  TP <- pnorm(sqrt(r/(1+r)^2*n) * delta_A/sigma - qnorm(1-alpha/2))
  return(data.frame(EP = EP, TP = TP))
}

#****************************************************************************#
# Exemplary call according to Example 3.4 (compare Table 3.3a)             #
#****************************************************************************#

power_expected(170, 10, 20, 0.001, 0.05, 1)

##          EP        TP
## 1 0.8143798 0.9031373

##############################################################################
# Function to calculate the sample size for a given expected power for a     #
# two-sided z-test and normal prior according to formulas (3.19) and (3.21) #
##############################################################################
```

```
#############################################################################
# Parameters to specify:
# EP      = expected Power
# delta_A = difference under the alternative or prior expectation
# sigma   = standard deviation
# pr_neg  = probability for delta < 0
# alpha   = significance level
# r       = allocation ratio of sample sizes (n_E/n_C)
#############################################################################

samplesize_expected_power <- function(EP, delta_A, sigma, pr_neg, alpha, r){
  n <- r + 1
  EP_it <- 0

  while(EP_it < EP){
    n <- n + r + 1
    n_E <- ceiling(r/(1+r) * n)
    n_C <- ceiling(1/(1+r) * n)
    n_total <- n_E + n_C
    actual_r <- n_E/n_C
    tau <- -delta_A / qnorm(pr_neg)

    EP_it <- 1 - pnorm(sqrt((1+actual_r)^2/(actual_r*n_total) * sigma^2 /
          ((1+actual_r)^2/(actual_r*n_total) * sigma^2 + tau^2)) *
          (qnorm(1-alpha/2) - sqrt(actual_r*n_total / (1+actual_r)^2) *
          delta_A/sigma))
    TP <- pnorm(sqrt(actual_r/(1+actual_r)^2*n_total) * delta_A/sigma -
          qnorm(1-alpha/2))
  }

    return(data.frame(n_total = n_total, n_E = n_E, n_C = n_C,
          actual_r = actual_r, actual_EP = EP_it, actual_TP = TP))
}

#****************************************************************************#
# Exemplary call according to Example 3.4 (compare Table 3.3b)              #
#****************************************************************************#

samplesize_expected_power(0.9, 10, 20, 0.001, 0.05, 1)

##   n_total n_E n_C actual_r actual_EP actual_TP
## 1     268 134 134        1 0.9006278 0.9835258
```

A.4   R Code for Chapter 4

```
# _____ SECTION 4.2 _____

##############################################################################
# Function to calculate the sample size for a two group comparison for    #
# continuous data with a non-parametric approach according to formula (4.3) #
##############################################################################

##############################################################################
# Parameters to specify:
# pi_A  = relative effect
# alpha = significance level
# beta  = type II error rate
# r     = allocation ratio of sample sizes (n_E/n_C)
##############################################################################

samplesize_Mann_Whitney <- function(pi_A, alpha, beta, r){
  n <- (1+r)^2/r * (qnorm(1-alpha/2) + qnorm(1-beta))^2 / (12*(pi_A-0.5)^2)
  n_E <- ceiling(r/(1+r) * n)
  n_C <- ceiling(1/(1+r) * n)
  n_total <- n_E + n_C
  actual_r <- n_E/n_C
  actual_power <- pnorm((pi_A-0.5) / (1+actual_r) *
                        sqrt(n_total*actual_r*12) - qnorm(1-alpha/2))

  return(data.frame(n_total = n_total, n_E = n_E, n_C = n_C,
                    actual_r = actual_r, actual_power = actual_power))
}

#****************************************************************************#
# Exemplary call according to Example 4.1                                   #
#****************************************************************************#

samplesize_Mann_Whitney(0.638, 0.05, 0.1, 1)

##   n_total n_E n_C actual_r actual_power
## 1     184  92  92        1    0.9001316

#
#
# _____ SECTION 4.3.1 _____

##############################################################################
# Function to calculate the sample size for a two group comparison for    #
# ordered categorical data with an assumption-free approach               #
# according to formula (4.7)                                              #
##############################################################################

##############################################################################
# Parameters to specify:
# p_C  = vector of probabilities p_C_i, for i in M categories
# p_E  = vector of probabilities p_E_i, for i in M categories
# alpha = significance level
# beta  = type II error rate
# r     = allocation ratio of sample sizes (n_E/n_C)
##############################################################################

#-----------------------------------------------------#
# Calculation of the relative effect pi_A             #
```

```r
# according to formula (4.4) (Auxiliary function)  #
#---------------------------------------------------#
relative_effect <- function(p_C, p_E){
  x <- 0
  p_C_cum <- cumsum(p_C)
  for (i in 2:length(p_E)){
    x <- x + p_E[i] * p_C_cum[i-1]
  }
  y <- 0
  for (i in 1:length(p_E)){
    y <- y + 0.5 * p_E[i] * p_C[i]
  }
  return(x+y)
}

samplesize_Afs <- function(p_C, p_E, alpha, beta, r){
  pi_A <- relative_effect(p_C, p_E)
  x <- 0
  for (i in 1:length(p_C)){
    x <- x + ((r*p_C[i] + p_E[i]) / (1 + r))^3
  }
  n <- (1+r)^2/r * (qnorm(1-alpha/2) + qnorm(1-beta))^2 / 12 *
    (1-x)/(pi_A - 0.5)^2
  n_E <- ceiling(r/(1+r) * n)
  n_C <- ceiling(1/(1+r) * n)
  n_total <- n_E + n_C
  actual_r <- n_E/n_C
  actual_power <- pnorm(abs(pi_A-0.5) / (1 + actual_r)*
                    sqrt(n_total * actual_r * 12/(1-x)) - qnorm(1-alpha/2))
  return(data.frame(n_total = n_total, n_E = n_E, n_C = n_C,
                    actual_r = actual_r, actual_power = actual_power))
}

#************************************************************************#
# Exemplary call according to Example 4.2                              #
#************************************************************************#

samplesize_Afs(c(0.01, 0.03, 0.05, 0.12, 0.55, 0.22, 0.02),
               c(0.00, 0.01, 0.02, 0.03, 0.31, 0.43, 0.20),
               0.05, 0.1, 2)

##    n_total n_E n_C actual_r actual_power
## 1       66  44  22        2    0.9013279

#
#
#_____ SECTION 4.3.2 _____

########################################################################
# Function to calculate the sample size for a two group comparison for  #
# ordered categorical data assuming proportional odds according         #
# to formula (4.8)                                                       #
########################################################################

########################################################################
# Parameters to specify:
# p_C  = vector of probabilities p_C_i, for i in M categories
# theta_A = common log odds ratio
```

```
# alpha = significance level
# beta  = type II error rate
# r     = allocation ratio of sample sizes (n_E/n_C)
#######################################################################

#----------------------------------------------------#
# Calculation of the probabilities for the           #
# experimental group p_E (Auxiliary function)        #
#----------------------------------------------------#
prop_experimental <- function(p_C, theta_A){
  Q_C <- cumsum(p_C)
  Q_E <- numeric(length(p_C))
  for (i in 1:length(p_C)){
    Q_E[i] <- Q_C[i] / (Q_C[i] + (1-Q_C[i]) * exp(-theta_A))
  }
  p_E <- numeric(length(p_C))
  p_E[1] <- Q_E[1]
  for (i in 2:length(p_C)){
    p_E[i] <- Q_E[i] - Q_E[i-1]
  }
  return(round(p_E, 3))
}

samplesize_po <- function(p_C, theta_A, alpha, beta, r){
  p_E <- prop_experimental(p_C, theta_A)
  x=0
  for (i in 1:length(p_C)){
    x = x + ((p_C[i] + r*p_E[i]) / (1 + r))^3
  }
  n <- (1+r)^2/r * 3*(qnorm(1-alpha/2) + qnorm(1-beta))^2 /
    (theta_A^2 * (1-x))
  n_E <- ceiling(r/(1+r) * n)
  n_C <- ceiling(1/(1+r) * n)
  n_total <- n_E + n_C
  actual_r <- n_E/n_C
  actual_power <- pnorm(sqrt(n_total * actual_r * theta_A^2 * (1-x)/
                      (3 * (1 + actual_r)^2)) - qnorm(1-alpha/2))
  return(data.frame(n_total = n_total, n_E = n_E, n_C = n_C,
                 actual_r = actual_r, actual_power = actual_power))
}

#*********************************************************************#
# Exemplary call according to Example 4.3                            #
#*********************************************************************#

samplesize_po(c(0.10, 0.20, 0.10, 0.15, 0.20, 0.25), 0.5878, 0.05, 0.1, 2)

##   n_total n_E n_C actual_r actual_power
## 1     426 284 142        2    0.9007518
```

A.5   R Code for Chapter 5

```
# _____ SECTION 5.2.1 _____

###############################################################################
# Functions to calculate the approximate (formula (5.8)) and exact          #
# (formula (5.9)) sample size for a normal approximation test and           #
# respective exact power (formula (5.9))                                    #
###############################################################################

###############################################################################
# Parameters to specify:
# p_EA  = event probability in experimental group
# p_CA  = event probability in control group
# alpha = significance level
# beta  = type II error rate
# r     = allocation ratio of sample sizes (n_E/n_C)
###############################################################################

#--------------------------------------------------#
# Exact power according to formula (5.9)           #
#--------------------------------------------------#
power_exact <- function(n_E, n_C, p_EA, p_CA, alpha){
  delta_A <- p_EA - p_CA
  i <- 0:n_C
  j <- 0:n_E
  p_E_hat <- j/n_E
  p_C_hat <- i/n_C
  p_0_hat <- outer(j, i, "+")/(n_E + n_C)
  U <- sqrt(n_C*n_E / (n_C+n_E)) * (outer(p_E_hat, p_C_hat, "-")) /
    sqrt(p_0_hat*(1-p_0_hat))
  if(delta_A > 0){  # Large rates refer to a positive result
  exact_power <- sum(outer(dbinom(j, n_E, p_EA),dbinom(i, n_C, p_CA), "*") *
        (U >= qnorm(1-alpha/2)), na.rm = TRUE)
  }
  else   # small rates refer to a positive result
  exact_power <- sum(outer(dbinom(j, n_E, p_EA),dbinom(i, n_C, p_CA), "*") *
        (U <= -qnorm(1-alpha/2)), na.rm = TRUE)

  return(exact_power)
}

#--------------------------------------------------#
# Approximate sample size according to             #
# formula (5.8)                                    #
#--------------------------------------------------#
samplesize_chisquare <- function(p_EA, p_CA, alpha, beta, r){

p_0 <- (p_CA + r*p_EA)/(1+r)
delta_A <- p_EA - p_CA
n <- (1+r)/r * (qnorm(1-alpha/2) * sqrt((1+r)*(1-p_0)*p_0) +
    qnorm(1-beta) * sqrt(r*p_CA*(1-p_CA) + p_EA*(1-p_EA)))^2 / delta_A^2
n_E <- ceiling(r/(1+r) * n)
n_C <- ceiling(1/(1+r) * n)
n_total <- n_E + n_C
actual_r <- n_E/n_C
actual_power <- pnorm((abs(delta_A) - (1+actual_r) *
            sqrt(p_0 * (1-p_0) / (n_total * actual_r)) * qnorm(1-alpha/2))/
            sqrt((1+actual_r) / n_total*(p_CA*(1-p_CA)+(1-p_EA)*p_EA/r)))
exact_power <- power_exact(n_E, n_C, p_EA, p_CA, alpha)
```

```
   return(data.frame(n_total = n_total, n_E = n_E, n_C = n_C,
                     actual_r = actual_r, actual_power = actual_power,
                     exact_power = exact_power))
}

#----------------------------------------------------#
# Exact sample size according to formula (5.9)      #
#----------------------------------------------------#
samplesize_exact_chisquare <- function(p_EA, p_CA, alpha, beta, r){
  exact_power <- 0
  n <- 2
  while(exact_power < 1-beta){
    n <- n+1
    n_E <- ceiling(r/(1+r) * n)
    n_C <- ceiling(1/(1+r) * n)
    exact_power <- power_exact(n_E, n_C, p_EA, p_CA, alpha)
  }

  return(data.frame(n_total = n_E + n_C, n_E = n_E, n_C = n_C,
                    exact_power = exact_power))
}

#*****************************************************************************#
# Exemplary call according to Example 5.1                                    #
#*****************************************************************************#

samplesize_chisquare(0.5, 0.3, 0.05, 0.2, 2)

##   n_total n_E n_C actual_r actual_power exact_power
## 1     212 141  71 1.985915    0.8034595   0.8064196

samplesize_exact_chisquare(0.5, 0.3, 0.05, 0.2, 2)

##   n_total n_E n_C exact_power
## 1     207 138  69    0.800901

#
#
# _____ SECTION 5.2.2 _____

##############################################################################
# Function to calculate the sample size for a test based on risk             #
# ratio according to formula (5.13)                                          #
##############################################################################

##############################################################################
# Parameters to specify:
# p_EA  = event probability in experimental group
# p_CA  = event probability in control group
# alpha = significance level
# beta  = type II error rate
# r     = allocation ratio of sample sizes (n_E/n_C)
##############################################################################

samplesize_RR <- function(p_EA, p_CA, alpha, beta, r){
  p_0 <- (p_CA + r*p_EA)/(1+r)
  RR_A <- p_EA/p_CA
```

```
  n <- (1+r)/r * (qnorm(1-alpha/2) * sqrt((1+r)*(1-p_0)/p_0) +
      qnorm(1-beta) * sqrt(r*(1-p_CA)/p_CA + (1-p_EA)/p_EA))^2 / log(RR_A)^2
  n_E <- ceiling(r/(1+r) * n)
  n_C <- ceiling(1/(1+r) * n)
  n_total <- n_E + n_C
  actual_r <- n_E/n_C
  if(RR_A>1){ # large rates refer to a positive result
  actual_power <- pnorm((sqrt(n_total * actual_r/(1+actual_r)) *
      log(RR_A) - qnorm(1-alpha/2) * sqrt((1+actual_r)*(1-p_0) / p_0)) /
      sqrt(actual_r*(1-p_CA)/p_CA + (1-p_EA)/p_EA))
  }
  else  # small rates refer to a positive result
  actual_power <- pnorm((sqrt(n_total * actual_r/(1+actual_r)) *
      -log(RR_A) - qnorm(1-alpha/2) * sqrt((1+actual_r)*(1-p_0) / p_0)) /
      sqrt(actual_r*(1-p_CA)/p_CA + (1-p_EA)/p_EA))

  return(data.frame(n_total = n_total, n_E = n_E, n_C = n_C,
                    actual_power = actual_power))
}

#***************************************************************************#
# Exemplary call according to Example 5.4                                  #
#***************************************************************************#

samplesize_RR(0.5, 0.3, 0.05, 0.2, 1)

##   n_total n_E n_C actual_power
## 1     188  94  94    0.8032498

#
#
# _____ SECTION 5.2.3 _____

##############################################################################
# Function to calculate the sample size for a test based on odds             #
# ratio according to formula (5.17)                                          #
##############################################################################

##############################################################################
# Parameters to specify:
# p_EA  = event probability in experimental group
# p_CA  = event probability in control group
# alpha = significance level
# beta  = type II error rate
# r     = allocation ratio of sample sizes (n_E/n_C)
##############################################################################

samplesize_OR <- function(p_EA, p_CA, alpha, beta, r){
  OR_A <- p_EA * (1-p_CA) / (p_CA * (1-p_EA))
  n <- (1+r)/r * (qnorm(1-alpha/2) + qnorm(1-beta))^2 *
    (1/(p_EA*(1-p_EA)) + r/(p_CA*(1-p_CA))) / log(OR_A)^2
  n_E <- ceiling(r/(1+r) * n)
  n_C <- ceiling(1/(1+r) * n)
  n_total <- n_E + n_C
  actual_r <- n_E/n_C
  if(OR_A>1){ # large rates refer to a positive result
  actual_power <- pnorm(log(OR_A) / sqrt((1+actual_r) /
    (actual_r * n_total) * (1/(p_EA*(1-p_EA)) + r/(p_CA*(1-p_CA)))) -
```

```r
    qnorm(1-alpha/2))
  }
  else  # small rates refer to a positive result
  actual_power <- pnorm(-log(OR_A) / sqrt((1+actual_r) /
    (actual_r * n_total) * (1/(p_EA*(1-p_EA)) + r/(p_CA*(1-p_CA)))) -
    qnorm(1-alpha/2))

  return(data.frame(n_total = n_total, n_E = n_E, n_C = n_C,
                    actual_r = actual_r, actual_power = actual_power))
}

#*************************************************************************#
# Exemplary call according to Example 5.5                               #
#*************************************************************************#

samplesize_OR(0.5, 0.3, 0.05, 0.2, 1)

##    n_total n_E n_C actual_r actual_power
## 1      192  96  96        1    0.8008457

#
#
#  _____ SECTION 5.2.4 _____

################################################################################
# Function to calculate the sample size for a logistic regression model     #
# with a binary confounder according to formula (5.22)                      #
################################################################################

################################################################################
# Parameters to specify:
# p_y2  = probability that the binary confounder assumes a value of 1
# p0_EA = event probability in the experimental group for stratum Y2=0
# p0_CA = event probability in the control group for stratum Y2=0
# p1_EA = event probability in the experimental for stratum Y2=1
# p1_CA = event probability in the control group for stratum Y2=1
# Note: exactly three event probabilities have to be specified for a
#       well-defined model
# alpha = significance level
# beta  = type II error rate
# r     = allocation ratio of sample sizes (n_E/n_C)
################################################################################

samplesize_logistic <- function(p_y2, p0_CA = NA, p1_CA = NA, p0_EA = NA,
                                p1_EA = NA, alpha, beta, r){

  # Check if exactly 3 response parameters are specified
  if (sum(is.na(c(p0_CA, p0_EA, p1_CA, p1_EA))) != 1)
    return("Response parameters are over/-underidentified, please insert
           exactly 3 parameters")

  # Calculate the missing fourth parameter and the conditional odds ratio of
  # the treatment effect
  if (is.na(p0_CA)){
    OR_cond <- p1_EA / (1-p1_EA) / (p1_CA/(1-p1_CA))
    p0_CA <- p0_EA / (1-p0_EA) / OR_cond / (1 + p0_EA/(1-p0_EA)/OR_cond)
  }
  if (is.na(p0_EA)){
```

```r
    OR_cond <- p1_EA / (1-p1_EA) / (p1_CA/(1-p1_CA))
    p0_EA <- p0_CA / (1-p0_CA) * OR_cond / (1 + p0_CA/(1-p0_CA)*OR_cond)
  }
  if (is.na(p1_CA)){
    OR_cond <- p0_EA / (1-p0_EA) / (p0_CA/(1-p0_CA))
    p1_CA <- p1_EA / (1-p1_EA) / OR_cond / (1 + p1_EA/(1-p1_EA)/OR_cond)
  }
  if (is.na(p1_EA)){
    OR_cond <- p0_EA / (1-p0_EA) / (p0_CA/(1-p0_CA))
    p1_EA <- p1_CA / (1-p1_CA) * OR_cond / (1 + p1_CA/(1-p1_CA)*OR_cond)
  }

  # Calculate marginal response in the control group
  pmarg_CA <- p0_CA * (1-p_y2) + p1_CA * p_y2

  # Calculate marginal response in the experimentalgroup
  pmarg_EA <- p0_EA * (1-p_y2) + p1_EA * p_y2

  # Calculate marginal odds ratio
  OR_marg <- pmarg_EA/(1-pmarg_EA)/(pmarg_CA/(1-pmarg_CA))

  # Calculate odds ratio of the confounder
  OR_Y2 <- (p1_CA/(1-p1_CA)/(p0_CA/(1-p0_CA)))

  # Define the variance components of sigma^2
  e <- p0_CA / (1-p0_CA)
  f <- r
  g <- OR_cond
  h <- OR_Y2
  a <- (e * (1-p_y2)) / ((1+e)^2 * (1+f))
  b <- (e*f*g*(1-p_y2)) / ((1+e*g)^2 * (1+f))
  c <- (e*f*g*h*p_y2) / ((1+e*g*h)^2 * (1+f))
  d <- (e*h*p_y2) / ((1+e*h)^2 * (1+f))
  sigma2 <- ((1/a + 1/b)*(1/c + 1/d))/(1/a + 1/b + 1/c + 1/d)

  n <- (qnorm(1-alpha/2) + qnorm(1-beta))^2 * sigma2 / log(OR_cond)^2
  n_E <- ceiling(n * r/(1+r))
  n_C <- ceiling(n * 1/(1+r))
  n_total <- n_E + n_C
  actual_r <- n_E/n_C
  actual_power <- pnorm(log(OR_cond) * sqrt(n_total) /
                        sqrt(sigma2) - qnorm(1-alpha/2))

  return(data.frame(n_total = n_total, n_E = n_E, n_C = n_C,
     actual_r = actual_r, actual_power = actual_power, OR_cond = OR_cond,
     OR_marg = OR_marg, p0_CA = p0_CA, p1_CA = p1_CA, pmarg_CA = pmarg_CA,
     p0_EA = p0_EA, p1_EA = p1_EA, pmarg_EA = pmarg_EA, OR_Y2 = OR_Y2))
}

#****************************************************************************#
# Exemplary call according to Example 5.6                                   #
#****************************************************************************#

samplesize_logistic(p_y2 = 0.75, p0_CA = 0.2, p1_CA = 0.333, p0_EA = 0.368,
                    alpha = 0.05, beta = 0.2, r = 2)
```

```
##    n_total n_E n_C actual_r actual_power OR_cond  OR_marg p0_CA p1_CA
## 1      228 152  76        2    0.8011771 2.329114 2.291953   0.2 0.333

##    pmarg_CA p0_EA     p1_EA  pmarg_EA    OR_Y2
## 1   0.29975 0.368 0.5376387 0.4952291 1.997001
```

```
#
#
#_____ SECTION 5.3.2 _____

#######################################################################
# Function to calculate exact sample size for Fisher-Boschloo test    #
# iteratively                                                         #
#######################################################################

#######################################################################
# Parameters to specify:
# p_EA     = event probability in the experimental group
# p_CA     = event probability in the control group
# gamma    = odds ratio non-inferiority margin (= 1 in case of superiority)
# alpha    = significance level
# beta     = type II error rate
# r        = allocation ratio of sample sizes (n_E/n_C)
# size_acc = accuracy of grid for determination of size maximum
#######################################################################

# development version: https://github.com/s-kilian/binary
devtools::install_github("s-kilian/binary@v0.1.0", quiet = TRUE)
library(binary)

#*********************************************************************#
# Exemplary call according to Example 5.7                           #
#*********************************************************************#

# Use of function specified in section 8.4.2.4 for non-inferiority with
# use of gamma = 1 for superiority

samplesize_exact_boschloo_NI(0.5, 0.3, 1, 0.025, 0.2, 2)

## $n_C
## [1] 71
##
## $n_E
## [1] 142
##
## $nom_alpha
## nom_alpha_mid
##    0.03393585
##
## $exact_power
##    0.3, 0.5
## 0.8036507
```

A.6   R Code for Chapter 6

```
# _____ SECTION 6.2 _____

################################################################################
# Function to calculate the sample size for a log-rank test with the        #
# approach of exponentially distributed time-to-event data according to     #
# formulas (6.9) and (6.11)                                                  #
################################################################################

################################################################################
# Parameters to specify:
# theta_A  = hazard ratio
# lambda_C = hazard rate for the control group
# lambda_E = hazard rate for the experimental group
# a = acrual period
# f = minimal follow-up period
# alpha = significance level
# beta  = type II error rate
# r     = ratio of number of events (d_E/d_C)
################################################################################

samplesize_exponential <- function(theta_A, lambda_C, lambda_E, a, f, alpha,
                                   beta, r){
  # Calculation of the number of events according to formula (6.9)
  d <- (1+r)^2/r * (qnorm(1-alpha/2) + qnorm(1-beta))^2 / log(theta_A)^2
  d_E <- ceiling(r/(1+r) * d)
  d_C <- ceiling(1/(1+r) * d)
  d_total <- d_E + d_C
  actual_r <- d_E/d_C
  actual_power <-  pnorm(sqrt(d_total * log(theta_A)^2 * actual_r /
                         (1 + actual_r)^2) - qnorm(1-alpha/2))
  # Calculation of the probability of an event according to formula (6.11)
  p_D <- 1 - 1/(a*(1+r)) *
    ((exp(-lambda_C * f) - exp(-lambda_C * (a + f))) / lambda_C +
      r/lambda_E * (exp(-lambda_E * f) - exp(-lambda_E * (a + f))))
  n_total <- ceiling(d_total/p_D)
  n_E <- ceiling(r/(1+r) * n_total)
  n_C <- ceiling(1/(1+r) * n_total)
  return(data.frame(n_total = n_total, n_E = n_E, n_C = n_C, p_D = p_D,
               d_total = d_total, d_E = d_E, d_C = d_C,
               actual_r = actual_r, actual_power = actual_power))
}

#*****************************************************************************#
# Exemplary call according to Example 6.3                                   #
#*****************************************************************************#

samplesize_exponential(0.769, log(2)/20, log(2)/26, 24, 36, 0.05, 0.15, 1)

##   n_total n_E n_C        p_D d_total d_E d_C actual_r actual_power
## 1     686 343 343 0.7610624     522 261 261        1    0.8509749

#
#
# _____ SECTION 6.3.1 _____

################################################################################
# Function to calculate the sample size for a log-rank test with the        #
# approach of Schoenfeld according to formulas (6.15) and (6.17)            #
```

```
##########################################################################
##########################################################################
# Parameters to specify:
# theta_A  = hazard ratio
# lambda_C = hazard rate for the control group
# a = acrual period
# f = minimal follow-up period
# alpha = significance level
# beta  = type II error rate
# r     = ratio of number of events (d_E/d_C)
##########################################################################

samplesize_Schoenfeld <- function(theta_A, lambda_C, a, f, alpha, beta, r){
  # Calculation of the number of events according to formula (6.15)
  d <- (1+r)^2/r * (qnorm(1-alpha/2) + qnorm(1-beta))^2 / log(theta_A)^2
  d_E <- ceiling(r/(1+r) * d)
  d_C <- ceiling(1/(1+r) * d)
  d_total <- d_E + d_C
  actual_r <- d_E/d_C
  actual_power <-  pnorm(sqrt(d_total * log(theta_A)^2 * actual_r /
                       (1 + actual_r)^2) - qnorm(1-alpha/2))
  # Calculation of the probability of an event according to formula (6.17)
  p_D <- 1 - 1/(6*(1+r)) *
            (exp(-lambda_C * f) + r * exp(-lambda_C * f)^theta_A
          + 4 * (exp(-lambda_C * (a/2 + f))
               + r * exp(-lambda_C*(a/2 + f))^theta_A )
          + exp(-lambda_C * (a + f)) + r * exp(-lambda_C*(a + f))^theta_A)
  n_total <- ceiling(d_total/p_D)
  n_E <- ceiling(r/(1+r) * n_total)
  n_C <- ceiling(1/(1+r) * n_total)
  return(data.frame(n_total = n_total, n_E = n_E, n_C = n_C, p_D = p_D,
            d_total = d_total, d_E = d_E, d_C = d_C,
            actual_r = actual_r, actual_power = actual_power))
}

#************************************************************************#
# Exemplary call according to Example 6.3                              #
#************************************************************************#

samplesize_Schoenfeld(0.769, log(2)/20, 24, 36, 0.05, 0.15, 1)

##    n_total n_E n_C      p_D d_total d_E d_C actual_r actual_power
## 1      686 343 343 0.7609855     522 261 261        1    0.8509749

#
#
# _____ SECTION 6.3.2 _____

##########################################################################
# Function to calculate the sample size for a log-rank test with the    #
# approach of Freedman according to formulas (6.18) and (6.20)          #
##########################################################################

##########################################################################
# Parameters to specify:
# theta_A  = hazard ratio
# lambda_C = hazard rate for the control group
```

```r
# a = acrual period
# f = minimal follow-up period
# alpha = significance level
# beta  = type II error rate
# r     = ratio of number of events (d_E/d_C)
################################################################################

samplesize_Freedman <- function(theta_A, lambda_C, a, f, alpha, beta, r){
  # Calculation of the number of events according to formula (6.18)
  d <- 1/r * (qnorm(1-alpha/2) + qnorm(1-beta))^2 *
    (theta_A * r + 1)^2 / (theta_A - 1)^2
  d_E <- ceiling(r/(1+r) * d)
  d_C <- ceiling(1/(1+r) * d)
  d_total <- d_E + d_C
  actual_r <- d_E/d_C
  actual_power <-  pnorm(sqrt(d_total * (theta_A - 1)^2/(theta_A + 1)^2
                         *actual_r) - qnorm(1-alpha/2))
  # Calculation of the probability of an event according to formula (6.20)
  p_D <- 1 - 1/(1+r) * (exp(-lambda_C*(a/2 + f)) +
                       r * exp(-lambda_C * (a/2 + f))^theta_A)
  p_D_C <- 1 - exp(-lambda_C * (a + f)) * (exp(-lambda_C * a) - 1) /
    (lambda_C * a)
  lambda_E <- log(2)/20
  p_D_E <- 1 - exp(-lambda_E * (a + f)) * (exp(-lambda_E * a) - 1) /
    (lambda_E * a)
  n_total <- ceiling(d_total/p_D)
  n_E <- ceiling(r/(1+r) * n_total)
  n_C <- ceiling(1/(1+r) * n_total)
  return(data.frame(n_total = n_total, n_E = n_E, n_C = n_C, p_D = p_D,
              d_total = d_total, d_E = d_E, d_C = d_C,
              actual_r = actual_r, actual_power = actual_power))
}

#***************************************************************************#
# Exemplary call according to Example 6.3                                   #
#***************************************************************************#

samplesize_Freedman(0.769, log(2)/20, 24, 36, 0.05, 0.15, 1)

##   n_total n_E n_C        p_D d_total d_E d_C actual_r actual_power
## 1     690 345 345 0.7661479     528 264 264        1    0.8509665
```

A.7   R Code for Chapter 7

```
# Inclusion of packages for chapter 7
library(multiCA)

# _____ SECTION 7.2 _____

##############################################################################
# Function to calculate the sample size for the F-test according to      #
# formula (7.3)                                                          #
##############################################################################

##############################################################################
# Parameters to specify:
# delta_mu = vector of differences between group mu_i and study average mu
# sigma = standard deviation
# alpha = significance level
# beta  = type II error rate
# r     = vector of allocation ratios (n_i/n_1)
##############################################################################

samplesize_Ftest <- function(delta_mu, sigma, alpha, beta, r){
  k <- length(r)
  c <- sum(delta_mu^2*r) / (sum(r)*sigma^2)
  n <- k +1
  nc <- n * c
  while(qf(1-alpha, k-1, n-k, 0) > qf(beta, k-1, n-k, nc)){
    n <- n + 1
    nc <- n * c
  }
  n_groups <- ceiling(r/sum(r) * n) # sample size per group
  n_total <- sum(n_groups)
  actual_r <- n_groups / n_groups[1]
  # calculation of actual power
  nc <- n_total* sum(delta_mu^2 * actual_r) / (sum(actual_r) * sigma^2)
  actual_power <- 1 - pf(qf(1-alpha, k-1, n_total-k), k-1, n_total-k, nc)
  return(list(n_total = n_total, n_groups = n_groups,
                    actual_r = actual_r, actual_power = actual_power))
}

#****************************************************************************#
# Exemplary call according to Example 7.1                                 #
#****************************************************************************#

samplesize_Ftest(c(5, 5, 0), 20, 0.05, 0.1, c(1, 1, 1))

## $n_total
## [1] 309
##
## $n_groups
## [1] 103 103 103
##
## $actual_r
## [1] 1 1 1
##
## $actual_power
## [1] 0.902266

#
#
```

```
#        SECTION 7.3

##############################################################################
# Function to calculate the sample size for the Kruskal-Wallis test          #
# according to formula (7.8)                                                  #
##############################################################################

##############################################################################
# Parameters to specify:
# pi_A  = matrix (pi_A)_{i,j} = Prob(X_i > X_j) + 0.5 * Prob(X_i=X_j)
# alpha = significance level
# beta  = type II error rate
# r     = vector of allocation ratios (n_i/n_1)
##############################################################################

samplesize_kruskal_wallis <- function(pi_A, alpha, beta, r){
  stopifnot(dim(pi_A)[1] == dim(pi_A)[2])
  k <- dim(pi_A)[1]
  sr <- sum(r)
  tau <- cnonct(qchisq(1-alpha, df = k-1), beta, k-1)
  sr <- sum(r)
  denom <- 0
  for (l in 1:k){
    denom <- denom + (r[l] / sr) *
      sum( r[1:k != l] * (pi_A[l, 1:k != l] - 0.5) / sr )^2
  }
  n <- tau / (12 * denom)
  n_groups <- ceiling(r/sum(r) * n) # sample size per group
  n_total <- sum(n_groups)
  actual_r <- n_groups/n_groups[1]

  # calculation of actual power
  nc <- 12 * denom * n_total^2 / (n_total +1)
  actual_power <- 1 - pchisq(qchisq(1-alpha, df = k-1), df = k-1, ncp = nc)
  return(list(n_total = n_total, n_groups = n_groups,
              actual_r = actual_r, actual_power = actual_power))
}

#****************************************************************************#
# Exemplary call according to Example 7.2                                    #
#****************************************************************************#

samplesize_kruskal_wallis(matrix(ncol = 3, data =
  c( 0.5,     0.570,  0.638,
     1-0.570, 0.5,    0.570,
     1-.638,  1-.570, 0.5)),
  alpha = 0.05, beta = 0.1, r = c(1,1,1))

## $n_total
## [1] 330
##
## $n_groups
## [1] 110 110 110
##
## $actual_r
## [1] 1 1 1
##
```

```
## $actual_power
## [1] 0.8999639

#
#
#_____ SECTION 7.4.1 _____

###############################################################################
# Function to calculate the sample size for a chi-square test according to  #
# formula (7.13)                                                            #
###############################################################################


###############################################################################
# Parameters to specify:
# p_A   = vector of event probability in different groups
# alpha = significance level
# beta  = type II error rate
# r     = vector of allocation ratios (n_i/n_1)
###############################################################################

samplesize_chisquare2 <- function(p_A, alpha, beta, r){
  k <- length(p_A)
  tau <- cnonct(qchisq(1-alpha, df = k-1), beta, k-1)
  p_A_bar <- sum(r*p_A)/sum(r)
  n <- tau * sum(r) * p_A_bar *(1-p_A_bar) / sum(r * (p_A - p_A_bar)^2)
  n_groups <- ceiling(r/sum(r) * n) # sample size per group
  n_total <- sum(n_groups)
  actual_r <- n_groups/n_groups[1]

  # calculation of actual power
  nc <- n_total * sum(actual_r * (p_A - p_A_bar)^2) /
    (sum(actual_r) * p_A_bar *(1-p_A_bar))
  actual_power <- 1 - pchisq(qchisq(1-alpha, df = k-1), df =  k-1, ncp = nc)

  return(list(n_total = n_total, n_groups = n_groups,
                    actual_r = actual_r, actual_power = actual_power))
}

#*****************************************************************************#
# Exemplary call according to Example 7.3                                   #
#*****************************************************************************#

samplesize_chisquare2(c(0.3, 0.4, 0.5), 0.05, 0.2, c(1, 1, 1))

## $n_total
## [1] 348
##
## $n_groups
## [1] 116 116 116
##
## $actual_r
## [1] 1 1 1
##
## $actual_power
## [1] 0.8013913

#
#
```

```
# _____ SECTION 7.4.2 _____

###############################################################################
# Function to calculate the sample size for a Cochran-Armitage test         #
# according to formula (7.18)                                               #
###############################################################################

###############################################################################
# Parameters to specify:
# p_A   = vector of event probability in different groups
# alpha = significance level
# beta  = type II error rate
# r     = vector of allocation ratios (n_i/n_1)
###############################################################################
samplesize_cochran_armitage <- function(p_A, s, alpha, beta, r){
  tau <- cnonct(qchisq(1-alpha, df = 1), beta, 1)
  p_A_bar <- sum(r * p_A) / sum(r)
  s_bar <- sum(r * s) / sum(r)
  n <- tau * sum(r)* p_A_bar * (1-p_A_bar) *(sum(r * (s - s_bar)^2)) /
    sum(r * p_A * (s - s_bar))^2
  n_groups <- ceiling(r / sum(r) * n)  # sample size per group
  n_total <- sum(n_groups)
  actual_r <- n_groups/n_groups[1]

  # calculation of actual power
  nc <- n_total * sum(r * p_A * (s - s_bar))^2 /
    (sum(r) * p_A_bar * (1-p_A_bar) * (sum(r * (s - s_bar)^2)))
  actual_power <- 1 - pchisq(qchisq(1-alpha, df = 1), df =  1, ncp = nc)

  return(list(n_total = n_total, n_groups = n_groups,
              actual_r = actual_r, actual_power = actual_power))
}

#*****************************************************************************#
# Exemplary call according to Example 7.4                                   #
#*****************************************************************************#

samplesize_cochran_armitage(c(0.1,0.2,0.3), c(1,2,3), 0.1, 0.1, c(1,1,1))

## $n_total
## [1] 207
##
## $n_groups
## [1] 69 69 69
##
## $actual_r
## [1] 1 1 1
##
## $actual_power
## [1] 0.9018205

#
#
```

```
# _____ SECTION 7.5 _____

###############################################################################
# Function to calculate the sample size for a log-rank test                  #
# according to formula (7.24)                                                #
###############################################################################

###############################################################################
# Parameters to specify:
# lambda_A = vector of hazard rates under the alternative
# a = acrual period
# f = minimal follow-up period
# alpha = significance level
# beta  = type II error rate
# r     = vector of allocation ratios
###############################################################################
samplesize_generalized_Schoenfeld <- function(lambda_A, a, f, alpha, beta, r){
  theta_A <- lambda_A / lambda_A[1]
  k <- length(lambda_A)
  tau <- cnonct(qchisq(1-alpha, df = k-1), beta, k-1)
  sr <- sum(r)
  denom <- 0
  for (i in 2:(k-1)){
    denom <- denom +
      r[i] * log(theta_A[i]) * sum(r[(i+1):k] * log(theta_A[(i+1):k]))
  }
  denom <- sum( (sr - r[-1]^2) * (log(theta_A[-1])^2) ) - 2 * denom
  d <- sum(r)^2 * tau / denom
  Pr_Di <- 1 - (exp(-lambda_A*f) - exp(-lambda_A * (a + f))) / (a*lambda_A)
  Pr_D <- sum(r*Pr_Di) / sum(r)
  d_groups <- ceiling(r/sum(r) * d)  # number of events per group
  d_total <- sum(d_groups)
  n <- d_total/Pr_D
  n_groups <- ceiling(r/sum(r) * n) # sample size per group
  n_total <- sum(n_groups)
  actual_r <- n_groups/n_groups[1]

  # calculation of actual power
  nc <- d_total*denom/sr^2
  actual_power <- 1 - pchisq(qchisq(1-alpha, df = k-1), df = k-1, ncp = nc)

  return(list(n_total = n_total, n_groups = n_groups, probevent = Pr_D,
              nevent_total = d_total, nevent_groups = d_groups,
              actual_r = actual_r, actual_power = actual_power))
}
```

```
#***********************************************************************#
# Exemplary call according to Example 7.5                              #
#***********************************************************************#

samplesize_generalized_Schoenfeld(log(2)/c(20, 26, 23),
                                  24, 36, 0.05, 0.15, c(1, 1, 1))
## $n_total
## [1] 1251
##
## $n_groups
## [1] 417 417 417
##
## $probevent
## [1] 0.7605279
##
## $nevent_total
## [1] 951
##
## $nevent_groups
## [1] 317 317 317
##
## $actual_r
## [1] 1 1 1
##
## $actual_power
## [1] 0.8500926
```

A.8   R Code for Chapter 8

```
# _____ SECTION 8.2.1 _____
##############################################################################
# Functions to calculate the sample size for normally distributed outcomes  #
# to test the difference of means for non-inferiority                       #
##############################################################################

##############################################################################
# Parameters to specify:
# delta_A = difference between groups
# sigma   = common standard deviation
# delta   = non-inferiority margin
# alpha   = significance level
# beta    = type II error rate
# r       = allocation ratio of sample sizes (n_E/n_C)
##############################################################################

#----------------------------------------------------#
# Exact sample size according to iteratively          #
# resolving formula (3.10) using nc given in (8.3)    #
#----------------------------------------------------#
samplesize_diff_NI_exact <- function(delta_A, sigma, delta, alpha, beta, r){
  power <- 0
  n <- 2
  while(power < 1 - beta){
    n <- n + 1
    n_E <- ceiling(r/(1+r) * n)
    n_C <- ceiling(1/(1+r) * n)
    actual_r <- n_E/n_C
    n_total <- n_E + n_C
    nc <- sqrt(actual_r/(1+actual_r)^2 * n_total) * (delta_A+delta)/sigma
    power <- 1 - pt(qt(1-alpha/2, n_total-2), n_total-2, nc)
  }

  return(data.frame(n_total = n_total, n_E = n_E, n_C = n_C,
            actual_r = actual_r, actual_power = power))
}

#----------------------------------------------------#
# Approximate sample size formula according           #
# to Guenther/Schouten (8.4)                          #
#----------------------------------------------------#
samplesize_diff_GS_NI <- function(delta_A, sigma, delta, alpha, beta, r){
  n <- (1+r)^2/r * (qnorm(1-alpha/2) + qnorm(1-beta))^2 *
    (sigma/(delta_A+delta))^2 + (qnorm(1-alpha/2))^2/2
  n_E <- ceiling(r/(1+r) * n)
  n_C <- ceiling(1/(1+r) * n)
  n_total <- n_E + n_C
  actual_r <- n_E/n_C
  actual_power <- pnorm(abs((delta_A+delta)/sigma)
                    * sqrt(actual_r)/(1+actual_r)
                    * sqrt(n_total - (qnorm(1-alpha/2))^2/2)
                    - qnorm(1-alpha/2))

  return(data.frame(n_total = n_total, n_E = n_E, n_C = n_C,
            actual_r = actual_r, actual_power = actual_power))
}
```

```
#***********************************************************************#
# Exemplary call according to Example 8.1                              #
#***********************************************************************#
samplesize_diff_NI_exact(0, 275, 100, 0.05, 0.1, 1)

##   n_total n_E n_C actual_r actual_power
## 1     320 160 160        1    0.9001945

samplesize_diff_GS_NI(0, 275, 100, 0.05, 0.1, 1)

##   n_total n_E n_C actual_r actual_power
## 1     320 160 160        1    0.9002054


########################################################################
# Functions to calculate the sample size to test for non-inferiority   #
# using an ANCOVA                                                      #
########################################################################


########################################################################
# Parameters to specify:
# delta_A  = difference between groups
# sigma = common standard deviation
# rho = correlation between outcome and covariate
# delta = non-inferiority margin
# alpha = significance level
# beta  = type II error rate
# r     = allocation ratio of sample sizes (n_E/n_C)
########################################################################


#--------------------------------------------------#
# Exact sample size according to iteratively        #
# resolving formula (3.15) using nc given in (8.6) #
#--------------------------------------------------#
samplesize_ANCOVA_NI_exact <- function(delta_A, sigma, rho, delta,
                                       alpha, beta, r){

  power <- 0
  n <- 2
  while(power < 1 - beta){
    n <- n + 1
    n_E <- ceiling(r/(1+r) * n)
    n_C <- ceiling(1/(1+r) * n)
    actual_r <- n_E/n_C
    n_total <- n_E + n_C
    nc <- sqrt(actual_r/(1+actual_r)^2 * n_total) * (delta_A+delta) /
      sqrt(1-rho^2) / sigma
    power <- 1 - pt(qt(1-alpha/2, n_total-3), n_total-3, nc)
  }

  return(data.frame(n_total = n_total, n_E = n_E, n_C = n_C,
              actual_r = actual_r, actual_power = power))
}


#--------------------------------------------------#
# Approximate sample size according                 #
# to formula (8.7)                                  #
#--------------------------------------------------#
samplesize_ANCOVA_NI_approx <- function(delta_A, sigma, rho, delta,
```

```
                                         alpha, beta, r){
  n <- (1+r)^2/r * (qnorm(1-alpha/2) + qnorm(1-beta))^2 *
    (sigma/(delta_A+delta))^2 * (1-rho^2) + (qnorm(1-alpha/2))^2/2
  n_E <- ceiling(r/(1+r) * n)
  n_C <- ceiling(1/(1+r) * n)
  n_total <- n_E + n_C
  actual_r <- n_E/n_C
  actual_power <- pnorm(abs((delta_A+delta)/sigma) / sqrt(1-rho^2) *
                        sqrt(actual_r)/(1+actual_r) *
                        sqrt(n_total - (qnorm(1-alpha/2))^2/2) -
                        qnorm(1-alpha/2))

  return(data.frame(n_total = n_total, n_E = n_E, n_C = n_C,
                 actual_r = actual_r, actual_power = actual_power))
}


#
#
#  _____ SECTION 8.2.2 _____

###############################################################################
# Function to calculate the sample size for normally distributed outcomes   #
# to test the ratio of means for non-inferiority                            #
###############################################################################

###############################################################################
# Parameters to specify:
# theta_A  = ratio of means (mu_e/mu_C)
# CV = coefficient of variation
# delta = non-inferiority margin
# alpha = significance level
# beta  = type II error rate
# r      = allocation ratio of sample sizes (n_E/n_C)
###############################################################################

#-----------------------------------------------------#
# Exact sample size according to iteratively          #
# resolving formula (3.10) using                      #
# nc given in (8.10b)                                 #
#-----------------------------------------------------#
samplesize_ratio_NI_exact <- function(delta_A, CV, delta, alpha, beta, r){
  power <- 0
  n <- 2
  while(power < 1 - beta){
    n <- n + 1
    n_E <- ceiling(r/(1+r) * n)
    n_C <- ceiling(1/(1+r) * n)
    actual_r <- n_E/n_C
    n_total <- n_E + n_C
    nc <- sqrt(actual_r/(1+actual_r)/(1+delta^2*actual_r) * n_total) *
      (delta_A-delta)/CV
    power <- 1 - pt(qt(1-alpha/2, n_total-2), n_total-2, nc)
  }

  return(data.frame(n_total = n_total, n_E = n_E, n_C = n_C,
              actual_r = actual_r, actual_power = power))
}
```

```
#-----------------------------------------------------#
# Approximate sample size formula according           #
# to Guenther/Schouten (8.11b)                        #
#-----------------------------------------------------#
samplesize_ratio_GS_NI <- function(theta_A, CV, delta, alpha, beta, r){
  n <- (1+r) * (1+delta^2*r)/r * (qnorm(1-alpha/2) + qnorm(1-beta))^2 *
       (CV/(theta_A-delta))^2 + qnorm(1-alpha/2)^2/2
  n_E <- ceiling(r/(1+r) * n)
  n_C <- ceiling(1/(1+r) * n)
  n_total <- n_E + n_C
  actual_r <- n_E/n_C
  actual_power <- pnorm(sqrt((n_total - qnorm(1-alpha/2)^2 / 2) /
                             ((1+actual_r)*(1+delta^2*actual_r)/actual_r))*
                        abs((theta_A-delta)/CV) - qnorm(1-alpha/2))
  return(data.frame(n_total = n_total, n_E = n_E, n_C = n_C,
                    actual_r = actual_r, actual_power = actual_power))
}

#***************************************************************************#
# Exemplary call according to Example 8.2                                   #
#***************************************************************************#
samplesize_ratio_GS_NI(1, 0.135, 0.9, 0.05, 0.10, 1)

##   n_total n_E n_C actual_r actual_power
## 1      72  36  36        1     0.903063

samplesize_ratio_NI_exact(1, 0.135, 0.9, 0.05, 0.10, 1)

##   n_total n_E n_C actual_r actual_power
## 1      72  36  36        1    0.9028414

#
#
#        SECTION 8.3

##############################################################################
# Function to calculate the sample size for continuous and ordered           #
# categorical outcomes to test the ratio of means for non-inferiority        #
# according to formula (8.13)                                                #
##############################################################################

##############################################################################
# Parameters to specify:
# pi_A = relative effect
# sigma_E = standard deviation for experimental group
# sigma_C = standard deviation for control group
# delta = non-inferiority margin
# alpha = significance level
# beta  = type II error rate
# r     = allocation ratio of sample sizes (n_E/n_C)
##############################################################################

samplesize_categorical_NI <- function(pi_A, sigma_E, sigma_C, delta,
                                       alpha, beta, r){
  n <- (1+r)/r * (qnorm(1-alpha/2) + qnorm(1-beta))^2 *
       (sigma_E^2 + r*sigma_C^2) / (pi_A - (0.5 - delta))^2
  n_E <- ceiling(r/(1+r) * n)
```

```
  n_C <- ceiling(1/(1+r) * n)
  n_total <- n_E + n_C
  actual_r <- n_E/n_C
  actual_power <- pnorm(sqrt(n_total * actual_r /(1+actual_r) *
                        (pi_A - (0.5 - delta))^2 / (sigma_E^2 + r*sigma_C^2)) -
                        qnorm(1-alpha/2))
  return(data.frame(n_total = n_total, n_E = n_E, n_C = n_C,
                    actual_r = actual_r, actual_power = actual_power))
}



#
#
# _____ SECTION 8.4.1.1 _____

################################################################################
# Function to calculate the sample size to test the difference of rates       #
# for non-inferiority according to formula (8.18)                             #
################################################################################

################################################################################
# Parameters to specify:
# p_EA    = event probability in the experimental group
# p_CA    = event probability in the control group
# delta   = non-inferiority margin (>0 if higher rates are better,
#           <0 if lower rates are better)
# alpha   = significance level (one-sided)
# beta    = type II error rate
# r       = allocation ratio of sample sizes (n_E/n_C)
################################################################################

samplesize_diff_rates_NI <- function(p_EA, p_CA, delta, alpha, beta, r){
  theta <- 1/r

  a <- 1 + theta
  b <- -(1 + theta + p_EA + theta * p_CA - delta * (theta + 2))
  c <- delta^2 - delta * (2*p_EA + theta + 1) + p_EA + theta * p_CA
  d <- p_EA * delta * (1 - delta)

  v <- b^3/(3*a)^3 - b*c/(6*a^2) + d/(2*a)
  u <- sign(v) * sqrt(b^2/(3*a)^2 - c/(3*a))
  w <- 1/3 * (pi + acos(v/u^3))

  # probabilities under the alternative
  p_E0 <- 2*u*cos(w) - b/(3*a)
  p_C0 <- p_E0 + delta

  n <- (1+r)/r *
    (qnorm(1-alpha) * sqrt(r * p_C0 * (1-p_C0) + p_E0 * (1-p_E0)) +
        qnorm(1-beta) * sqrt(r * p_CA * (1-p_CA) + p_EA * (1-p_EA)))^2 /
    (p_EA - p_CA + delta)^2

  n_C <- ceiling(1/(1+r) * n)
  n_E <- ceiling(r/(1+r) * n)
  n_total <- n_E + n_C
  actual_r <- n_E/n_C
  actual_power <- pnorm((sqrt(n_total * actual_r/(1+actual_r) *
```

```
                              (p_EA - p_CA + delta)^2) - qnorm(1-alpha) *
                     sqrt(r * p_C0 * (1-p_C0) + p_E0 * (1-p_E0))) /
                     sqrt(r * p_CA * (1-p_CA) + p_EA * (1-p_EA)))
  return(data.frame(n_total = n_total, n_E = n_E, n_C = n_C,
                    actual_r = actual_r, actual_power = actual_power))
}

#*****************************************************************************#
# Exemplary call according to Example 8.3                                    #
#*****************************************************************************#
samplesize_diff_rates_NI(0.4, 0.4, -0.105, 0.025, 0.15, 2)

##    n_total n_E n_C actual_r actual_power
## 1      857 571 286 1.996503    0.8504379

#
#
#_____ SECTION 8.4.1.2 _____


##############################################################################
# Function to calculate the sample size to test the ratio of rates for       #
# non-inferiority according to formula (8.24)                                #
##############################################################################


##############################################################################
# Parameters to specify:
# p_EA   = event probability in the experimental group
# p_CA   = event probability in the control group
# delta  = non-inferiority margin (>1 if higher rates are better,
#          <1 if lower rates are better)
# alpha  = significance level (one-sided)
# beta   = type II error rate
# r      = allocation ratio of sample sizes (n_E/n_C)
##############################################################################

samplesize_ratio_rates_NI <- function(p_EA, p_CA, delta, alpha, beta, r){

  theta <- 1/r

  a <- 1 + theta
  b <- -(delta*(1 + theta*p_CA) + theta + p_EA)
  c <- delta * (p_EA + theta * p_CA)

  # probabilities under the alternative
  p_E0 <- (-b - sqrt(round(b^2 - 4*a*c,10)))/(2*a)
  p_C0 <- p_E0 / delta

  n <- (1+r)/r *
    (qnorm(1-alpha) * sqrt(r * delta^2 * p_C0 * (1-p_C0) + p_E0 * (1-p_E0)) +
       qnorm(1-beta) * sqrt(r * delta^2 * p_CA * (1-p_CA) + p_EA *
                              (1-p_EA)))^2 / (p_EA - delta * p_CA)^2

  n_C <- ceiling(1/(1+r) * n)
  n_E <- ceiling(r/(1+r) * n)
  n_total<- n_E + n_C
  actual_r <- n_E/n_C
  actual_power <- pnorm((sqrt(n_total * actual_r / (1 + actual_r) *
                          (p_EA - delta * p_CA)^2) -
```

```
                                qnorm(1-alpha)*sqrt(actual_r * delta^2 * p_C0 *
                                                    (1-p_C0) + p_E0*(1-p_E0))) /
                        sqrt(actual_r * delta^2 * p_CA * (1-p_CA) +
                                p_EA * (1-p_EA)))
  return(data.frame(n_total = n_total, n_E = n_E, n_C = n_C,
                    actual_r = actual_r, actual_power = actual_power))
}

#****************************************************************************#
# Exemplary call according to Example 8.4                                   #
#****************************************************************************#
samplesize_ratio_rates_NI(0.0385, 0.0550, 1.25, 0.025, 0.08, 1)

##    n_total  n_E  n_C actual_r actual_power
## 1     2826 1413 1413        1    0.9200239

#
#
#        SECTION 8.4.1.3 _____

################################################################################
# Function to calculate the sample size to test the odds ratio for            #
# non-inferiority according to formula (8.27)                                 #
################################################################################

################################################################################
# Parameters to specify:
# p_EA    = event probability in the experimental group
# p_CA    = event probability in the control group
# gamma   = non-inferiority margin for the odds ratio (<1 if higher rates are
#           better, >1 if lower rates are better)
# alpha   = significance level (one-sided)
# beta    = type II error rate
# r       = allocation ratio of sample sizes (n_E/n_C)
################################################################################

samplesize_OR_NI <- function(p_EA, p_CA, gamma, alpha, beta, r){
  delta <- log(gamma)
  OR_A <- p_EA * (1-p_CA) / (p_CA * (1-p_EA))
  n <- (1+r)/r * (qnorm(1-alpha) + qnorm(1-beta))^2 *
    (1/(p_EA * (1-p_EA)) + r/(p_CA * (1-p_CA))) / (log(OR_A) - delta)^2
  n_C <- ceiling(1/(1+r) * n)
  n_E <- ceiling(r/(1+r) * n)
  n_total <- n_E + n_C
  actual_r <- n_E/n_C
  actual_power <- pnorm(sqrt(n_total * actual_r / (1+actual_r) *
                        (log(OR_A) - delta)^2 / (1/(p_EA * (1-p_EA)) +
                        actual_r/(p_CA * (1-p_CA)))) - qnorm(1-alpha))
  return(data.frame(n_total = n_total, n_E = n_E, n_C = n_C,
                    actual_r = actual_r, actual_power = actual_power))
}

#****************************************************************************#
# Exemplary call according to Example 8.5                                   #
#****************************************************************************#
samplesize_OR_NI(0.8, 0.8, 0.7, 0.025, 0.2, 1)
```

```
##   n_total n_E n_C actual_r actual_power
## 1    1544 772 772        1    0.8004015

#
#
#  _____ SECTION 8.4.2.2 _____

################################################################################
# Function to calculate exact sample size to test the difference of rates    #
# for non-inferiority                                                         #
################################################################################


################################################################################
# Parameters to specify:
# p_EA      = event probability in the experimental group
# p_CA      = event probability in the control group
# delta     = risk difference non-inferiority margin (>0 if higher rates are
#               better, <0 if lower rates are better)
# alpha     = significance level (one-sided)
# beta      = type II error rate
# r         = allocation ratio of sample sizes (n_E/n_C)
# size_acc  = accuracy of grid for determination of size maximum
# method    = "RD" for test of risk difference
# better    = "low" or "high" depending on whether lower or higher rates
#               are advantageous
################################################################################


# development version: https://github.com/s-kilian/binary
devtools::install_github("s-kilian/binary@v0.1.0", quiet = TRUE)
library(binary)


#*****************************************************************************#
# Exemplary call according to Example 8.6                                    #
#*****************************************************************************#


samplesize_exact(p_EA = 0.4, p_CA = 0.4, delta = -0.105, alpha = 0.025,
                 beta = 0.15, r = 2, size_acc = 2, method = "RD",
                 better = "low")

## $n_C
## [1] 290
##
## $n_E
## [1] 580
##
## $crit.val
## $crit.val$crit.val.mid
## [1] -1.981404
##
##
## $exact_power
## 0.4, 0.4
## 0.8514205

#
#
```

```
# _____ SECTION 8.4.2.3 _____

#############################################################################
# Function to calculate exact sample size to test the ratio of rates       #
# for non-inferiority                                                        #
#############################################################################

#############################################################################
# Parameters to specify:
# p_EA      = event probability in the experimental group
# p_CA      = event probability in the control group
# delta     = risk ratio non-inferiority margin (<1 if higher rates are
#             better, >1 if lower rates are better)
# alpha     = significance level (one-sided)
# beta      = type II error rate
# r         = allocation ratio of sample sizes (n_E/n_C)
# size_acc  = accuracy of grid for determination of size maximum
# method    = "RR" for test of risk ratio
# better    = "low" or "high" depending on whether lower or higher rates
#                 are advantageous
#############################################################################

# development version: https://github.com/s-kilian/binary
devtools::install_github("s-kilian/binary@v0.1.0", quiet = TRUE)
library(binary)

#***************************************************************************#
# Exemplary call according to Example 8.7                                   #
#***************************************************************************#

samplesize_exact(p_EA = 0.0385, p_CA = 0.055, delta = 1.25, alpha = 0.025,
                 beta = 0.08, r = 1, size_acc = 3, method = "RR",
                 better = "low")

## $n_C
## [1] 1431
##
## $n_E
## [1] 1431
##
## $crit.val
## $crit.val$crit.val.mid
## [1] -2.015836
##
##
## $exact_power
## 0.055, 0.0385
##    0.9200947

#
#
# _____ SECTION 8.4.2.4 _____

#############################################################################
# Function to calculate exact sample size to test the odds ratio           #
# for non-inferiority using the Fisher-Boschloo test                        #
#############################################################################

#############################################################################
# Parameters to specify:
```

```
# p_EA      = event probability in the experimental group
# p_CA      = event probability in the control group
# gamma     = odds ratio non-inferiority margin
# alpha     = significance level (one-sided)
# beta      = type II error rate
# r         = allocation ratio of sample sizes (n_E/n_C)
# size_acc = accuracy of grid for determination of size maximum
###########################################################################

# development version: https://github.com/s-kilian/binary
devtools::install_github("s-kilian/binary@v0.1.0", quiet = TRUE)
library(binary)

#***********************************************************************#
# Exemplary call according to Example 8.8                              #
#***********************************************************************#

samplesize_exact_boschloo_NI(p_EA = 0.8, p_CA = 0.8, gamma = 0.7,
                    alpha = 0.025, beta = 0.2, r = 1, size_acc = 2)

## $n_C
## [1] 785
##
## $n_E
## [1] 785
##
## $nom_alpha
## nom_alpha_mid
##      0.02771634
##
## $exact_power
##  0.8, 0.8
## 0.8002347

#
#
# _____ SECTION 8.5 _____

###########################################################################
# Function to calculate the sample size for time-to-event outcomes to test  #
# for non-inferiority according to formula (8.33)                           #
###########################################################################
```

```
###############################################################################
# Parameters to specify:
# delta = non-inferiority margin
# a     = acrual period
# f     = minimal follow-up period
# s     = 1-year survival rate
# alpha = significance level
# beta  = type II error rate
# r     = ratio of number of events (d_E/d_C)
###############################################################################

samplesize_surv_ni <- function(delta, a, f, s, alpha, beta, r){

  # Calculation of the number of events according to formula (8.33)
  d <- (1+r)^2 / r * ((qnorm(1-alpha/2) * sqrt(delta) + qnorm(1-beta) *

                          (1+delta*r) / (1+r))^2) / (delta-1)^2
  d_E <- ceiling(r / (1+r) * d)
  d_C <- ceiling(1 / (1+r) * d)
  d_total <- d_E + d_C
  actual_r <- d_E/d_C
  actual_power <- pnorm((1+actual_r) / (1+delta*actual_r) *
                          (sqrt(d_total * actual_r) *
                              abs((delta-1) / (1+actual_r))-
                              (qnorm(1-alpha/2) * sqrt(delta))))

  # Calculation of the probability of an event according to formula (6.10)
  lambda <- -log(s)/12
  p_D <- 1 - 1/(a * lambda) * (exp(-lambda * f) - exp(-lambda * (a+f)))
  n_total <- ceiling(d_total / p_D)
  n_E <- ceiling(r / (1+r) * n_total)
  n_C <- ceiling(1 / (1+r) * n_total)

  return(data.frame(n_total = n_total, n_E = n_E, n_C = n_C, p_D = p_D,
                    d_total = d_total, d_E = d_E, d_C = d_C,
                    actual_r = actual_r,
                    actual_power = actual_power))
}

#*****************************************************************************#
# Exemplary call according to Example 8.9                                     #
#*****************************************************************************#

samplesize_surv_ni(1.43, 18, 12, 0.7, 0.05, 0.2, 1)

##   n_total n_E n_C       p_D d_total d_E d_C actual_r actual_power
## 1     538 269 269 0.4578881     246 123 123        1    0.8013332
```

A.9   R Code for Chapter 9

```r
# Inclusion of packages for chapter 9
library(mvtnorm)

# _____ SECTION 9.2 _____

################################################################################
# Functions to calculate the sample size for the net effect approach for a     #
# given allocation ratio according to formulas (9.6) and (9.7)                 #
################################################################################

################################################################################
# Parameters to specify:
# Delta_ER, Delta_EP = differences under the alternative
# sigma = standard deviation
# delta = non-inferiority margin > 0
# alpha = significance level
# beta  = type II error rate
# r_R   = allocation ratio n_R/n_P
# r_E   = allocation ratio n_E/n_P
################################################################################

# For hypothesis given in (9.3)
samplesize_net_effect_2 <- function(Delta_ER, Delta_EP, sigma, delta, alpha,
                                    beta, r_R, r_E){
  n_P <- 1
  power <- 0

  while(power <= 1 - beta){
    n_P <- n_P + 1
    n_R <- ceiling(r_R * n_P)
    n_E <- ceiling(r_E * n_P)
    actual_r_R <- n_R/n_P
    actual_r_E <- n_E/n_P
    rho <- sqrt(actual_r_R / ((1 + actual_r_E) * (actual_r_E + actual_r_R)))
    cor <- matrix(c(1, rho, rho, 1), 2, 2)

    c_EP <- qnorm(1 - alpha/2) - sqrt(actual_r_E / (1 + actual_r_E) * n_P) *
      Delta_EP/sigma
    c_ER <- qnorm(1 - alpha/2) - sqrt(actual_r_E * actual_r_R /
            (actual_r_E + actual_r_R) * n_P) * (Delta_ER + delta)/sigma

    power <- pmvnorm(lower = c(c_EP, c_ER), upper = c(Inf, Inf), corr = cor)
  }

  n_total <- n_P + n_E + n_R

  return(data.frame(n_total = n_total, n_P = n_P, n_E = n_E, n_R = n_R,
    actual_r_R = actual_r_R, actual_r_E = actual_r_E, actual_power = power))
}

# For hypothesis given in (9.4)
################################################################################
# additional parameter to specify:
# Delta_RP = difference under the alternative
################################################################################

samplesize_net_effect_3 <- function(Delta_ER, Delta_EP, Delta_RP, sigma,
```

```
                                          delta, alpha, beta, r_R, r_E){
  n_P <- 1
  power <- 0

  while(power <= 1 - beta){
    n_P <- n_P + 1
    n_R <- ceiling(r_R * n_P)
    n_E <- ceiling(r_E * n_P)
    actual_r_R <- n_R/n_P
    actual_r_E <- n_E/n_P
    rho_12 <- sqrt(actual_r_R / ((1 + actual_r_E) * (actual_r_E +
                                                     actual_r_R)))
    rho_13 <- sqrt(actual_r_E * actual_r_R / ((1 + actual_r_E) *
                                              (1 + actual_r_R)))
    rho_23 <- -sqrt(actual_r_E / ((actual_r_R + actual_r_E) *
                    (1 + actual_r_R)))
    cor <- matrix(c(1, rho_12, rho_13, rho_12, 1, rho_23, rho_13,
                    rho_23, 1), 3, 3)

    c_EP <- qnorm(1 - alpha/2) - sqrt(actual_r_E / (1 + actual_r_E) * n_P) *
      Delta_EP / sigma
    c_ER <- qnorm(1 - alpha/2) - sqrt(actual_r_E * actual_r_R /
              (actual_r_E + actual_r_R) * n_P) * (Delta_ER + delta)/sigma
    c_RP <- qnorm(1 - alpha/2) - sqrt(actual_r_R / (1+actual_r_R) * n_P) *
      (Delta_RP)/sigma

    power <- pmvnorm(lower = c(c_EP, c_ER, c_RP), upper = c(Inf, Inf, Inf),
                     corr = cor)
  }

  n_total <- n_P + n_E + n_R

return(data.frame(n_total = n_total, n_P = n_P, n_E = n_E, n_R = n_R,
    actual_r_R = actual_r_R, actual_r_E = actual_r_E, actual_power = power))
}

#*****************************************************************************#
# Exemplary call according to Example 9.1                                    #
#*****************************************************************************#

samplesize_net_effect_2(0, 4, 6, 2, 0.05, 0.2, 1, 1)

##   n_total n_P n_E n_R actual_r_R actual_r_E actual_power
## 1     426 142 142 142          1          1    0.8019773

samplesize_net_effect_3(0, 4, 4, 6, 2, 0.05, 0.2, 1, 1)

##   n_total n_P n_E n_R actual_r_R actual_r_E actual_power
## 1     426 142 142 142          1          1    0.8018604

##############################################################################
# Functions to calculate the minimal sample size for the net effect         #
# approach by determining the optimal allocation ratio by a grid seach       #
# according to Stucke and Kieser (2012) for hypothesis given in (9.3)        #
##############################################################################

##############################################################################
# Parameters to specify:
```

```
# Delta_ER, Delta_EP = differences under the alternative
# sigma = standard deviation
# delta = non-inferiority margin > 0
# alpha = significance level
# beta  = type II error rate
####################################################################

#------------------------------------------------------#
# Calculation of power for net effect grid search  #
# according to formula (9.6) (Auxiliary function)  #
#------------------------------------------------------#
power_net_effect_2 <- function(Delta_ER, Delta_EP, sigma, delta, alpha, n_E,
                                n_R, n_P){

  gamma12 <- sqrt(n_R * n_P) / sqrt((n_R + n_E) * (n_P + n_E))
  r12 <- matrix(c(1, gamma12, gamma12, 1), ncol = 2)

  t1 <- Delta_EP / (sigma * sqrt(1/(n_E) + 1/(n_P))) - qnorm(1-alpha/2)
  t2 <- (Delta_ER+delta) / (sigma * sqrt(1/n_E + 1/(n_R))) - qnorm(1-alpha/2)

  p12 <- pmvnorm(lower = c(-Inf, -Inf), upper = c(t1, t2), mean = c(0, 0),
                 sigma = r12)
  return(p12)
}


# Grid search to find the minimum sample size under restriction r_E >= 1,
# r_R >= 1, and r_E >= r_R
samplesize_grid_2 <- function(Delta_ER, Delta_EP, sigma, delta, alpha, beta){

  Ngesmin <- 10000
  NPmin <- 10000

  for(i in 1:100){
    CR <- i/100 # CR = n_R/n_E as defined in Stucke and Kieser (2012)

    for(j in 1:i) {
      CP <- j/100 # CP = n_P/n_E as defined in Stucke and Kieser (2012)
      n_E <- 10
      power <- 0

      while(power < 1-beta & n_E < 10000){
        n_R <- ceiling(n_E*CR)
        n_P <- ceiling(n_E*CP)
        power <- power_net_effect_2(Delta_ER, Delta_EP, sigma, delta, alpha,
                                     n_E, n_R, n_P)
        n_E <- n_E + 1
      }

      NE <- n_E - 1
      NR <- ceiling(CR * NE)
      NP <- ceiling(CP * NE)
      Nges <- NE + NR + NP

      if((Nges < Ngesmin) | (Nges == Ngesmin && NP < NPmin)){
        Ngesmin <- Nges
        NPmin <- NP
```

```
          NRmin <- NR
          NEmin <- NE
        }
      }
  }

  n_total <- NPmin + NRmin + NEmin

return(data.frame(n_total = n_total, n_P = NPmin, n_E = NEmin,
                  n_R = NRmin, opt_r_R = NRmin/NPmin, opt_r_E = NEmin/NPmin,
                  actual_power = power))
}


# Grid search to find the minimum sample size for r_E = r_R
samplesize_grid_equal_2 <- function(Delta_ER, Delta_EP, sigma, delta, alpha,
                                     beta){
  Ngesmin <- 10000
  NPmin <- 10000
  CR <- 1 # CR = n_R/n_E as defined in Stucke and Kieser (2012)

  for(j in 1:100) {
    CP <- j/100 # CP = n_P/n_E as defined in Stucke and Kieser (2012)
    n_E <- 10
    power <- 0

    while(power < 1-beta & n_E < 10000){
      n_R <- ceiling(n_E * CR)
      n_P <- ceiling(n_E * CP)
      power <- power_net_effect_2(Delta_ER, Delta_EP, sigma, delta, alpha,
                                  n_E, n_R, n_P)
      n_E <- n_E + 1
    }

    NE <- n_E - 1
    NR <- ceiling(CR * NE)
    NP <- ceiling(CP * NE)
    Nges <- NE + NR + NP

    if((Nges < Ngesmin) | (Nges == Ngesmin && NP < NPmin)){
      Ngesmin <- Nges
      NPmin<- NP
      NRmin<- NR
      NEmin<- NE
    }
  }

  n_total <- NPmin + NRmin + NEmin

return(data.frame(n_total = n_total, n_P = NPmin, n_E = NEmin,
                  n_R = NRmin, opt_r_R = NRmin/NPmin, opt_r_E = NEmin/NPmin,
                  actual_power = power))
}

#***************************************************************************#
# Exemplary call according to Example 9.1                                  #
#***************************************************************************#
```

```
samplesize_grid_2(0, 4, 6, 2, 0.05, 0.2)

##    n_total n_P n_E n_R  opt_r_R  opt_r_E actual_power
## 1      338  43 152 143 3.325581 3.534884    0.8019773

samplesize_grid_equal_2(0, 4, 6, 2, 0.05, 0.2)

##    n_total n_P n_E n_R  opt_r_R  opt_r_E actual_power
## 1      338  46 146 146 3.173913 3.173913    0.8019773

###########################################################################
# Functions to calculate the minimal sample size for the net effect      #
# approach by determining the optimal allocation ratio by a grid seach for #
# hypothesis given in (9.4)                                               #
###########################################################################

###########################################################################
# Parameters to specify:
# Delta_ER, Delta_EP, Delta_RP = differences under the alternative
# sigma = standard deviation
# delta = non-inferiority margin > 0
# alpha = significance level
# beta  = type II error rate
###########################################################################

#--------------------------------------------------------------#
# Calculation of power for net effect grid search              #
# functions (Auxiliary function)                               #
#--------------------------------------------------------------#
power_net_effect_3 <- function(Delta_ER, Delta_EP, Delta_RP, sigma, delta,
                               alpha, n_E, n_R, n_P){

  gamma12 <- sqrt(n_R * n_P) / (sqrt((n_R + n_E) * (n_P + n_E)))
  gamma13 <- sqrt(n_R * n_E) / (sqrt((n_R + n_P) * (n_E + n_P)))
  gamma23 <- -sqrt(n_E * n_P) / (sqrt((n_E + n_R) * (n_R + n_P)))
  r <- matrix(c(1, gamma12, gamma13, gamma12, 1, gamma23, gamma13, gamma23,
                1), ncol = 3)

  t1 <- Delta_EP / (sigma * sqrt(1/(n_E) + 1/(n_P))) - qnorm(1-alpha/2)
  t2 <- (Delta_ER+delta) / (sigma * sqrt(1/n_E + 1/(n_R))) - qnorm(1-alpha/2)
  t3 <- Delta_RP / (sigma * sqrt(1/(n_R) + 1/(n_P))) - qnorm(1-alpha/2)

  p123 <- pmvnorm(lower = c(-Inf, -Inf, -Inf), upper = c(t1, t2, t3),
                  mean = c(0, 0, 0), sigma = r)
  return(p123)
}


# Grid search to find the minimum sample size under restriction r_E>=1,
# r_R>=1, and r_E>=r_R
samplesize_grid_3 <- function(Delta_ER, Delta_EP, Delta_RP, sigma, delta,
                              alpha, beta){
  Ngesmin <- 10000
  NPmin <- 10000

  for(i in 1:100){
    CR <- i/100 # CR = n_R/n_E as defined in Stucke and Kieser (2012)
```

```
  for(j in 1:i) {
    CP <- j/100 # CP = n_P/n_E as defined in Stucke and Kieser (2012)
    n_E <- 10
    power <- 0

    while(power < 1-beta & n_E < 10000){
      n_R <- ceiling(n_E * CR)
      n_P <- ceiling(n_E * CP)
      power <- power_net_effect_3(Delta_ER, Delta_EP, Delta_RP, sigma,
                                  delta, alpha, n_E, n_R, n_P)
      n_E <- n_E + 1
    }

    NE <- n_E - 1
    NR <- ceiling(CR * NE)
    NP <- ceiling(CP * NE)
    Nges <- NE + NR + NP

    if((Nges < Ngesmin) | (Nges == Ngesmin && NP < NPmin)){
      Ngesmin <- Nges
      NPmin <- NP
      NRmin <- NR
      NEmin <- NE
    }
  }
}

n_total <- NPmin + NRmin + NEmin

return(data.frame(n_total = n_total, n_P = NPmin, n_E = NEmin,
                  n_R = NRmin, opt_r_R = NRmin/NPmin, opt_r_E = NEmin/NPmin,
                  actual_power = power))
}

# Grid search to find the minimum sample size for r_E = r_R
samplesize_grid_equal_3 <- function(Delta_ER, Delta_EP, Delta_RP, sigma,
                                    delta, alpha, beta){
  Ngesmin <- 10000
  NPmin <- 10000
  CR <- 1

  for(j in 1:100) {
    CP <- j/100
    n_E <- 10
    power <- 0

    while(power < 1-beta & n_E < 10000){
      n_R <- ceiling(n_E * CR)
      n_P <- ceiling(n_E * CP)
      power <- power_net_effect_3(Delta_ER, Delta_EP, Delta_RP, sigma, delta,
                                  alpha, n_E, n_R, n_P)
      n_E <- n_E + 1
    }

    NE <- n_E - 1
    NR <- ceiling(CR * NE)
```

```
    NP <- ceiling(CP * NE)
    Nges <- NE + NR + NP

    if((Nges < Ngesmin) | (Nges == Ngesmin && NP < NPmin)){
      Ngesmin <- Nges
      NPmin <- NP
      NRmin <- NR
      NEmin <- NE
    }
  }

  n_total <- NPmin + NRmin + NEmin

  return(data.frame(n_total = n_total, n_P = NPmin, n_E = NEmin,
                    n_R = NRmin, opt_r_R = NRmin/NPmin,
                    opt_r_E = NEmin/NPmin, actual_power = power))
}


#**************************************************************************#
# Exemplary call according to Example 9.1                                 #
#**************************************************************************#

samplesize_grid_3(0, 4, 4, 6, 2, 0.05, 0.2)

##   n_total n_P n_E n_R opt_r_R opt_r_E actual_power
## 1     348  48 150 150   3.125   3.125    0.8018608

samplesize_grid_equal_3(0, 4, 4, 6, 2, 0.05, 0.2)

##   n_total n_P n_E n_R opt_r_R opt_r_E actual_power
## 1     348  48 150 150   3.125   3.125    0.8018533

#
#
# _____ SECTION 9.3 _____

###############################################################################
# Functions to calculate the sample size for the fraction effect approach     #
# for a given allocation ratio according to formula (9.13) (exact) and        #
# formula (9.14) (approximate)                                                #
###############################################################################

###############################################################################
# Parameters to specify:
# fr = Delta_EP / Delta_RP = ratio of differences
# st_eff_RP = standardized effect (delta_RP/sigma)
# gamma = non-inferiority margin as defined in (9.9)
# alpha = significance level
# beta  = type II error rate
# r_R   = allocation ratio n_R/n_P
# r_E   = allocation ratio n_E/n_P
###############################################################################

# exact calculation according to formula (9.13)
samplesize_exact_fraction_effect <- function(fr, st_eff_RP, gamma, alpha,
                                             beta, r_R, r_E) {
  n <- 10
```

```
  eq <- FALSE
  while(eq == FALSE){
    n <- n + 1

    n_P <- ceiling(n / (r_E + r_R + 1))
    n_E <- ceiling(n / (r_E + r_R + 1) * r_E)
    n_R <- ceiling(n / (r_E + r_R + 1) * r_R)
    n_total <- n_P + n_E + n_R
    actual_r_R <- n_R/n_P
    actual_r_E <- n_E/n_P

    nc <- sqrt(actual_r_E * actual_r_R / (actual_r_R + gamma^2 * actual_r_E +
      (1-gamma)^2 * actual_r_E * actual_r_R) * n_P) * (fr - gamma) * st_eff_RP
    eq <- qt(1-alpha/2, n_total-3, 0) <= qt(0.2, n_total-3, nc)
  }

  actual_power <- 1- pt(qt(1-alpha/2, n_total-3), n_total-3, nc)

  return(data.frame(n_total = n_total, n_P = n_P, n_E = n_E, n_R = n_R,
                    actual_r_R = actual_r_R, actual_r_E = actual_r_E,
                    actual_power = actual_power))
}

# approximation formula according to formula (9.14)
samplesize_fraction_effect <- function(fr, st_eff_RP, gamma, alpha, beta,
                                       r_R, r_E) {
  n <- (1 + r_E + r_R) * (r_R + gamma^2 * r_E + (1-gamma)^2 * r_E * r_R) /
    (r_E * r_R) * (qnorm(1-alpha/2) + qnorm(1-beta))^2 *
    (1/(st_eff_RP * (fr-gamma)))^2

  n_P <- ceiling(n / (r_E + r_R + 1))
  n_E <- ceiling(n / (r_E + r_R + 1) * r_E)
  n_R <- ceiling(n / (r_E + r_R + 1) * r_R)
  n_total <- n_P + n_E + n_R

  actual_r_R <- n_R/n_P
  actual_r_E <- n_E/n_P
  power <- pnorm(sqrt(n_total / (1 + actual_r_R + actual_r_E) *
    (actual_r_R * actual_r_E) / (actual_r_R + gamma^2*actual_r_E +
    (1-gamma)^2*actual_r_E*actual_r_R)) * (fr - gamma)*st_eff_RP -
    qnorm(1-alpha/2))

return(data.frame(n_total = n_total, n_P = n_P, n_E = n_E, n_R = n_R,
  actual_r_R = actual_r_R, actual_r_E = actual_r_E, actual_power = power))
}
#***************************************************************************#
# Exemplary call according to Example 9.2                                  #
#***************************************************************************#

samplesize_fraction_effect(1, 2/3, 0.5, 0.05, 0.2, 1, 1)

##   n_total n_P n_E n_R actual_r_R actual_r_E actual_power
## 1     318 106 106 106          1          1    0.8001485

samplesize_exact_fraction_effect(1, 2/3, 0.5, 0.05, 0.2, 1, 1)

##   n_total n_P n_E n_R actual_r_R actual_r_E actual_power
## 1     321 107 107 107          1          1     0.801454
```

A.10   R Code for Chapter 10

```r
# Inclusion of packages for chapter 10
library(mvtnorm)

# _____ SECTION 10.2 _____

##############################################################################
# Function to calculate the sample size for normally distributed outcomes   #
# to test the difference of means for equivalence                           #
##############################################################################

##############################################################################
# Parameters to specify:
# delta_A  = mean difference
# sigma = common standard deviation
# delta = non-inferiority margin
# alpha = significance level
# beta  = type II error rate
# r     = allocation ratio of sample sizes (n_E/n_C)
##############################################################################

#--------------------------------------------------#
# Exact sample size according to                   #
# iteratively resolving formula (10.4)             #
#--------------------------------------------------#
samplesize_diff_eq_exact <- function(delta_A, sigma, delta, alpha, beta, r){
  power <- 0
  n <- 2
  while(power < 1 - beta){
    n <- n + 1
    n_E <- ceiling(r/(1+r) * n)
    n_C <- ceiling(1/(1+r) * n)
    actual_r <- n_E/n_C
    n_total <- n_E + n_C
    nc_L <- sqrt(actual_r/(1+actual_r)^2 * n_total) * (delta_A+delta) / sigma
    nc_U <- sqrt(actual_r/(1+actual_r)^2 * n_total) * (delta_A-delta) / sigma
    power <- pmvt(lower = c(-Inf,qt(1-alpha, n_total-2)),
                  upper = c(-qt(1-alpha, n_total-2), Inf),
                  delta = c(nc_U, nc_L), df = n_total-2, corr = matrix(1,2,2))
  }

  return(data.frame(n_total = n_total, n_E = n_E, n_C = n_C,
                    actual_r = actual_r, actual_power = power))
}

#--------------------------------------------------#
# Approximate sample size formula according        #
# to Guenther/Schouten (10.5, 10.6)                #
#--------------------------------------------------#
samplesize_diff_eq_GS <- function(delta_A, sigma, delta, alpha, beta, r){
  n <- ifelse(
        delta_A == 0,
        (1+r)^2 / r * (qnorm(1-alpha) + qnorm(1-beta/2))^2 *
              (sigma/delta)^2 + (qnorm(1-alpha)^2)/2,
```

```
         (1+r)^2 / r * (qnorm(1-alpha) + qnorm(1-beta))^2 *
              (sigma/(delta - abs(delta_A)))^2 + (qnorm(1-alpha)^2)/2
  )

  n_E <- ceiling(r/(1+r) * n)
  n_C <- ceiling(1/(1+r) * n)
  n_total <- n_E + n_C
  actual_r <- n_E/n_C
  actual_power <- ifelse(
                  delta_A == 0,
                  1 - 2 * (1 - pnorm(abs(delta/sigma) *
                              sqrt(actual_r) / (1+actual_r) *
                              sqrt(n_total - (qnorm(1-alpha)^2)/2) -
                              qnorm(1-alpha))),
                  pnorm(sqrt((n_total - (qnorm(1-alpha)^2)/2) *
                          (delta - abs(delta_A))^2/sigma^2 *
                          actual_r / (1+actual_r)^2) -
                      qnorm(1-alpha)))

  return(data.frame(n_total = n_total, n_E = n_E, n_C = n_C,
                  actual_r = actual_r, actual_power = actual_power))
}

#************************************************************************#
# Exemplary call according to Example 10.1                             #
#************************************************************************#

samplesize_diff_eq_exact(0, 0.2442/sqrt(2), log(1.25), 0.05, 0.1, 1)

##   n_total n_E n_C actual_r actual_power
## 1      28  14  14        1    0.9074943

samplesize_diff_eq_GS(0, 0.2442/sqrt(2), log(1.25), 0.05, 0.1, 1)

##   n_total n_E n_C actual_r actual_power
## 1      28  14  14        1    0.9090791

#
#
#_____ SECTION 10.3 _____

################################################################################
# Function to calculate the sample size for normally distributed outcomes    #
# to test the ratio of means for equivalence                                 #
################################################################################

################################################################################
# Parameters to specify:
# theta_A  = ratio of means
# CV = coefficient of variation
# delta = non-inferiority margin
# alpha = significance level
# beta  = type II error rate
# r     = allocation ratio of sample sizes (n_E/n_C)
################################################################################

#----------------------------------------------------#
# Exact sample size according to                     #
```

```r
# iteratively resolving formula (10.4)              #
#----------------------------------------------------#
samplesize_ratio_eq_exact <- function(theta_A, CV, delta, alpha, beta, r){
  power <- 0
  n <- 2
  while(power < 1 - beta){
    n <- n + 1
    n_E <- ceiling(r/(1+r) * n)
    n_C <- ceiling(1/(1+r) * n)
    actual_r <- n_E/n_C
    n_total <- n_E + n_C
    nc_L <- sqrt(actual_r / (1+actual_r) / (1+delta^2*r) * n_total) *
      (theta_A-delta) / CV
    nc_U <- sqrt(actual_r / (1+actual_r) / (1+1/delta^2*r) * n_total) *
      (theta_A-1/delta) / CV
    power <- pmvt(lower = c(-Inf,qt(1-alpha, n_total-2)),
                  upper = c(-qt(1-alpha, n_total-2), Inf),
                  delta = c(nc_U, nc_L), df = n_total-2, corr = matrix(1,2,2))
  }

  return(data.frame(n_total = n_total, n_E = n_E, n_C = n_C,
                    actual_r = actual_r, actual_power = power))
}


#----------------------------------------------------#
# Approximate sample size according                 #
# to formulas (10.13a)-(10.13c)                     #
#----------------------------------------------------#
samplesize_ratio_eq <- function(theta_A, CV, delta, alpha, beta, r){

  # Calculation of the sample size according to formula (10.13a)
  if (theta_A == 1){
    if (r<1){
      n <- (1+r) * (1 + delta^2*r)/r * (qnorm(1-alpha) + qnorm(1-beta/2))^2 *
           (CV / (1-delta))^2 + qnorm(1-alpha)^2/2
    }
    else{
      n <- (1+r) * (1 + 1/(delta^2)*r)/r * (qnorm(1-alpha) +
                                            qnorm(1-beta/2))^2 *
           (CV / (1 - 1/delta))^2 + qnorm(1-alpha)^2/2
    }
  }

  # Calculation of the sample size according to formula (10.13b)
  else if(theta_A < 1){
    n <- (1+r)*(1+delta^2*r)/r * (qnorm(1-alpha) + qnorm(1-beta))^2 *
         (CV/(theta_A-delta))^2 + qnorm(1-alpha)^2/2
  }

  # Calculation of the sample size according to formula (10.13c)
  else {
    n <- (1+r)*(1+(1/delta^2)*r)/r * (qnorm(1-alpha) + qnorm(1-beta))^2 *
         (CV/(1/delta-theta_A))^2 + qnorm(1-alpha)^2/2
  }

  n_E <- ceiling(r/(1+r) * n)
  n_C <- ceiling(1/(1+r) * n)
```

```
  n_total <- n_E + n_C
  actual_r <- n_E/n_C

  if (theta_A == 1){
    if (actual_r < 1){
      actual_power <- 1 - 2*(1 - pnorm(sqrt((n_total - qnorm(1-alpha)^2/2) *
                      actual_r / ((1+actual_r) * (1+delta^2*actual_r))) *
                        abs((1-delta)/CV) - qnorm(1-alpha)))
    }
    else{
      actual_power <- 1 - 2*(1 - pnorm(sqrt((n_total - qnorm(1-alpha)^2/2) *
                      actual_r / ((1+actual_r) * (1+1/delta^2*actual_r))) *
                        abs((1-1/delta)/CV) - qnorm(1-alpha)))
    }
  }
  else if(theta_A < 1){
    actual_power <- pnorm(sqrt((n_total - qnorm(1-alpha)^2/2) *
                    actual_r / ((1+actual_r) * (1+delta^2*actual_r))) *
                      abs((theta_A - delta) / CV) - qnorm(1-alpha))
  }
  else {
    actual_power <- pnorm(sqrt((n_total - qnorm(1-alpha)^2/2)*
                    actual_r / ((1+actual_r) * (1+1/delta^2*actual_r))) *
                      abs((1/delta - theta_A) / CV) - qnorm(1-alpha))
  }
  return(data.frame(n_total = n_total, n_E = n_E, n_C = n_C,
                    actual_r = actual_r, actual_power = actual_power))
}

#***********************************************************************#
# Exemplary call according to Example 10.2                             #
#***********************************************************************#

samplesize_ratio_eq_exact(1, 0.363, 0.55, 0.025, 0.10, 1)

##   n_total n_E n_C actual_r actual_power
## 1      26  13  13        1    0.9275896

samplesize_ratio_eq(1, 0.363, 0.55, 0.025, 0.10, 1)

##   n_total n_E n_C actual_r actual_power
## 1      24  12  12        1    0.9008725
```

A.11   R Code for Chapter 11

```
# Inclusion of packages for chapter 11
library(mvtnorm)

# _____ SECTION 11.3.2 _____

###############################################################################
# Function to calculate the sample size for the "all or none" success        #
# criterion according to formula (11.6)                                       #
###############################################################################

###############################################################################
# Parameters to specify:
# st_eff  = standardized effect (delta_A/sigma_A)
# rho     = matrix of correlation coefficients
# k       = number of endpoints
# ptol, maxiter = paramters passed to the stochastic root-finding algorithm
#                 in qmvnorm()
# alpha   = significance level
# 1-beta  = conjunctive power
# r       = allocation ratio of sample sizes (n_E/n_C)
###############################################################################

samplesize_all_or_none <- function(st_eff, rho, k, ptol, maxiter, alpha,
                      beta, r){
  q1 <- qmvnorm(1-beta, tail = "lower.tail", mean = 0, corr = rho,
                ptol = ptol, maxiter = maxiter)$quantile
  n <- (1+r)^2/r * (qnorm(1-alpha/2) + q1)^2 * (1/st_eff)^2
  n_E <- ceiling(r/(1+r) * n)
  n_C <- ceiling(1/(1+r) * n)
  n_total <- n_E + n_C
  actual_r <- n_E/n_C
  actual_power <- pmvnorm(lower = rep(-Inf, k), upper =
    sqrt(n_total * actual_r) * st_eff / (1 + actual_r) - qnorm(1-alpha/2),
    mean = 0, corr = rho)
  return(data.frame(n_total = n_total, n_E = n_E, n_C = n_C,
                    actual_r = actual_r, actual_power = actual_power))
}

#*****************************************************************************#
# Exemplary call according to Example 11.1                                   #
#*****************************************************************************#

samplesize_all_or_none(0.2, matrix(c(1,0.3,0.3,1),2,2), 2, 0.0001, 1000,
                       0.05, 0.1, 1)

##   n_total n_E n_C actual_r actual_power
## 1    1274 637 637        1    0.9004714

###############################################################################
# Function to calculate the sample size for the "all or none" success        #
# criterion according to the method proposed by Varga et al. (2017)          #
###############################################################################

###############################################################################
# Parameters to specify:
# st_eff  = standardized effect (delta_A/sigma_A)
# k       = number of endpoints
# alpha   = significance level
```

```
# 1-beta  = conjunctive power
# r       = allocation ratio of sample sizes (n_E/n_C)
#           (rounding works only for whole numbers)
###########################################################################

samplesize_all_or_none_Varga <- function(st_eff, k, alpha, beta, r){

  # Step 1:
  ind_pow <- (1-beta)^(1/k)
  n <- (1+r)^2/r * (qnorm(ind_pow) + qnorm(1-alpha/2))^2 * (1/max(st_eff))^2
  n <- ceiling(n/(r+1))*(r+1)
  ind_pow <- pnorm(sqrt(n)*st_eff/2 - qnorm(1-alpha/2))

  # Repeat Step 2 until conjunctive power is at least 1-beta
  while(prod(ind_pow) < 1-beta){
    n <- ceiling(n + r + 1)
    ind_pow <- pnorm(sqrt(n)*st_eff/2 - qnorm(1-alpha/2))
  }

  return(list(n_total = n, n_E = r/(1+r) * n, n_C = 1/(1+r) * n,
              ind_power = ind_pow))
}

#*************************************************************************#
# Exemplary call according to Example 11.1                               #
#*************************************************************************#

samplesize_all_or_none_Varga(c(0.2, 0.2), 2, 0.05, 0.1, 1)

## $n_total
## [1] 1292
##
## $n_E
## [1] 646
##
## $n_C
## [1] 646
##
## $ind_power
## [1] 0.9489205 0.9489205

###########################################################################
# Function to calculate the sample size for the "at least one" success   #
# criterion according to formula (11.10)                                 #
###########################################################################

###########################################################################
# Parameters to specify:
# st_eff = standardized effect (delta_A/sigma_A)
# rho    = matrix of correlation coefficients
# k      = number of endpoints
# ptol, maxiter = paramters passed to the stochastic root-finding algorithm
#                 in qmvnorm()
# alpha  = significance level
# 1-beta = conjunctive power
# r      = allocation ratio of sample sizes (n_E/n_C)
###########################################################################
```

```
samplesize_at_least_one <- function(st_eff, rho, k, ptol, maxiter, alpha,
                                    beta, r){
  q1 <- qmvnorm(beta, tail = "lower.tail", mean = 0, corr = rho, ptol = ptol,
                maxiter = maxiter)$quantile
  n <- (1+r)^2/r * (qnorm(1-alpha/(2*k)) - q1)^2 * (1/st_eff)^2
  n_E <- ceiling(r/(1+r) * n)
  n_C <- ceiling(1/(1+r) * n)
  n_total <- n_E + n_C
  actual_r <- n_E/n_C
  actual_power <- 1-pmvnorm(lower = rep(-Inf, k),
    upper = -sqrt(n_total*actual_r)*st_eff/(1+actual_r) +
      qnorm(1-alpha/(2*k)), mean = 0, corr = rho)
  return(data.frame(n_total = n_total, n_E = n_E, n_C = n_C,
                    actual_r = actual_r, actual_power = actual_power))
}

#*****************************************************************************#
# Exemplary call according to Example 11.2                                   #
#*****************************************************************************#

samplesize_at_least_one(0.2,  matrix(c(1,0.3,0.3,1),2,2), 2, 0.0001, 1000,
                        0.05, 0.1, 1)

##   n_total n_E n_C actual_r actual_power
## 1     838 419 419        1    0.9006748

#
#
# _____ SECTION 11.4.2 _____

##############################################################################
# Function to calculate the sample size for the "all or none" success        #
# criterion for more than two groups using the Dunnett test                  #
# according to formula (11.14)                                               #
##############################################################################

##############################################################################
# Parameters to specify:                                                     #
# st_eff = standardized effect (delta_A/sigma)
#          (assumed to be equal for each group)
# k      = number of treatment groups (without control)
# ptol, maxiter = paramters passed to the stochastic root-finding algorithm
#                 in qmvnorm()
# alpha  = significance level
# 1-beta = conjunctive power
# r      = allocation ratio of sample sizes (n_i/n_0)
#          (assumed to be equal for i=1,...,k)
##############################################################################

samplesize_all_or_none_dunnett <- function(st_eff, k, ptol, maxiter, alpha,
                                           beta, r){
  rho <- matrix(rep(r/(1+r), k^2), k, k)
  diag(rho) <- rep(1, k)
  n <- (1+r)*(1+r*k)/r * (qmvnorm(1-alpha/2, tail = "lower.tail", mean = 0,
    corr = rho, ptol = ptol, maxiter = maxiter)$quantile + qmvnorm(1-beta,
    tail = "lower.tail", mean = 0, corr = rho, ptol = ptol,
    maxiter = maxiter)$quantile)^2 * 1/st_eff^2
  n_0 <- ceiling(n/(1+k*r))
```

```
  n_i <- ceiling(n/(1/r+k))
  n_per_group <- c(n_0, rep(n_i, k))
  n_total <- sum(n_per_group)
  actual_r <- n_i/n_0
  actual_power <- pmvnorm(lower = rep(-Inf, k),
    upper = sqrt(n_total*actual_r / ((1 + actual_r) * (1 + actual_r*k))) *
    st_eff - qmvnorm(1-alpha/2, tail = "lower.tail", mean = 0, corr = rho,
    ptol = ptol, maxiter = maxiter)$quantile, mean = 0, corr = rho)[1]

  return(list(n_total = n_total, n_per_group = n_per_group,
                   actual_r = actual_r, actual_power = actual_power))
}

#*************************************************************************#
# Exemplary call according to Example 11.3                              #
#*************************************************************************#

samplesize_all_or_none_dunnett(0.5, 2, 0.0001, 1000, 0.05, 0.1, 1/sqrt(2))

## $n_total
## [1] 341
##
## $n_per_group
## [1] 141 100 100
##
## $actual_r
## [1] 0.7092199
##
## $actual_power
## [1] 0.9023376

#########################################################################
# Function to calculate the sample size for the "at least one" success  #
# criterion for more than two groups using the Dunnett test             #
# according to formula (11.16)                                          #
#########################################################################

#########################################################################
# Parameters to specify:
# st_eff = standardized effect (delta_A/sigma)
#          (assumed to be equal for each group)
# k      = number of treatment groups (without control)
# ptol, maxiter = paramters passed to the stochastic root-finding algorithm
#                 in qmvnorm()
# alpha  = significance level
# 1-beta = conjunctive power
# r      = allocation ratio of sample sizes (n_i/n_0)
#          (assumed to be equal for i = 1,...,k)
#########################################################################
```

```
samplesize_at_least_one_dunnett <- function(st_eff, k, ptol, maxiter, alpha,
                                            beta, r){
  rho <- matrix(rep(r/(1+r), k^2), k, k)
  diag(rho) <- rep(1, k)
  n <- (1+r)*(1+r*k)/r * (qmvnorm(1-alpha/2, tail = "lower.tail", mean = 0,
    corr = rho, ptol = ptol )$quantile -
    qmvnorm(beta, tail = "lower.tail", mean = 0, corr = rho, ptol = ptol,
           maxiter = maxiter)$quantile)^2 * 1/st_eff^2

  n_0 <- ceiling(n/(1+k*r))
  n_i <- ceiling(n/(1/r+k))
  n_per_group <- c(n_0, rep(n_i, k))
  n_total <- sum(n_per_group)
  actual_r <- n_i/n_0
  actual_power <- 1 - pmvnorm(lower = rep(-Inf, k), upper =
    qmvnorm(1-alpha/2, tail = "lower.tail", mean = 0, corr = rho)$quantile -
    sqrt(n_total*actual_r/((1+actual_r)*(1+actual_r*k))) * st_eff,
    mean = 0, corr = rho)[1]
  return(list(n_total = n_total, n_per_group = n_per_group,
              actual_r = actual_r, actual_power = actual_power))
}

#***************************************************************************#
# Exemplary call according to Example 11.3                                 #
#***************************************************************************#

samplesize_at_least_one_dunnett(0.5, 2, 0.0001, 1000, 0.05, 0.1, 1/sqrt(2))

## $n_total
## [1] 202
##
## $n_per_group
## [1] 84 59 59
##
## $actual_r
## [1] 0.702381
##
## $actual_power
## [1] 0.9008463
```

A.12   R Code for Chapter 12

```
# _____ SECTION 12.2 _____
################################################################################
# Function to calculate the sample size for the exact binomial test           #
# according to formula (12.2)                                                  #
################################################################################

################################################################################
# Parameters to specify:
# p_A   = event probability under the alternative
# p_0   = event probability under the null hypothesis
# alpha = significance level
# beta  = type II error rate
################################################################################
samplesize_exact_binomial <- function(p_A, p_0, alpha, beta){
  power <- 0
  n <- 1
  while(power < 1 - beta){
    n <- n + 1
    i <- 1:n
    pr_0 <- dbinom(i, n, p_0, log = FALSE)
    k_L <- max(which(cumsum(pr_0) <= alpha/2), 0)
    k_U <- min(which(rev(cumsum(rev(pr_0))) <= alpha/2), n+1)

    pr_A <- dbinom(i, n, p_A, log = FALSE)
    power <- ifelse(k_L > 0, sum(pr_A[1:k_L]), 0) +
      ifelse(k_U < n+1, sum(pr_A[k_U:n]), 0)
  }

  return(data.frame(n = n, actual_power = power))
}

#*****************************************************************************#
# Exemplary call according to Table 12.1                                     #
#*****************************************************************************#

samplesize_exact_binomial(0.25, 0.3, 0.05, 0.2)

##     n actual_power
## 1 633    0.8026741
################################################################################
# Function to calculate the sample size for the approximate score test        #
# according to formula (12.3)                                                 #
################################################################################

################################################################################
# Parameters to specify:
# p_A   = event probability under the alternative
# p_0   = event probability under the null hypothesis
# alpha = significance level
# beta  = type II error rate
################################################################################

samplesize_approximate_score <- function(p_A, p_0, alpha, beta){
  n <- ceiling((qnorm(1-alpha/2) * sqrt(p_0 * (1-p_0)) + qnorm(1-beta) *
    sqrt(p_A * (1-p_A)))^2 / (p_A - p_0)^2)
```

```
approx_power <- pnorm((sqrt(n) * abs(p_A-p_0) -
    qnorm(1-alpha/2) * sqrt(p_0 * (1-p_0))) / sqrt(p_A * (1-p_A)))

# exact power (according to formula 12.2) for approximate sample size
i <- 1:n
pr_0 <- dbinom(i, n, p_0, log = FALSE)
k_L <- max(which(cumsum(pr_0) <= alpha/2), 0)
k_U <- min(which(rev(cumsum(rev(pr_0))) <= alpha/2), n+1)
pr_A <- dbinom(i, n, p_A, log = FALSE)
exact_power <- ifelse(k_L > 0, sum(pr_A[1:k_L]), 0) +
    ifelse(k_U < n+1, sum(pr_A[k_U:n]), 0)

return(data.frame(n = n, approx_power = approx_power,
                    exact_power = exact_power))
}


#**************************************************************************#
# Exemplary call according to Table 12.1                                  #
#**************************************************************************#

samplesize_approximate_score(0.25, 0.3, 0.05, 0.2)

##     n approx_power exact_power
## 1 638    0.8002146   0.7954096

#
#
# _____ SECTION 12.3 _____

###########################################################################
# Function to calculate the expected width of the Clopper-Pearson- and    #
# Wilson score confidence interval according to formula (12.4)            #
###########################################################################

###########################################################################
# Parameters to specify:
# p     = event probability
# n     = sample size
# alpha = significance level
###########################################################################
#-----------------------------------------------------------#
# This function also acts as an auxiliary function          #
# for samplesize_clopper_pearson() and samplesize_wilson()  #
#-----------------------------------------------------------#
exp_confidence_width <- function(p, n, alpha){
  k <- 0:n
  W_CP <- qbeta(1-alpha/2, k+1, n-k) - qbeta(alpha/2, k, n-k+1)
  z <- qnorm(1-alpha/2)
  W_Wilson <- z * sqrt(z^2 + 4*n*k/n*(1-k/n)) / (n+z^2)
  EW_CP <- sum(dbinom(k, n, p, log = FALSE) * W_CP)
  EW_Wilson <- sum(dbinom(k, n, p, log = FALSE) * W_Wilson)

  return(data.frame(EW_CP = EW_CP, EW_Wilson = EW_Wilson))
}

#**************************************************************************#
# Exemplary call according to Example 12.1                               #
```

```
#***********************************************************************#

exp_confidence_width(0.01, 3300, 0.05)

##         EW_CP    EW_Wilson
## 1 0.007096106 0.006855392

################################################################################
# Functions to calculate the sample size to assure that the expected width  #
# of either the Clopper-Pearson or the Wilson score confidence interval is  #
# not larger than a pre-specified amount according to formula (12.4)        #
################################################################################


################################################################################
# Parameters to specify:
# p     = event probability
# width = width of confidence interval
# alpha = significance level
################################################################################


# Clopper-Pearson interval
samplesize_clopper_pearson <- function(p, width, alpha){
  n <- 1
  w <- exp_confidence_width(p, n, alpha)$EW_CP
  while(w > width){
    n <- n + 1
    w <- exp_confidence_width(p, n, alpha)$EW_CP
  }

  return(data.frame(n = n, actual_expected_width = w))
}

# Wilson score interval
samplesize_wilson <- function(p, width, alpha){
  n <- 1
  w <- exp_confidence_width(p, n, alpha)$EW_Wilson
  while(w > width){
    n <- n + 1
    w <- exp_confidence_width(p, n, alpha)$EW_Wilson
  }

  return(data.frame(n = n, actual_expected_width = w))
}

#***********************************************************************#
# Exemplary call according to Table  12.2                              #
#***********************************************************************#

samplesize_clopper_pearson(0.2, 0.164, 0.05)

##     n actual_expected_width
## 1 100             0.1639844

samplesize_wilson(0.2, 0.155, 0.05)

##     n actual_expected_width
## 1 100             0.1543218
```

```
#
#
#  _____ SECTION 12.4 _____

###############################################################################
# Function to calculate sample size required to observe an event at least    #
# once according to formula (12.5)                                           #
###############################################################################

###############################################################################
# Parameters to specify:
# p        = event probability
# 1-beta = probability to observe the event at least once
###############################################################################

samplesize_event_occurence <- function(p, beta){
  n <- ceiling(log(beta)/log(1-p))
  # probability to observe an event at least once for sample size n
  actual_prob <- 1-exp(n * log(1-p))
return(data.frame(n = n, actual_prob = actual_prob))
}

#***************************************************************************#
# Exemplary call according to Table 12.3a                                  #
#***************************************************************************#

samplesize_event_occurence(0.005, 0.2)

##     n actual_prob
## 1 322    0.800918

###############################################################################
# Function to calculate probability to observe at least one event for given #
# sample size according to formula (12.5)                                   #
###############################################################################

###############################################################################
# Parameters to specify:
# p = event probability
# n = sample size
###############################################################################

prob_atLeastOne_Event <- function(p, n){
  pr <- round(1-(1-p)^n,4)
return(pr)
}

#***************************************************************************#
# Exemplary call according to Table 12.3b                                  #
#***************************************************************************#

prob_atLeastOne_Event(0.005, 100)

## [1] 0.3942
```

A.13   R Code for Chapter 13

```
# _____ SECTION 13.2 _____

##########################################################################
# Function to calculate the sample size for the t-test iteratively in the  #
# setting of cluster-randomized trials according to formula (13.11)        #
##########################################################################


##########################################################################
# Parameters to specify:
# delta_A = difference between groups
# sigma = standard deviation
# m     = sample size per cluster
# ICC   = intra-cluster correlation coefficient (see formula (13.8))
# alpha = significance level
# beta  = type II error rate
# r     = allocation ratio of sample sizes (K_E/K_C)
##########################################################################

samplesize_tTest_cluster <- function(delta_A, sigma, m, ICC, alpha, beta, r){
  DE <- 1 + (m-1) * ICC # design effect according to forumula (13.9)
  power <- 0
  K <- 2
  while(power < 1-beta){
    K <- K+1
    K_E <- ceiling(r/(1+r) * K) # number of clusters in experimental group
    K_C <- ceiling(1/(1+r) * K) # number of clusters in control group
    K <- K_E + K_C
    actual_r <- K_E/K_C
    nc <- sqrt(actual_r/(1+actual_r)^2 * K) * delta_A / sigma / sqrt(DE/m)
    power <- 1- pt(qt(1-alpha/2, K-2), K-2, nc)
  }

  return(data.frame(DE = DE, K_total = K, K_E = K_E,
            K_C = K_C, n_total = K*m, n_E = K_E*m,
            n_C = K_C*m, actual_r = actual_r,
            actual_power = round(power, 4)))
}


#************************************************************************#
# Exemplary call according to Example 13.1                             #
#************************************************************************#

samplesize_tTest_cluster(10, 20, 12, 0.01, 0.05, 0.1, 1)

##    DE K_total K_E K_C n_total n_E n_C actual_r actual_power
## 1 1.11      18   9   9     216 108 108        1       0.9052
```

A.14   R Code for Chapter 14

```r
# Inclusion of packages for chapter 14
library(mvtnorm)

# _____ SECTION 14.2 _____

###############################################################################
# Method 1, unconditional probability                                         #
# Function to calculate portion of the total sample size w_s                  #
# required for the specified region according to Formula (14.6)               #
###############################################################################

###############################################################################
# Parameters to specify:
# alpha = two-sided significance level
# 1-beta = Power to show effect delta in the entire study population
# 1-beta_c = Power consistency condition, unconditional probability
# gamma = effect fraction
###############################################################################

w_uncond_prob_M1 <- function(beta_c, gamma, alpha, beta){
  w <- (qnorm(1-beta_c)^2) / (((1-gamma)^2) *
                              ((qnorm(1-alpha/2) + qnorm(1-beta))^2) +
                              (qnorm(1-beta_c)^2) * gamma * (2-gamma) )

  return(data.frame(w_s = w))
}

#*****************************************************************************#
# Exemplary call (compare Figure 14.1)                                       #
#*****************************************************************************#

w_uncond_prob_M1(0.2, 0.5, 0.05, 0.2)

##         w_s
## 1 0.2840731

###############################################################################
# Method 1, conditional probability of obtaining consistent results          #
# Function to calculate conditional probability according to                 #
# Formula (14.7) and (14.8)                                                   #
###############################################################################

###############################################################################
# Parameters to specify:
# w = sample size portion of the specified region w_s
# gamma = effect fraction
# alpha = two-sided significance level
# 1-beta = Power to show effect delta in the entire study population
# r = allocation ratio
###############################################################################

prob_cond_M1<-function(w, gamma, alpha, beta, r){
  corr <- ((1-gamma) * sqrt(w)) / (sqrt(1 + gamma*(gamma-2)*w))
  sigma <- matrix(c(1, corr, corr, 1), nrow = 2)
  c1 <- -corr * (qnorm(1-alpha/2) + qnorm(1-beta))
  c2 <- -qnorm(1-beta)
  psig <- 1-beta
  pjoint <- pmvnorm(lower = c(c1, c2), upper = c(Inf, Inf), mean = c(0, 0)),
```

```
                        sigma = sigma)
  p_cond <- pjoint/psig

  return(data.frame(p_cond = p_cond))
}

#*********************************************************************#
# Exemplary call (compare Figure 14.2)                               #
#*********************************************************************#

prob_cond_M1(0.5, 0.5, 0.05, 0.2, 1)

##       p_cond
## 1 0.9303356

#
#
#_____ SECTION 14.3 _____

#####################################################################
# Method 2, unconditional probability                               #
# Function to calculate unconditional probability according to      #
# Formula (14.11)                                                   #
#####################################################################

#####################################################################
# Parameters to specify:
# w1 = smallest region, sample size portion
# K = number of regions
# alpha = two-sided significance level
# 1-beta = power
#####################################################################

prob_uncond_M2 <- function(w1, K, alpha, beta){
  q <- qnorm(1-alpha/2) + qnorm(1-beta)
  p_uncond <- ((pnorm(sqrt(w1)*q))^(K-1)) * pnorm(sqrt(1-(K-1)*w1) * q)

  return(data.frame(p_uncond = p_uncond))
}

#*********************************************************************#
# Exemplary call (compare Figure 14.3)                               #
#*********************************************************************#

prob_uncond_M2(0.2, 2, 0.05, 0.2)

##    p_uncond
## 1 0.8894139
```

```
###############################################################################
# Method 2, conditional probability for K=3 regions                           #
# Function to calculate conditional probability according to                  #
# Formula (14.12)                                                             #
###############################################################################

###############################################################################
# Parameters to specify:
# w1, w2, w3 = proportion sample size for region 1, 2, 3
# alpha = two-sided significance level
# 1-beta = Power to demonstrate a significant overall result
###############################################################################

prob_cond_M2<-function(w1, w2, w3, alpha, beta){

  qa <- qnorm(1-alpha/2)
  sigma <- matrix(c(1,0,0,sqrt(w1),
                    0,1,0,sqrt(w2),
                    0,0,1,sqrt(w3),
                    sqrt(w1),sqrt(w2),sqrt(w3),1),
                  nrow=4)
  qb <- qnorm(1-beta)
  c1 <- -(qa+qb) * sqrt(w1)
  c2 <- -(qa+qb) * sqrt(w2)
  c3 <- -(qa+qb) * sqrt(w3)
  call <- -qb

  psig <- 1-beta
  pjoint <- pmvnorm(lower = c(c1,c2,c3,call), upper = c(Inf,Inf,Inf,Inf),
                    mean=c(0,0,0,0)), sigma = sigma)

  p_cond <- pjoint/psig

  return(data.frame(p_cond = p_cond))
}

#***************************************************************************#
# Exemplary call (compare Table 14.3, K=3)                                  #
#***************************************************************************#

prob_cond_M2(1/3, 1/3, 1/3, 0.05, 0.2)

##      p_cond
## 1 0.9312545
```

A.15   R Code for Chapter 15

```
# Inclusion of packages for chapter 15
library(mvtnorm)

# _____ SECTION 15.2 _____

###############################################################################
# Functions to solve the optimization problem of choosing the optimal       #
# sample size in phase II/III drug development programs                      #
# given cost and benefit constraints                                        #
###############################################################################

###############################################################################
# Parameters to specify:
# Delta = standardized true treatment group difference
# tau2  = variance for prior distribution of standardized treatment effect
# kappa = threshold value for decision rule after phase II
# n2    = total sample size in phase II
# fixed = set TRUE or FALSE when modelling the true assumed treatment
#         effect as fixed or by a prior distribution
# step1, step2 = lower and upper boundary for effect size categories
# c02, c03     = fixed costs for phase II and III
# c2, c3       = variable per-patient costs for phase II and III
# steps1, stepm1, stepl1 = lower boundaries of effect size categories
# b1, b2, b3             = benefit categories
# n2min, n2max, stepn2   = parameters to define optimization region for n2
# kappamin, kappamax, stepkappa = parameters to define optimization
#                                 region for kappa
# alpha = one-sided significance level
# beta  = type II error rate
# r     = allocation ratio of sample sizes (n_E/n_C)
###############################################################################

# prior distribution for Delta (compare Figure 15.1)
prior <- function(x, Delta, tau2){
  dnorm(x, Delta, sqrt(tau2))
}

# (expected) probability to go to phase III
pgo <- function(kappa, n2, Delta, tau2, r, fixed){

  nC2 <- n2/(r+1)
  nE2 <- n2/(1/r+1)
  if(fixed){
    return(
      pnorm((Delta-kappa)/sqrt((nC2+nE2)/(nC2*nE2)))
    )
  }else{
    return(
      integrate(function(x){
        sapply(x, function(x){
          pnorm((x-kappa)/sqrt((nC2+nE2)/(nC2*nE2))) *
            prior(x, Delta, tau2)
        })
      }, - Inf, Inf)$value
    )
  }
}
```

```
# expected sample size for phase III
# when going to phase III (compare formula 15.1)
En3 <- function(kappa, n2, alpha, beta, Delta, tau2, r, fixed){
  nC2 <- n2/(r+1)
  nE2 <- n2/(1/r+1)

  if(fixed){
    return(
      integrate(function(y){
        ((((1+r)^2/r)*(qnorm(1-alpha)+qnorm(1-beta))^2)/y^2) *
          dnorm(y,
                mean = Delta,
                sd = sqrt((nC2+nE2)/(nC2*nE2)))
      }, kappa, Inf)$value
    )
  }else{
    return(
      integrate(function(x){
        sapply(x, function(x){
          integrate(function(y){
            ((((1+r)^2/r)*(qnorm(1-alpha)+qnorm(1-beta))^2)/y^2) *
              dnorm(y,
                    mean = x,
                    sd = sqrt((nC2+nE2)/(nC2*nE2)))*
              prior(x, Delta, tau2)
          }, kappa, Inf)$value
        })
      }, - Inf, Inf)$value
    )
  }
}


# (expected) probability of a successful program
POSP <- function(kappa, n2, alpha, beta, step1, step2, Delta, tau2, r, fixed)
{

  nC2 <- n2/(r+1)
  nE2 <- n2/(1/r+1)
  c = (qnorm(1 - alpha) + qnorm(1 - beta))^2

  if(fixed){
    return(
      integrate(function(y){
        ( pnorm(qnorm(1 - alpha) + step2/sqrt(y^2/c),
                mean = Delta/sqrt(y^2/c),
                sd = 1) -
            pnorm(qnorm(1 - alpha) + step1/sqrt(y^2/c),
                  mean = Delta/sqrt(y^2/c),
                  sd = 1) ) *
          dnorm(y,
                mean = Delta,
                sd = sqrt((nC2+nE2)/(nC2*nE2)))
      }, kappa, Inf)$value
    )
  }else{
    return(
```

```
    integrate(function(x){
      sapply(x, function(x){
        integrate(function(y){
          ( pnorm(qnorm(1 - alpha) + step2/sqrt(y^2/c),
                  mean = x/sqrt(y^2/c),
                  sd = 1) -
              pnorm(qnorm(1 - alpha) + step1/sqrt(y^2/c),
                    mean = x/sqrt(y^2/c),
                    sd = 1) ) *
            dnorm(y,
                  mean = x,
                  sd = sqrt((nC2+nE2)/(nC2*nE2))) *
            prior(x, Delta, tau2)
        }, kappa, Inf)$value
      })
    }, - Inf, Inf)$value
  )
}

}

# utility function
utility <- function(n2, kappa, Delta, tau2, r,
                    alpha, beta,
                    c2, c3, c02, c03,
                    steps1, stepm1, stepl1,
                    b1, b2, b3, fixed){
  steps2 <- stepm1
  stepm2 <- stepl1
  stepl2 <- Inf

  n3   <- En3(kappa = kappa, n2 = n2, alpha = alpha, beta = beta,
             Delta = Delta, tau2 = tau2, r = r, fixed = fixed)

  pg   <- pgo(kappa = kappa, n2 = n2,
             Delta = Delta, tau2 = tau2, r = r, fixed = fixed)

  n3_go <- ceiling(n3/pg) # compare formula 15.5

  K2   <- c02 + c2 * n2        # cost phase II
  K3   <- c03 * pg + c3 * n3   # cost phase III

  # (expected) probability of a successful program;
  # small, medium, large effect size
  prob1 <- POSP(kappa = kappa, n2 = n2, alpha = alpha, beta = beta,
               step1 = steps1, step2 = steps2,
               Delta = Delta, tau2 = tau2, r = r, fixed = fixed)
  prob2 <- POSP(kappa = kappa, n2 = n2, alpha = alpha, beta = beta,
               step1 = stepm1, step2 = stepm2,
               Delta = Delta, tau2 = tau2, r = r, fixed = fixed)
  prob3 <- POSP(kappa = kappa, n2 = n2, alpha = alpha, beta = beta,
               step1 = stepl1, step2 = stepl2,
               Delta = Delta, tau2 = tau2, r = r, fixed = fixed)
  # (expected) probability of a successful program
  SP    <- prob1 + prob2 + prob3
  # gain
```

```r
    G      <-  b1 * prob1 + b2 * prob2 + b3 * prob3
    # expected utility
    EU     <-  - K2 - K3 + G

    return(c(EU, n3_go, SP, pg, K2, K3, prob1, prob2, prob3))

}

# function to find optimal threshold value and sample size
optimal <- function(Delta, tau2, r,
                     n2min, n2max, stepn2,
                     kappamin, kappamax, stepkappa,
                     alpha, beta,
                     c2, c3, c02, c03,
                     steps1 = 0, stepm1 = 3/8, stepl1 = 5/8,
                     b1, b2, b3,
                     fixed = FALSE){

  KAPPA <- seq(kappamin, kappamax, stepkappa)
  N2    <- seq(n2min, n2max, stepn2)

  ufkt <- spfkt <- pgofkt <- K2fkt <- K3fkt <-
    sp1fkt <- sp2fkt <- sp3fkt <- n2fkt <- n3fkt <- matrix(0, length(N2),
                                                           length(KAPPA))

  for(j in 1:length(KAPPA)){

    kappa <- KAPPA[j]
    result <- sapply( N2, utility, kappa, Delta, tau2, r,
                        alpha, beta,
                        c2, c3, c02, c03,
                        steps1, stepm1, stepl1,
                        b1, b2, b3, fixed)

    ufkt[, j]    <- result[1, ]
    n3fkt[, j]   <- result[2, ]
    spfkt[, j]   <- result[3, ]
    pgofkt[, j]  <- result[4, ]
    K2fkt[, j]   <- result[5, ]
    K3fkt[, j]   <- result[6, ]
    sp1fkt[, j]  <- result[7, ]
    sp2fkt[, j]  <- result[8, ]
    sp3fkt[, j]  <- result[9, ]
  }

  ind   <- which(ufkt == max(ufkt), arr.ind <- TRUE)

  I <- as.vector(ind[1, 1])
  J <- as.vector(ind[1, 2])

  Eud   <- ufkt[I, J]
  n3_go <- n3fkt[I, J]
  prob  <- spfkt[I, J]
  pg    <- pgofkt[I, J]
  k2    <- K2fkt[I, J]
  k3    <- K3fkt[I, J]
  prob1 <- sp1fkt[I, J]
```

```
  prob2 <- sp2fkt[I, J]
  prob3 <- sp3fkt[I, J]

  if(fixed){
    result <-  data.frame(Delta = Delta, b1 = b1, b2 = b2, b3 = b3,
                          Kappa = KAPPA[J], n2 = N2[I], n3_go = n3_go,
                          pgo = round(pg,2), sProg = round(prob,2),
                          u = round(Eud,1))
  }else{
    result <-  data.frame(Delta = Delta, tau2 = tau2,
                          b1 = b1, b2 = b2, b3 = b3,
                          Kappa = KAPPA[J], n2 = N2[I], n3_go = n3_go,
                          pgo = round(pg,2), sProg = round(prob,2),
                          u = round(Eud,1))
  }
  return(result)
}

#***************************************************************************#
# Exemplary call according to Example 15.1                                 #
#***************************************************************************#

# Table 15.1 (line 3):
n3 <- En3(kappa=2/8, n2=100, alpha=0.025,
               beta=0.1, Delta=4/8, tau2=0, r=1, fixed=TRUE)
pg <- pgo(kappa=2/8, n2=100, Delta=4/8, tau2=0, r=1, fixed=TRUE)

data.frame(
  n_2    = 100,
  En3_go = ceiling(n3 / pg),
  pgo    = pg,
  posp   = POSP(kappa=2/8, n2=100, alpha=0.025, beta=0.1, step1=0,
               step2=Inf, Delta=4/8, tau2=0, r=1, fixed=TRUE)
)

##    n_2 En3_go      pgo      posp
## 1 100    195 0.8943502 0.7337882
```

```
# Table 15.2 (line 5):
opt <- optimal(Delta = 4/8, tau2 = 0, r=1,
               n2min = 20, n2max = 120, stepn2 = 2,
               kappamin = 0.8/8, kappamax = 2/8, stepkappa = 0.01,
               alpha = 0.025, beta = 0.1,
               c2 = 0.675, c3 = 0.72, c02 = 15, c03 = 20,
               steps1 = 0,
               stepm1 = 3/8,
               stepl1 = 5/8,
               b1 = 500, b2 = 1000, b3 = 10000,
               fixed = TRUE
               )

data.frame(
  kappa = opt$Kappa * 8,
  n_2   = opt$n2,
  En_3go = opt$n3_go,
  pgo   = opt$pgo,
  posp  = opt$sProg,


  Eun   = opt$u
)

##   kappa n_2 En_3go  pgo posp Eun
## 1   1.6  86    221 0.92 0.75 235

# The link https://web.imbi.uni-heidelberg.de/drugdevelopR/ provides open
# source access to user friendly R Shiny applications and an R package, which
# ease optimal phase II/III drug development planning by the introduced
# framework. Various scenarios in drug development, e.g. scenarios with
# time-to-event, binary or normally distributed endpoint, where the true
# assumed treatment effect is a fixed value or modelled by a prior
# distribution, with multiple arms in phase II and III, with multiple trials
# in phase III, with bias adjustment of the treatment effect estimate, with
# constraints on sample size and/or the expected costs, with the option to
# skip phase II or to model different population structures in phase II and
# III, can be investigated.
```

A.16   R Code for Chapter 20

```
#  _____  CHAPTER 20  _____

###########################################################################
# Compute characteristics of blinded sample size recalculation           #
# for a normally distributed outcome in a superiority trial               #
###########################################################################

# development version: https://github.com/imbi-heidelberg/blindrecalc
# blindrecalc version 0.1.0 was used for the examples of this book
devtools::install_github("imbi-heidelberg/blindrecalc@v0.1.0", quiet = TRUE)
library(blindrecalc)

# define the setting
design <- setupStudent(alpha = .025, beta = .2, r = 1, delta = 3.5,
                       delta_NI = 0, alternative = "greater", n_max = 156)

# define the nuisance parameter sequence
sigma <- c(4, 5.5, 7)

# compute the type I error rate
toer(design = design, n1 = 20, nuisance = sigma, recalculation = TRUE, iters =
1e5, seed = 2020)

## [1] 0.02473 0.02458 0.02553

# compute the power
pow(design = design, n1 = 20, nuisance = sigma, recalculation = TRUE, iters =
1e5, seed = 2020)

## [1] 0.79981 0.78589 0.76866

# compute some characteristics of the sample size distribution
n_dist(design = design, n1 = 20, nuisance = sigma, summary = TRUE, plot = FALS
E,
       iters = 1e5, seed = 2020)

##     sigma = 4        sigma = 5.5       sigma = 7
##  Min.   : 20.00   Min.   : 20.0    Min.   : 23.0
##  1st Qu.: 39.00   1st Qu.: 67.0    1st Qu.:103.0
##  Median : 48.00   Median : 83.0    Median :130.0
##  Mean   : 49.81   Mean   : 86.1    Mean   :125.2
##  3rd Qu.: 59.00   3rd Qu.:103.0    3rd Qu.:156.0
##  Max.   :154.00   Max.   :156.0    Max.   :156.0
```

A.17   R Code for Chapter 21

```
# _____ CHAPTER 21 _____

##############################################################################
# Compute characteristics of blinded sample size recalculation               #
# for a binary outcome in a superiority trial                                #
##############################################################################

# development version: https://github.com/imbi-heidelberg/blindrecalc
devtools::install_github("imbi-heidelberg/blindrecalc@v0.1.0", quiet = TRUE)
library(blindrecalc)

# define the setting
design <- setupChiSquare(alpha = 0.025, beta = 0.2, r = 1, delta = -0.15,
  alternative = "smaller")

# compute the actual level of the fixed design and the recalculation design
# for different values of the nuisance parameter
nuisance <- seq(from = 0.1, to = 0.5, by = 0.1)
toer(design = design, n1 = 314, nuisance = nuisance, recalculation = FALSE)

## [1] 0.02528920 0.02533138 0.02478511 0.02462797 0.02403954

toer(design = design, n1 = 158, nuisance = nuisance, recalculation = TRUE)

## [1] 0.02563106 0.02497733 0.02524282 0.02536103 0.02354299

# compute the actual level of the recalculation design for different
# values of n1
n1 <- seq(from = 100, to = 300, by = 50)
toer(design = design, n1 = n1, nuisance = 0.345, recalculation = TRUE)

## [1] 0.02506655 0.02512955 0.02510210 0.02512273 0.02521349

# compute power of the fixed design and the recalculation design for
# different values of the nuisance parameter
pow(design = design, n1 = 314, nuisance = nuisance, recalculation = FALSE)

## [1] 0.9980262 0.9224842 0.8306453 0.7756085 0.7555553

pow(design = design, n1 = 158, nuisance = nuisance, recalculation = TRUE)

## [1] 0.9304756 0.7948725 0.7983968 0.8002579 0.7917329

# compute adjusted alpha for the fixed design and the recalculation design
adjusted_alpha(design = design, n1 = 314, nuisance = nuisance,
  precision = 0.001, nuis_ass = 0.345, recalculation = FALSE)

## [1] 0.024

adjusted_alpha(design = design, n1 = 158, nuisance = nuisance,
  precision = 0.001, recalculation = TRUE)

## [1] 0.024

# compute some characteristics of the sample size distribution
n_dist(design = design, n1 = 158, nuisance = nuisance, summary = TRUE,
  plot = FALSE)

##          p = 0.1   p = 0.2   p = 0.3   p = 0.4   p = 0.5
## Min.    158.0000 158.0000 202.0000 270.0000 318.0000
```

```
## 1st Qu. 158.0000 202.0000 278.0000 326.0000 344.0000
## Median  158.0000 224.0000 290.0000 334.0000 346.0000
## Mean    159.0582 220.5748 289.7078 331.5282 345.2945
## 3rd Qu. 158.0000 240.0000 304.0000 340.0000 348.0000
## Max.    214.0000 300.0000 342.0000 348.0000 348.0000
```

A.18   R Code for Chapter 22

```
# _____ CHAPTER 22 _____

##############################################################################
# Compute characteristics of blinded sample size recalculation              #
# for a normally distributed outcome in a non-inferiority trial             #
##############################################################################

# current development version: https://github.com/imbi-heidelberg/blindrecalc
# blindrecalc version 0.1.0 was used for the examples of this book
devtools::install_github("imbi-heidelberg/blindrecalc@v0.1.0", quiet = TRUE)
library(blindrecalc)

# define the setting
design <- setupStudent(alpha = .025, beta = .2, r = 1, delta = 0,
                       delta_NI = 1.5, n_max = 848)

# define the nuisance parameter sequence
sigma <- c(2, 5.5, 9)

# compute the type I error rate
toer(design = design, n1 = 20, nuisance = sigma, recalculation = TRUE, iters =
1e5, seed = 2020)

## [1] 0.02921 0.02579 0.02501

# compute the power
pow(design = design, n1 = 20, nuisance = sigma, recalculation = TRUE, iters =
1e5, seed = 2020)

## [1] 0.78431 0.77169 0.65978

# compute some characteristics of the sample size distribution
n_dist(design = design, n1 = 20, nuisance = sigma, summary = TRUE, plot = FALS
E,
       iters = 1e5, seed = 2020)

##     sigma = 2         sigma = 5.5       sigma = 9
## Min.    : 20.00   Min.    : 69.0   Min.    :185.0
## 1st Qu.: 43.00    1st Qu.:325.0    1st Qu.:848.0
## Median : 54.00    Median :408.0    Median :848.0
## Mean    : 56.38   Mean    :422.6   Mean    :812.1
## 3rd Qu.: 67.00    3rd Qu.:506.0    3rd Qu.:848.0
## Max.    :163.00   Max.    :848.0   Max.    :848.0

# compute the adjusted level of significance
adjusted_alpha(design = design, n1 = 20, nuisance = sigma, tol = 1e-4, iters =
1e5, seed = 2020)

## [1] 0.02091981
```

A.19   R Code for Chapter 23

```
# _____ CHAPTER 23 _____

################################################################################
# Compute characteristics of blinded sample size recalculation            #
# for a binary outcome in a non-inferiority trial                         #
################################################################################

# development version: https://github.com/imbi-heidelberg/blindrecalc
devtools::install_github("imbi-heidelberg/blindrecalc@v0.1.0", quiet = TRUE)
library(blindrecalc)

# define the setting
design <- setupFarringtonManning(alpha = 0.025, beta = 0.2, r = 1,
  delta = 0, delta_NI = 0.15)

# compute the actual level of the fixed design and the recalculation design
# for different values of the nuisance parameter
nuisance <- seq(from = 0.1, to = 0.5, by = 0.1)
toer(design = design, n1 = 244, nuisance = nuisance, recalculation = FALSE)

## [1] 0.02198404 0.02532780 0.02568273 0.02571317 0.02807810

toer(design = design, n1 = 100, nuisance = nuisance, recalculation = TRUE)

## [1] 0.02526287 0.02590192 0.02571951 0.02538209 0.02407422

# compute the actual level of the recalculation design for different
# values of n1
n1 <- seq(from = 50, to = 200, by = 50)
toer(design = design, n1 = n1, nuisance = 0.78, recalculation = TRUE)

## [1] 0.02608315 0.02587430 0.02578084 0.02541980

# compute power of the fixed design and the recalculation design for
# different values of the nuisance parameter
pow(design = design, n1 = 244, nuisance = nuisance, recalculation = FALSE)

## [1] 0.9628712 0.8322862 0.7314467 0.6769343 0.6729255

pow(design = design, n1 = 100, nuisance = nuisance, recalculation = TRUE)

## [1] 0.8326088 0.8046838 0.7996089 0.7969354 0.7910965

# compute adjusted alpha for the fixed design and the recalculation design
adjusted_alpha(design = design, n1 = 244, nuisance = nuisance,
  precision = 0.001, nuis_ass = 0.78, recalculation = FALSE)

## [1] 0.022

adjusted_alpha(design = design, n1 = 100, nuisance = nuisance,
  precision = 0.001, recalculation = TRUE)

## [1] 0.024

# compute some characteristics of the sample size distribution
n_dist(design = design, n1 = 100, nuisance = nuisance,
  summary = TRUE, plot = FALSE)
```

```
##          p = 0.1  p = 0.2  p = 0.3  p = 0.4  p = 0.5
## Min.    100.0000 122.0000 182.0000 250.0000 298.0000
## 1st Qu. 130.0000 206.0000 276.0000 322.0000 340.0000
## Median  148.0000 230.0000 292.0000 332.0000 344.0000
## Mean    147.8524 227.8357 289.8365 328.1864 341.2734
## 3rd Qu. 166.0000 250.0000 306.0000 338.0000 344.0000
## Max.    244.0000 314.0000 344.0000 344.0000 344.0000
```

A.20   R Code for Chapter 27

```
# _____ CHAPTER 27 _____

###########################################################################
# Compute a group sequential design                                       #
###########################################################################

# adoptr version 0.3.2 was used in this book and can be installed via
# more recent versions may be on CRAN
devtools::install_github("kkmann/adoptr@0.3.2", quiet = TRUE)
library(adoptr)

###########################################################################
# Function to compute a group sequential design for usage with adoptr     #
###########################################################################

###########################################################################
# Parameters to specify:
# dist   = data distribution
# delta  = standardized alternative effect size
# alpha  = significance level
# beta   = type II error rate
# cf     = futility stop boundary
# order  = order for numerical integration
###########################################################################

gs_design <- function(dist, delta, alpha, beta, cf, order = 7L) {

    pow  <- Power(dist, PointMassPrior(delta, 1)) # power
    toer <- Power(dist, PointMassPrior(0, 1)) # type I error rate

    # fixed sample size
    n1 <- 2*((qnorm(1 - alpha) + qnorm(1 - beta))^2) / delta^2

    # inverse normal combination test
    omega_1 <- sqrt(0.5)
    c2_in <- function(z) (qnorm(1 - alpha) - omega_1 * z) /
      sqrt(1 - omega_1^2)

    # early-efficacy boundary
    critical <- function(c) {
```

```
        d <- GroupSequentialDesign(n1, cf, c, n1, 2, order)
        d@c2_pivots <- sapply(adoptr:::scaled_integration_pivots(d), c2_in)
        return(alpha - evaluate(toer, d))
    }
    ce <- uniroot(critical, c(1, 3.5))$root
    d  <- GroupSequentialDesign(n1, cf, ce, n1, 2, order)
    d@c2_pivots <- sapply(adoptr:::scaled_integration_pivots(d), c2_in)

    # decrease sample size to fulfill power constraint
    while(evaluate(pow, d) > 1 - beta) {
        d@n1 <- d@n1 - 1
        d@n2_pivots <- d@n1
    }
    return(d)
}
#*************************************************************************#
# Exemplary call according to Example 29.1                              #
#*************************************************************************#

gsd <- gs_design(Normal(), 6.5/12, 0.025, 0.1, 0)

# check the error rates
evaluate(Power(Normal(), PointMassPrior(0, 1)), gsd)

## [1] 0.025

evaluate(Power(Normal(), PointMassPrior(6.5/12, 1)), gsd)

## [1] 0.8994674
```

A.21   R Code for Chapter 28

```
# _____ CHAPTER 28 _____

###########################################################################
# Compute an adaptive design based on conditional power                   #
###########################################################################

# adoptr version 0.3.2 was used in this book and can be installed via
# more recent versions may be on CRAN
devtools::install_github("kkmann/adoptr@0.3.2", quiet = TRUE)
library(adoptr)


###########################################################################
# Function to compute a conditional power based design with adoptr        #
###########################################################################

###########################################################################
# Parameters to specify:
# dist    = data distribution
# Delta   = alternative mean difference
# sigma   = standard deviation
# alpha   = significance level
# beta    = type II error rate
# beta_cp = 1 - beta_cp is the desired conditional power
# n1      = first-stage sample size
# cf      = futility stop boundary
# effect  = on which effect size is the conditional power constraint desired?
# epsilon = prior belief in negative effect size
# order   = order for numerical integration
###########################################################################

cp_design <- function(dist, Delta, sigma, alpha, beta, beta_cp, n1, cf,
                      effect = c("interim", "initial", "posterior_mean"),
                      epsilon = NULL, order = 99L) {

  pow  <- Power(dist, PointMassPrior(Delta / sigma, 1)) # power
  toer <- Power(dist, PointMassPrior(0, 1)) # type I error rate

  # inverse normal combination test
  c_alpha <- qnorm(1 - alpha)
  omega_1 <- sqrt(0.5)
  c2_in <- function(z) (c_alpha - omega_1 * z) / sqrt(1 - omega_1^2)

  # early-efficacy boundary
  critical <- function(c) {
    d <- TwoStageDesign(1, cf, c, 1, 2, order)
    d@c2_pivots <- sapply(adoptr:::scaled_integration_pivots(d), c2_in)
    return(alpha - evaluate(toer, d))
  }
  ce <- uniroot(critical, c(1, 3.5))$root
  d  <- TwoStageDesign(1, cf, ce, 1, 2, order)
  d@c2_pivots <- sapply(adoptr:::scaled_integration_pivots(d), c2_in)


    # stage-two sample size
    n2_cp <- function(z, n1) {
```

```
        if (effect == "interim") del <- sigma * z / sqrt(n1 / 2)
        else if (effect == "initial") del <- Delta
        else if (effect == "posterior_mean") {
            tau <- Delta / qnorm(epsilon)
            del <- (Delta / tau^2 + (sigma * z / sqrt(n1 / 2)) /
                        (2 * sigma^2 / n1)) /
                (1 / tau^2 + 1 / (2 * sigma^2 / n1))
        }
        return(2 * (c2_in(z) + qnorm(1 - beta_cp))^2 / (del / sigma)^2)
    }

    if (is.null(n1)) {
        # increase n1 to fulfill the overall power constraint
        while(evaluate(pow, d) < 1 - beta) d@n1 <- d@n1 + 1
    } else {
        d@n1 <- n1 / 2 # adoptr uses per-group sample sizes
    }

    d@n2_pivots <- sapply(adoptr:::scaled_integration_pivots(d),
                          function(z) n2_cp(z, d@n1))

    return(d)
}

#***************************************************************************#
# Exemplary call according to Example 28.1                                 #
#***************************************************************************#

design1 <- cp_design(Normal(), 6.5, 12, 0.025, 0.1, 0.1, 72, 0.5, "interim")
design2 <- cp_design(Normal(), 6.5, 12, 0.025, 0.1, 0.1, 72, 0.5, "initial")
design3 <- cp_design(Normal(), 6.5, 12, 0.025, 0.1, 0.1, 72, 0.5,
                     "posterior_mean", 0.1)
design4 <- cp_design(Normal(), 6.5, 12, 0.025, 0.1, 0.1, 72, 0.5,
                     "posterior_mean", 0.2)


# check the sample sizes
# note that adoptr uses group-wise sample sizes and one has to multiply by 2
# for the computation of further characteristics see chapter 29
2 * n2(design1, c(1, 2))

## [1] 672  76

2 * n2(design2, c(1, 2))

## [1] 128  58

2 * n2(design3, c(1, 2))

## [1] 392  70

2 * n2(design4, c(1, 2))

## [1] 504  74
```

A.22   R Code for Chapter 29

```
# _____ CHAPTER 29 _____

###########################################################################
# Compute an optimal adaptive two-stage design                            #
###########################################################################

# adoptr version 0.3.2 was used in this book and can be installed via
devtools::install_github("kkmann/adoptr@0.3.2", quiet = TRUE)
library(adoptr)

# choose the data distribution
datadist <- Normal()

# define the hypotheses
H_0 <- PointMassPrior(theta = 0, mass = 1)
H_1 <- PointMassPrior(theta = 6.5/12, mass = 1)

# define scores for the optimization
ess  <- ExpectedSampleSize(datadist, H_1)
toer <- Power(datadist, H_0)
pow  <- Power(datadist, H_1)

# define a 'starting design' for optimization
init <- get_initial_design(theta = 6.5/12, alpha = 0.025, beta = 0.1,
                           type = "two-stage", dist = datadist, order = 7L)

# optimize
opt_design <- minimize(ess, subject_to(toer <= 0.025, pow >= 0.9),
                       initial_design = init)$design

# evaluate the expected sample size per group of this design
evaluate(ess, opt_design)

## [1] 54.11791
```

References

1. Aalen, O. O. (1989). Linear regression model for the analysis of life times. *Statistics in Medicine, 8,* 907–925.
2. Aaronson, N. K., Ahmedzai, S., Bergman, B., et al. (1993). The European Organization for Research and Treatment of Cancer QLQ-C30: A quality-of-life instrument for use in international clinical trials in oncology. *Journal of the National Cancer Institute, 85,* 365–376.
3. Abdulatif, M., Mukhtar, A., & Obayah, G. (2015). Pitfalls in reporting sample size calculation in randomized controlled trials published in leading anaesthesia journals: A systematic review. *British Journal of Anaesthesia, 115,* 699–707.
4. Abel, U., Jensen, K., Karapanagiotou-Schenkel, I., & Kieser, M. (2015). Some issues of sample size calculation for time-to-event endpoints using the Freedman and Schoenfeld formulas. *Journal of Biopharmaceutical Statistics, 25,* 1285–1311.
5. Acion, L., Peterson, J. J., Temple, S., & Arndt, S. (2006). Probabilistic index: An intuitive non-parametric approach to measuring the size of treatment effects. *Statistics in Medicine, 25,* 591–602.
6. Adams, G., Gulliford, M. C., Ukoumunne, O. C., Eldridge, S., Chinn, S., & Campbell, M. J. (2004). Patterns of intra-cluster correlation from primary care research to inform study design and analysis. *Journal of Clinical Epidemiology, 57,* 785–794.
7. Agresti, A. (2002). *Categorical data analysis* (2nd ed.). Hoboken: Wiley.
8. Ahnn, S., & Anderson, S. J. (1995). Sample size determination for comparing more than two survival distributions. *Statistics in Medicine, 14,* 2273–2282.
9. Andersen, P. K., & Gill, R. D. (1982). Cox's regression model for counting processes: A large sample study. *Annals of Statistics, 10,* 1100–1120.
10. Anderson, K. M. (2014). Timing and frequency of interim analyses in confirmatory trials. In W. He, J. Pinheiro, & O. M. Kuznetsova (Eds.), *Practical considerations for adaptive trial design and implementation* (pp. 115–123). New York: Springer.
11. Antonijevic, Z., Gallo, P., Chuang-Stein, C., Dragalin, V., Loewy, J., Menon, S., et al. (2013). Views on emerging issues pertaining to data monitoring committees for adaptive trials. *Therapeutic Innovation & Regulatory Science, 47,* 495–502.
12. Armitage, P. (1955). Test for linear trends in proportions and frequencies. *Biometrics, 11,* 375–386.
13. Arnold, B. F., Hogan, D. R., Colford, J. M., & Hubbard, A. E. (2011). Simulation methods to estimate design power: An overview for applied research. *BMC Medical Research Methodology, 11,* 1–10.
14. Arrowsmith, J., & Miller, P. (2013). Trial watch: Phase II and phase III attrition rates 2011–2012. *Nature Reviews Drug Discovery, 12,* 569.

15. Bang, H., Jung, S.-H., & George, S. L. (2005). Sample size calculation for simulation-based multiple-testing procedures. *Journal of Biopharmaceutical Statistics, 15,* 957–967.

16. Bariani, G. M., Ferrari, A. C. R. C., Hoff, P. M., Arai, R., Precivale, M., & Riechelmann, R. P. (2012). The quality of sample size calculation (ssc) reporting in cancer clinical trials. *Annals of Oncology, 23* Supplement 9, ix450.

17. Barnard, G. A. (1947). Significance test for 2x2 tables. *Biometrika, 34,* 123–138.

18. Bath, P. M. W., Lees, K. R., Schellinger, P. D., Altman, H., Bland, M., Hogg, C., et al. (2012). Statistical analysis of the primary outcome in acute stroke trials. *Stroke, 43,* 1171–1178.

19. Bauer, P. (1989). Multistage testing with adaptive designs. *Biometrie und Informatik in Medizin und Biologie, 20,* 130–148.

20. Bauer, P., Bretz, F., Dragalin, V., König, F., & Wassmer, G. (2016). 25 years of confirmatory adaptive designs: Opportunities and pitfalls. *Statistics in Medicine, 35,* 325–347.

21. Bauer, P., & Kieser, M. (1999). Combining different phases in the development of medical treatments within a single trial. *Statistics in Medicine, 18,* 1833–1848.

22. Bauer, P., & Köhne, K. (1994). Evaluation of experiments with adaptive interim analyses. *Biometrics, 50,* 1029–1041 (correction in 1996 *Biometrics, 52,* 380).

23. Bauer, P., & König, F. (2006). The reassessment of trial perspectives from interim data—A critical view. *Statistics in Medicine, 25,* 23–36.

24. Bauer, P., & Röhmel, J. (1995). An adaptive method for establishing a dose-response relationship. *Statistics in Medicine, 14,* 1595–1607.

25. Berger, R. L. (1982). Multiparameter hypothesis testing and acceptance sampling. *Technometrics, 24,* 295–300.

26. Berger, V. W. (2000). Pros and cons of permutation tests in clinical trials. *Statistics in Medicine, 19,* 1319–1328.

27. Birkett, M. A., & Day, S. J. (1994). Internal pilot studies for estimating sample size. *Statistics in Medicine, 19,* 1319–1328.

28. Bishop, Y. M. M., Fienberg, S. E., & Holland, P. W. (1975). *Discrete Multivariate Analysis.* Cambridge, MA: MIT Press.

29. Boschloo, R. D. (1970). Raised conditional level of significance for the 2×2-table when testing the equality of two probabilities. *Statistica Neerlandica, 24,* 1–35.

30. Brannath, W., Posch, M., & Bauer, P. (2002). Recursive combination tests. *Journal of the American Statistical Association, 97,* 236–244.

31. Brenner, C., Adrion, C., Grabmaier, U., Theisen, D., von Ziegler, F., Leber, A., et al. (2016). Sitagliptin plus granulocyte colony-stimulating factor in patients suffering from acute myocardial infarction: A double-blind, randomized placebo-controlled trial of efficacy and safety (SITAGRAMI trial). *International Journal of Cardiology, 205,* 26–30.

32. Bretz, F., Maurer, W., & Hommel, G. (2011). Test and power considerations for multiple endpoint analyses using sequentially rejective graphical procedures. *Statistics in Medicine, 30,* 1489–1501.

33. Brittain, E., & Schlesselman, J. J. (1982). Optimal allocation for the comparison of proportions. *Biometrics, 38,* 1003–1009.

34. Brown, B. W., Herson, J., Atkinson, E. N., & Rozell, M. E. (1987). Projection from previous studies: a Bayesian and frequentist compromise. *Controlled Clinical Trials, 8,* 29–44.

35. Brunner, E., & Munzel, U. (2000). The nonparametric Behrens-Fisher problem: Asymptotic theory and a small-sample approximation. *Biometrical Journal, 42,* 17–25.

36. Brunoni, A. R., Sampaio-Junior, B., Moffa, A. H., Borrione, L., Nogueira, B. S., Aparicio, L. V. M., et al. (2015). The escitalopram versus electric current therapy for treating depression clinical study (ELECT-TDCS): rationale and study design of a non-inferiority, triple-arm, placebo-controlled clinical trial. *Sao Paulo Medical Journal, 133,* 252–263.

37. Brunoni, A. R., Moffa, A. H., Sampaio-Junior, B., Borrione, L., Moreno, M. L., Fernandes, R. A., et al. (2017). Trial of electrical direct-current therapy versus Escitalopram for depression. *New England Journal of Medicine, 376,* 2523–2533.

38. Brutti, P., De Santis, F., & Gubbiotti, S. (2014). Bayesian-frequentist sample size determination: A game of two priors. *Metron, 72,* 133–151.

39. Burton, A., Altman, D. G., Royston, P., & Holder, R. L. (2006). The design of simulation studies in medical statistics. *Statistics in Medicine, 25*, 4279–4292.
40. Campbell, M. J., & Walters, S. J. (2014). *How to design, analyse and report cluster Randomised trials in medicine and health related research.* Chichester: Wiley.
41. Chakraborti, S., Hong, B., & van de Wiel, M. A. (2006). A note on sample size determination for a nonparametric test of location. *Technometrics, 48*, 88–94.
42. Chan, A.-W., Tetzlaff, J. M., Altman, D. G., Laupacis, A., Gøtzsche, P. C., Krleza-Jeric, K., et al. (2013). SPIRIT 2013 statement: defining standard protocol items for clinical trials. *Annals of Internal Medicine, 158*, 200–207.
43. Chan, I. S. F. (1998). Exact tests of equivalence and efficacy with a non-zero lower bound for comparative studies. *Statistics in Medicine, 17*, 1403–1413.
44. Chan, I. S. F. (2002). Power and sample size determination for non inferiority trials using an exact method. *Journal of Biopharmaceutical Statistics, 12*, 457–469.
45. Chan, I. S. F. (2003). Proving non-inferiority or equivalence of two treatments with dichotomous endpoints using exact methods. *Statistical Methods in Medical Research, 12*, 37–58.
46. Charles, P., Giraudeau, B., Dechartres, A., Baron, G., & Ravaud, P. (2009). Reporting of sample size calculation in randomised controlled trials: Review. *British Medical Journal, 338*, 1–6.
47. Chen, H., Zhang, N., Lu, X., & Chen, S. (2013). Caution regarding the choice of standard deviations to guide sample size calculations in clinical trials. *Clinical Trials, 10*, 522–529.
48. Chen, X., Lu, N., Nair, R., Xu, Y., Kang, C., Huang, Q., et al. (2012). Decision rules and associated sample size planning for regional approval utilizing multiregional clinical trials. *Journal of Biopharmaceutical Statistics, 22*, 1001–1018.
49. Chen, Y. H., DeMets, D. L., & Lan, K. K. G. (2004). Increasing the sample size when the unblinded interim result is promising. *Statistics in Medicine, 23*, 1023–1038.
50. Chow, S. C., Corey, R., & Lin, M. (2012). On the independence of data monitoring committee in adaptive design clinical trials. *Journal of Biopharmaceutical Statistics, 22*, 853–867.
51. Chow, S. C., Shao, J., & Wang, H. (2003). *Sample size calculation in clinical research.* New York: Chapman & Hall/CRC.
52. Chowdhury, S., Tiwari, R., & Ghosh, S. (2019). Non-inferiority testing for risk ratio, odds ratio and number needed to treat in three-arm trial. *Computational Statistics and Data Analysis, 132*, 70–83.
53. Ciarleglio, M. M., Arendt, C. D., Makuch, R. W., & Pedduzi, P. N. (2015). Selection of the treatment effect for sample size determination in a superiority clinical trial using a hybrid classical and Bayesian procedure. *Contemporary Clinical Trials, 41*, 160–171.
54. Ciarleglio, M. M., Arendt, C. D., & Peduzzi, P. N. (2016). Selection of the effect size for sample size determination for a continuous response in a superiority clinical trial using a hybrid classical and Bayesian procedure. *Clinical Trials, 13*, 275–285.
55. Ciarleglio, M. M., & Arendt, C. D. (2017). Sample size determination for a binary response in a superiority clinical trial using a hybrid classical and Bayesian procedure. *Trials, 18*, 83.
56. Clark, T., Berger, U., & Mansmann, U. (2013). Sample size determinations in original research protocols for randomised clinical trials submitted to UK research ethics committees: review. *British Medical Journal, 346*, f1135.
57. Clopper, C., & Pearson, E. S. (1934). The use of confidence or fiducial limits illustrated in the case of the binomial. *Biometrika, 26*, 404–413.
58. Cochran, W. G. (1954). Some methods for strengthening the common χ^2 test. *Biometrics, 10*, 417–451.
59. Coffey, C. S., & Muller, K. E. (1999). Exact test size and power of a Gaussian error linear model for an internal pilot study. *Statistics in Medicine, 18*, 1199–1214.
60. Collett, D. (2003). *Modelling survival data in medical research* (2nd ed.). Boca Raton: Chapman & Hall/CRC.
61. Cook, J. A., Bruckner, T., MacLennan, G. S., & Seiler, C. M. (2012). Clustering in surgical trials—database of intracluster correlations. *Trials, 13*, 2.

62. Cook, J. A., Hislop, J., Adewuyi, T. E., Harrild, K., Altman, D. G., Ramsay, C. R., Fraser, C., Buckley, B., Fayers, P., Harvey, I., Briggs, A. H., Norrie, J. D., Fergusson, D., Ford, I., & Vale, L. D. (2014). Assessing methods to specify the target difference for a randomised controlled trial: DELTA (Difference ELicitation in TriAls) review. *Health Technology Assessment, 18*, v–vi, 1–175.

63. Cook, J. A., Hislop, J., Altman, D. G., Fayers, P., Briggs, A. H., Ramsay, C. R., et al. (2015). Specifying the target difference in the primary outcome for a randomised controlled trial: guidance for researchers. *Trials, 16*, 12.

64. Cook, J. A., Julious, S. A., Sones, W., Hampson, L. V., Hewitt, C., Berlin, J. A., et al. (2018). DELTA^2 guidance on choosing the target difference and undertaking and reporting the sample size calculation for a randomised controlled trial. *British Medical Journal, 363*, k3750.

65. Cook, T., & Zea, R. (2020). Missing data and sensitivity analysis for binary data with implications for sample size and power of randomized clinical trials. *Statistics in Medicine, 39*, 192–204.

66. Copsey, B., Thompson, J. Y., Vadher, K., Ali, U., Dutton, S. J., Fitzpatrick, R., et al. (2018). Sample size calculations are poorly conducted and reported in many randomized trials of hip and knee osteoarthritis: Results of a systematic review. *Journal of Clinical Epidemiology, 104*, 52–61.

67. Cox, D. R. (1972). Regression models and life tables (with discussion). *Journal of the Royal Statistical Society, B, 74*, 187–220.

68. CPMP. (2000). Points to consider on switching between superiority and non-inferiority. European Agency for the Evaluation of Medicinal Products. http://www.emea.europa.eu/docs/en_GB/document_library/Scientific_guideline/2009/09/WC500003658.pdf. Accessed October 13, 2020.

69. Cui, L., Hung, H. M. J., & Wang, S.-J. (1999). Modification of sample size in group sequential clinical trials. *Biometrics, 55*, 853–857.

70. D'Agostino, R. B., Massaro, J. M., & Sullivan, L. M. (2003). Non-inferiority trials: design concepts and issues—The encounters of academic consultants in statistics. *Statistics in Medicine, 22*, 169–186.

71. Dallow, N., Best, N., & Montague, T. H. (2018). Better decision making in drug development through adoption of formal prior elicitation. *Pharmaceutical Statistics, 17*, 301–316.

72. Demchuk, A. M., Goyal, M., Menon, B. K., Eesa, M., Ryckborst, K. J., Kamal, N., et al. (2015). Endovascular treatment for small core and anterior circulation proximal occlusion with Emphasis on minimizing CT to recanalization times (ESCAPE) trial: methodology. *Stroke, 10*, 429–438.

73. Demidenko, E. (2007). Sample size determination for logistic regression revisited. *Statistics in Medicine, 26*, 3385–3397.

74. Denne, J. S. (2001). Sample size recalculation using conditional power. *Statistics in Medicine, 20*, 2645–2660.

75. Denne, J. S., & Jennison, C. (1999). Estimating the sample size for a *t*-test using an internal pilot. *Statistics in Medicine, 18*, 1575–1585.

76. Diener, M. K., Bruckner, T., Contin, P., Halloran, C., Glanemann, M., Schlitt, H. J., Mössner, J., Kieser, M., Werner, J., Büchler, M. W., & Seiler, C. M. (2010). ChroPac-trial: Duodenum-preserving pancreatic head resection versus pancreatoduodenectomy for chronic pancreatitis. Trial protocol of a randomised controlled multicentre trial. *Trials, 11*, 47.

77. Diener, M. K., Hüttner, F. J., Kieser, M., Knebel, P., Dörr-Harim, C., Distler, M., et al. (2017). Partial pancreatoduodenectomy versus duodenum-preserving pancreatic head resection in chronic pancreatitis: The multicentre, randomised, controlled, double-blind ChroPac trial. *Lancet, 390*, 1027–1037.

78. Diener, M. K., Hüttner, F. J., Kieser, M., Knebel, P., Dörr-Harim, C., Distler, M., Grützmann, R., Wittel, U. A., Schirren, R., Hau, H.-M., Kleespies, A., Heidecke, C.-D., Tomazic, A., Halloran, C. M., Wilhelm, T. J., Bahra, M., Beckurts, T., Börner, T., Glanemann, M., Steger, U., Treitschke, F., Staib, L., Thelen, K., Bruckner, T., Mihaljevic, A.

L., Werner, J., Ulrich, A., Hackert, T., & Büchler, M. W. (2017b). Partial pancreato-duodenectomy versus duodenum-preserving pancreatic head resection in chronic pancreatitis: the multicentre, randomised, controlled, double-blind ChroPac trial—Supplementary appendix. *Lancet* https://www.thelancet.com/cms/10.1016/S0140-6736(17)31960-8/attach ment/1c195581-8f67-4107-9a3b-fa1a42374d7f/mmc1.pdf. Accessed October 13, 2020.

79. Dmitrienko, A., Tamhane, A. C., & Bretz, F. (2010). *Multiple testing problems in pharmaceutical statistics.* Boca Raton: Chapman & Hall/CRC.

80. Dmitrienko, A., D'Agostino, R., Sr., & Huque, M. (2013). Key multiplicity issues in clinical drug development. *Statistics in Medicine, 32,* 1079–1111.

81. Donner, A., Birkett, N., & Buck, C. (1981). Randomization by cluster- sample size requirements and analysis. *American Journal of Epidemiology, 114,* 906–914.

82. Donner, A., & Klar, N. (2000). *Design and analysis of cluster randomization trials in health research.* London: Arnold.

83. Dumville, J. C., Hahn, S., Miles, J. N. V., & Torgerson, D. J. (2006). The use of unequal randomisation ratios in clinical trials: A review. *Contemporary Clinical Trials, 27,* 1–12.

84. Dunnett, C. W. (1955). A multiple comparison procedure for comparing several treatments with a control. *Journal of the American Statistical Association, 50,* 1096–1121.

85. Dupont, C., Campagne, A., & Constant, F. (2014). Efficacy and safety of a magnesium sulfate-rich natural mineral water for patients with functional constipation. *Clinical Gastroenterology and Hepatology, 12,* 1280–1287.

86. Durrett, R. (2010). *Probability: Theory and examples* (4th ed.). New York: Cambridge University Press.

87. Duru, G., & Fantino, B. (2008). The clinical relevance of changes in the Montgomery-Asberg depression rating scale using the minimum clinically important difference approach. *Current Medical Research and Opinion, 24,* 1329–1335.

88. Eldridge, S. M., Ashby, D., & Kerry, S. (2006). Sample size for cluster randomized trials: Effect of coefficient of variation of cluster size and analysis method. *International Journal of Epidemiology, 35,* 1292–1300.

89. Eldridge, S., & Kerry, S. (2012). *A practical guide to cluster randomised trials in health services research.* Chichester: Wiley.

90. Ellenberg, S. S., Fleming, T. R., & DeMets, D. L. (2019). *Data monitoring committees in clinical trials* (2nd ed.). Hoboken: Wiley.

91. Emerson, S. S., Levin, G. P., & Emerson, S. C. (2011). Comments on 'Adaptive increase in sample size when interim results are promising: A practical guide with examples'. *Statistics in Medicine, 30,* 3285–3301.

92. Engels, Y. (2005). Testing a European set of indicators for the evaluation of the management of primary care practices. *Family Practice, 23,* 137–147.

93. European Medicines Agency (EMA)/Committee for Medicinal Products for Human Use (CHMP). (2005). *Guideline on data monitoring committees.* https://www.ema.europa.eu/ en/documents/scientific-guideline/guideline-data-monitoring-committees_en.pdf. Accessed October 13, 2020.

94. European Medicines Agency (EMA)/Committee for Medicinal Products for Human Use (CHMP). (2007). *Reflection paper on methodological issues in confirmatory clinical trials planned with an adaptive design.* http://www.ema.europa.eu/docs/en_GB/document_library/ Scientific_guideline/2009/09/WC500003616.pdf. Accessed: October 13, 2020.

95. European Medicines Agency (EMA)/Committee for Medicinal Products for Human Use (CHMP). (2008). *Guideline on medicinal products for the treatment of Alzheimer's disease and other dementias.* http://www.ema.europa.eu/docs/en_GB/document_library/Scientific_g uideline/2009/09/WC500003562.pdf. Accessed: October 13, 2020.

96. European Medicines Agency (EMA)/Committee for Medicinal Products for Human Use (CHMP). (2015). *Guideline for adjustment of baseline covariates in clinical trials.* http://www.ema.europa.eu/docs/en_GB/document_library/Scientific_guideline/2015/ 03/WC500184923.pdf. Accessed: October 13, 2020.

97. European Medicines Agency (EMA)/Committee for Medicinal Products for Human Use (CHMP). (2016). *Guideline on multiplicity issues in clinical trials. Draft.* http://www.ema.europa.eu/docs/en_GB/document_library/Scientific_guideline/2017/03/WC500224998.pdf. Accessed: October 13, 2020.

98. Eypasch, E., Lefering, R., Kum, C. K., & Troidl, H. (1995). Probability of adverse events that have not yet occurred: a statistical reminder. *British Medical Journal, 311,* 619–620.

99. Fagerland, M. F., Lydersen, S., & Laake, P. (2017). *Statistical analysis of contingency tables.* Boca Raton: CRC Press.

100. Fahn, S., & Elton, R. (1987). Members of the updrs development committee. In: S. Fahn, C. D. Marsden, D. B. Calne, & M. Goldstein (Hrsg.), *Recent developments in Parkinson's Disease, Vol 2.* (S. 153–163 und S. 293–304). Florham Park, NJ: Macmillan Health Care Information.

101. Fan, C., Zhang, D., & Zhang, C.-H. (2011). On sample size of the Kruskal-Wallis Test with application to a mouse peritoneal cavity study. *Biometrics, 67,* 213–224.

102. Farrington, C. P., & Manning, G. (1990). Test statistics and sample size formulae for comparative binomial trials with null hypothesis of non-zero risk difference or non-unity relative risk. *Statistics in Medicine, 9,* 1447–1454.

103. Fedgchin, M., Trivedi, M., Daly, E. J., Melkote, R., Lane, R., Lim, P., et al. (2019). Efficacy and safety of fixed-dose esketamine nasal spray combined with a new oral antidepressant in treatment-resistant depression: Results of a randomized, double-blind, active-controlled study (TRANSFORM-1). *International Journal of Neuropsychopharmacology, 22,* 616–630.

104. Feldman, T. E., Reardon, M. J., Rajagopal, V., Makar, R. R., Bajwa, T. K., Kleiman, N. S., et al. (2018). Effect of mechanically expanded versus Self-expanding transcatheter aortic valve replacement on mortality and major adverse clinical events in high-risk patients with aortic Stenosis The REPRISE III randomized clinical trial. *JAMA, 319,* 27–37.

105. Fisher, L. D., Dixon, D. O., Herson, J., Frankowski, R. K., Hearon, M. S., & Pearce, K. E. (1990). Intention to treat in clinical trials. In K. E. Pearce (Ed.), *Statistical issues in drug research and development* (pp. 331–350). New York: Marcel Dekker.

106. Fleming, T. R., & Harrington, D. P. (1981). A class of hypothesis tests for one and two sample censored survival data. *Communication in Statistics—Theory and Methods, 10,* 763–794.

107. Food and Drug Administration (FDA). (2006). *Guidance for clinical trial sponsors: Establishment and operation of clinical trial data monitoring committees.* https://www.fda.gov/media/75398/download. Accessed: October 13, 2020.

108. Food and Drug Administration (FDA). (2013). *Draft guidance for industry: Alzheimer's Disease: Developing drugs for the treatment of early stage disease.* https://isctm.org/public_access/FDAGuidance_AD_Developing_Drugs_Early_Stage_Treatment.pdf. Accessed: October 13, 2020.

109. Food and Drug Administration (FDA). (2017). *Draft guidance for industry: Multiple endpoints in clinical trials.* https://www.fda.gov/downloads/drugs/guidancecomplianceregulatoryinformation/guidances/ucm536750.pdf. Accessed: October 13, 2020.

110. Food and Drug Administration (FDA). (2019). *Guidance for industry: Adaptive design for clinical trials of drugs and biologics.* https://www.fda.gov/media/78495/download. Accessed: October 13, 2020.

111. Food and Drug Administration (FDA). (2019). *Draft guidance for industry: Adjusting for covariates in randomized clinical trials for drugs and biologics with continuous outcomes.* https://www.fda.gov/media/123801/download. Accessed: October 13, 2020.

112. Fossler, M. J., Collins, D. A., Ino, H., Sarai, N., Ravindranath, R., Bowen, C. L., et al. (2015). Evaluation of bioequivalence of five 0.1 mg dutasteride capsules compared to one 0.5 mg dutasteride capsule: A randomized study in healthy male volunteers. *Journal of Drug Assessment, 4,* 24–29.

113. Freedman, L. S. (1982). Tables of the number of patients required in clinical trials using the logrank test. *Statistics in Medicine, 1,* 121–129.

114. Friede, T. (2000). *Methods for the determination of the sample size in clinical studies with internal pilot study* (in German). Dissertation, University of Heidelberg.

115. Friede, T., & Kieser, M. (2000). Re-calculating the sample size in internal pilot study designs with control of the type I error rate. *Statistics in Medicine, 19,* 901–911.

116. Friede, T., & Kieser, M. (2001). A comparison of methods for adaptive sample size adjustment. *Statistics in Medicine, 20,* 3861–3873.

117. Friede, T., & Kieser, M. (2002). On the inappropriateness of an EM algorithm based procedure for blinded sample size re-estimation. *Statistics in Medicine, 21,* 165–176.

118. Friede, T., & Kieser, M. (2003). Blinded sample size reassessment in non-inferiority and equivalence trials. *Statistics in Medicine, 22,* 995–1007.

119. Friede, T., & Kieser, M. (2004). Sample size recalculation for binary data in internal pilot study designs. *Pharmaceutical Statistics, 3,* 269–279.

120. Friede, T., & Kieser, M. (2006). Sample size recalculation in internal pilot study designs. *Biometrical Journal, 48,* 537–555.

121. Friede, T., & Kieser, M. (2011). Blinded sample size recalculation for clinical trials with normal data and baseline adjusted analysis. *Pharmaceutical Statistics, 10,* 8–13.

122. Friede, T., & Kieser, M. (2011b). Sample size reassessment in non-inferiority trials: internal pilot study designs with ANCOVA. *Methods of Information in Medicine, 50,* 237–243.

123. Friede, T., & Kieser, M. (2013). Blinded sample size re-estimation in superiority and noninferiority trials: bias versus variance in variance estimation. *Pharmaceutical Statistics, 12,* 141–146.

124. Friede, T., Mitchell, C., & Müller-Velten, G. (2007). Blinded sample size reestimation in non-inferiority trials with binary endpoints. *Biometrical Journal, 49,* 903–916.

125. Friede, T., & Schmidli, H. (2010). Blinded sample size reestimation with count data: methods and applications in multiple sclerosis. *Statistics in Medicine, 29,* 1145–1156.

126. Friede, T., Pohlmann, H., & Schmidli, H. (2019). Blinded sample size reestimation in event-driven clinical trials: Methods and an application in multiple sclerosis. *Pharmaceutical Statistics, 18,* 351–365.

127. Frison, L., & Pocock, S. J. (1999). Repeated measures in clinical trials: analysis using mean summary statistics and its implications for design. *Statistics in Medicine, 11,* 1685–1704.

128. Fürnkranz, A., Brugada, J., Albenque, J.-P., Tondo, C., Bestehorn, K., Wegscheider, K., et al. (2014). Rationale and design of FIRE AND ICE: a multicenter randomized trial comparing efficacy and safety of pulmonary vein isolation using a cryoballoon versus radiofrequencs ablation with 3D-reconstruction. *Journal of Cardiovascular Electrophysiology, 25,* 1314–1320.

129. Gaffney, M., & Ware, J. H. (2017). An evaluation of increasing sample size based on conditional power. *Journal of Biopharmaceutical Statistics, 27,* 797–808.

130. Galasko, D., Bennett, D., Sano, M., Ernesto, C., Thomas, R., Grundmann, M., et al. (1997). An inventory to assess activities of daily living for clinical trials in Alzheimer's disease. The Alzheimer's disease cooperative study. *Alzheimer Disease & Associated Disorders, 11,* 33–39.

131. Gallo, P. (2006). Operational challenges in adaptive design implementation. *Pharmaceutical Statistics, 5,* 119–124.

132. Gallo, P., DeMets, D., & LaVange, L. (2014). Considerations for interim analyses in adaptive trials, and perspectives on the use of DMCs. In W. He, J. Pinheiro, & O. M. Kuznetsova (Eds.), *Practical considerations for adaptive trial design and implementation* (pp. 259–272). New York: Springer.

133. Gan, H. K., You, B., Pond, G. R., & Chen, E. X. (2012). Assumptions of expected benefits in randomized phase III trials evaluating systematic treatments for cancer. *Journal of the National Cancer Institute, 104,* 590–598.

134. Gangnon, R., & Kosorok, M. (2004). Sample-size formula for clustered survival data using weighted log-rank statistics. *Biometrika, 91,* 263–275.

135. Gao, P., Ware, J. H., & Mehta, C. R. (2008). Sample size re-estimation for adaptive sequential design in clinical trials. *Journal of Biopharmaceutical Statistics, 18,* 1184–1196.

136. George, S. L., & Desu, M. M. (1974). Planning the size and duration of a clinical trial studying the time to some critical event. *Journal of Chronic Diseases, 27,* 15–24.

137. Gibbons, J. D., & Chakraborti, S. (2003). *Nonparametric statistical inference* (4th ed.). New York: Marcel Dekker.
138. Glimm, E. (2012). Comments on Adaptive increase in sample size when interim results are promising: A practical guide with examples by CR Mehta and SJ Pocock. *Statistics in Medicine, 31,* 98–99.
139. Glimm, E., & Läuter, J. (2013). Some notes on blinded sample size re-estimation. arXiv: 1301.4167v1 [stat.ME].
140. Glimm, E., Yau, L., & Woehling, H. (2019). Blinded sample size re-estimation in equivalence testing. arXiv: 1908.04695v1 [stat.AP].
141. Glöckler, T. (2015). Mehr Verordnungen, sinkende Umsätze. *Pharmazeutische Zeitung* 06.05.2015. https://www.pharmazeutische-zeitung.de/ausgabe-192015/mehr-verordnungen-sinkende-umsaetze/. Accessed: October 13, 2020.
142. Golkowski, D., Friede, T., & Kieser, M. (2014). Blinded sample size re-estimation in crossover bioequivalence trials. *Pharmaceutical Statistics, 13,* 157–162.
143. Gómez-Barrena, E., Padilla-Eguiluz, N. G., Avendano-Solá, C., Payares-Herrera, C., Velasco-Iglesias, A., Torres, F., Rosset, P., Gebhard, F., Baldini, N., Rubio-Suarez, J. C., Garciá-Rey, E., Cordero-Ampuero, J., Vaquero-Martin, J., Chana, F., Marco, F., García-Coiradas, J., Caba-Dessoux, P., de la Duadra, P., Hernigou, P., Flouzat-Lachaniette, C.-H., Gouin, F., Mainard, D., Laffosse, J. M., Kalbitz, M., Marzi, I., Südkamp, N., Stöckle, Ul., Ciapetti, G., Donati, D. M., Zagra, L., Pazzaglia, U., Zarattini, G., Capanna, R., & Catani, F. (2018). A multicentric, open-label, randomized, comparative clinical trial of two different doses of expanded hBM-MSCs plus biomaterial versus iliac crest autograft, for bone healing in nonunions after long bone fractures: Study protocol. *Stem Cells International* 2018, Article ID 6025918, 13 pages.
144. Good, P. (2013). *Permutation tests: A Practical guide to resampling methods for testing hypotheses* (2nd ed.). New York: Springer.
145. Gould, A. L. (1992). Interim analyses for monitoring clinical trials that do not materially affect the type I error rate. *Statistics in Medicine, 11,* 55–66.
146. Gould, A. L., & Shih, W. J. (1992). Sample size re-estimation without unblinding for normally distributed outcomes with unknown variance. *Communications in Statistics—Theory and Methods, 21,* 2833–2853.
147. Goyal, M., Demchuk, A. M., Menon, B. K., Eesa, M., Rempel, J. L., Thornton, J., et al. (2015). Randomized assessment of rapid endovascular treatment of ischemic stroke. *New England Journal of Medicine, 372,* 1019–1030.
148. Green, R., Schneider, L. S., Amato, D. A., Beelen, A. P., Wilcock, G., Swabb, E. A., et al. (2009). Effect of tarenflubil on cognitive decline and activities of daily living in patients with mild Alzheimer disease: A randomized controlled trial. *Journal of the American Medical Association, 302,* 2557–2564.
149. Guenther, W. C. (1981). Sample size formulas for normal theory t-tests. *American Statistician, 35,* 243–244.
150. Gutjahr, G., Brannath, W., & Bauer, P. (2011). An approach to the conditional error rate principle with nuisance parameters. *Biometrics, 67,* 1039–1046.
151. Hadziyannis, S. J., Sette, H., Jr., Morgan, T. R., et al. (2004). Peginterferon-alpha2a and ribavirin combination therapy in chronic hepatitis C: A randomized study of treatment duration and ribavirin dose. *Annals of Internal Medicine, 140,* 346–355.
152. Halabi, S., & Singh, B. (2004). Sample size determination for comparing several survival curves with unequal allocations. *Statistics in Medicine, 23,* 1793–1815.
153. Hamasaki, T., Sugimoto, T., Evans, S. R., & Sozu, T. (2013). Sample size determination for clinical trials with co-primary outcomes. Exponential event-times. *Pharmaceutical Statistics, 12,* 28–34.
154. Hamasaki, T., Evans, S. R., & Asakura, K. (2018). Design, data monitoring, and analysis of clinical trials with co-primary endpoints: A review. *Journal of Biopharmaceutical Statistics, 28,* 25–51.
155. Hamilton, M. (1960). A rating scale for depression. *Journal of Neurology, Neurosurgery and Psychiatry, 23,* 56–62.

156. Han, D., Chen, Z., & Hou, Y. (2018). Sample size for noninferiority clinical trial with time-to-event data in the presence of competing risks. *Journal of Biopharmaceutical Statistics, 28,* 797–807.

157. Hardin, J. W., & Hilbe, J. M. (2012). *Generalized estimation equations* (2nd ed.). Boca Raton: CRC Press.

158. Hasegawa, T. (2014). Sample size determination for the weighted log-rank test with the Fleming-Harrington class of weights in cancer vaccine studies. *Pharmaceutical Statistics, 13,* 128–135.

159. Hauck, W. W., & Donner, A. (1977). Wald's test as applied to hypotheses in logit analysis. *Journal of the American Statistical Association, 72,* 851–853.

160. Hauschke, D., Kieser, M., Diletti, E., & Burke, M. (1999). Sample size determination for proving equivalence based on the ratio of two means for normally distributed data. *Statistics in Medicine, 18,* 93–105.

161. Hauschke, D., & Pigeot, I. (2005). Establishing efficacy of a new experimental treatment in the 'gold standard' design. *Biometrical Journal, 47,* 782–786.

162. Hauschke, D., Steinijans, V., & Pigeot, I. (2007). *Bioequivalence studies in drug development.* New York: Wiley.

163. Hayes, R. J., & Moulton, L. H. (2009). *Cluster randomised trials.* Boca Raton: CRC Press.

164. Herrmann, C., Pilz, M., Kieser, M., & Rauch, G. (2020). A new conditional performance score for the evaluation of adaptive group sequential designs with sample size recalculation. *Statistics in Medicine, 39,* 2067–2100.

165. Hilton, J. F., & Mehta, C. R. (1993). Power and sample size calculations for exact conditional tests with ordered categorical data. *Biometrics, 49,* 609–616.

166. Hislop, J., Adewuyi, T. E., Vale, L. D., Harrild, K., Fraser, C., Gurung, T., et al. (2014). Methods for specifying the target difference in a randomised controlled trial: the Difference ELicitation in TriAls (DELTA) systematic review. *PloS Medicine, 11,* e1001645.

167. Hochberg, Y., & Tamhane, A. C. (1987). *Multiple comparison procedures.* New York: Wiley.

168. Hoffmann, B., Müller, V., Rochon, J., Gondan, M., Müller, B., Albay, Z., et al. (2014). Effects of a team-based assessment and intervention on patient safety culture in general practice: An open randomised controlled trial. *BMJ Quality & Safety, 23,* 35–46.

169. Holm, S. (1979). A simple sequentially rejective multiple test procedure. *Scandinavian Journal of Statistics, 6,* 65–70.

170. Hoover, D. R., & Blackwelder, W. C. (2001). Allocation of subjects to test null relative risks smaller than one. *Statistics in Medicine, 20,* 3071–3082.

171. Horn, M., & Vollandt, R. (1998). Sample sizes for comparisons of k treatments with a control based on different definitions of the power. *Biometrical Journal, 40,* 589–612.

172. Horn, M., & Vollandt, R. (2000). A survey of sample size formulas for pairwise and many-one multiple comparisons in the parametric, nonparametric and binomial case. *Biometrical Journal, 42,* 27–44.

173. Horn, M., & Vollandt, R. (2001). A manual for the determination of sample sizes for multiple comparisons—formulas and tables. *Informatik, Biometrie und Epidemiologie in Medizin und Biologie, 32,* 1–28.

174. Hsieh, F. Y. (1992). Comparing sample size formulae for trials with unbalanced allocation using the logrank test. *Statistics in Medicine, 11,* 1091–1098.

175. Hsieh, F. Y., Block, D. A., & Larsen, M. D. (1998). A simple method of sample size calculation for linear and logistic regression. *Statistics in Medicine, 17,* 1623–1634.

176. Huang, W.-S., Hung, H.-N., Hamasaki, T., & Hsiao, C.-F. (2017). Sample size determination for a specific region in multiregional clinical trials with multiple co-primary endpoints. *PLoS ONE, 12,* e0180405.

177. Hung, H.-M. J. (2006). Adaptive clinical trial designs: Ready for prime time? Discussion presented at 2004 Harvard-MIT division of health science and technology workshop. *Statistics in Medicine, 25,* 3313–3314.

178. ICH. (1998). *Topic E9: Statistical principles for clinical trials.* European Agency for the Evaluation of Medicinal Products. https://www.ema.europa.eu/documents/scientific-guidel ine/ich-e-9-statistical-principles-clinical-trials-step-5_en.pdf. Accessed: October 13, 2020.

179. ICH. (1998). *Topic E5(R1): Ethnic factors in the acceptability of foreign clinical data.* https://www.ema.europa.eu/en/documents/scientific-guideline/ich-e-5-r1-ethnic-factors-acceptability-foreign-clinical-data-step-5_en.pdf. Accessed: October 13, 2020.

180. ICH. (2001). *Topic E10: Choice of control group in clinical trials.* European Agency for the Evaluation of Medicinal Products. https://www.ema.europa.eu/documents/scientific-guideline/ich-e-10-choice-control-group-clinical-trials-step-5_en.pdf. Accessed: October 13, 2020.

181. ICH. (2006). *Q&A for the ICH E5(R1) Guideline on ethnic factors in the acceptability of foreign clinical data.* https://www.ema.europa.eu/en/documents/scientific-guideline/ich-e-5-r1-questions-answers-ethnic-factors-acceptability-foreign-clinical-data_en.pdf. Accessed: October 13, 2020.

182. Ikeda, K., & Bretz, F. (2010). Sample size and proportion of Japanese patients in multi-regional trials. *Pharmaceutical Statistics, 9,* 207–216.

183. Ingel, K., & Jahn-Eimermacher, A. (2014). Sample-size calculation and reestimation for a semiparametric analysis of recurrent event data taking robust standard errors into account. *Biometrical Journal, 56,* 631–648.

184. IQWiG. (2017). Allgemeine Methoden. Version 5.0, 10.07.2017, Technical Report. https://www.iqwig.de/de/methoden/methodenpapier.3020.html. Accessed: October 13, 2020.

185. Irwin, J. O. (1935). Tests of significance for differences between percentages based on small numbers. *Metron, 12,* 83–94.

186. Jaeschke, R., Singer, J., & Guyatt, G. H. (1989). Measurement of health status: Ascertaining the minimal clinically important difference. *Controlled Clinical Trials, 10,* 407–415.

187. James, S. (1991). Approximate multinormal probabilities applied to correlated multiple endpoints in clinical trials. *Statistics in Medicine, 10,* 1123–1135.

188. Jankovic, J., Watts, R. L., Martin, W., & Boroojerdi, B. (2007). Transdermal Rotigotine. *Archives of Neurology, 64,* 676–682.

189. Jennison, C., & Turnbull, B. W. (2000). *Group sequential methods with applications to clinical trials.* Boca Raton: Chapman & Hall/CRC.

190. Jennison, C., & Turnbull, B. W. (2003). Mid-course sample size modification in clinical trials based on the observed effect. *Statistics in Medicine, 22,* 971–993.

191. Jennison, C., & Turnbull, B. W. (2015). Adaptive sample size modification in clinical trials: Start small then ask for more? *Statistics in Medicine, 34,* 3793–3810.

192. Jung, S.-H., Kang, S. J., McCall, L. M., & Blumenstein, B. (2005). Sample size computation for two-sample noninferiority log-rank test. *Journal of Biopharmaceutical Statistics, 15,* 969–979.

193. Jung, S.-H., Kim, C., & Chow, S.-C. (2008). Sample size calculation for the log-rank test for multi-arm trials with a control. *Journal of the Korean Statistical Society, 37,* 11–22.

194. Jung, S.-H., & Chow, S.-C. (2012). On sample size calculation for comparing survival curves under general hypothesis testing. *Journal of Biopharmaceutical Statistics, 22,* 485–495.

195. Kawai, N., Chuang-Stein, C., Komiyama, O., & Ii, Y. (2008). An approach to rationalize partitioning sample size into individual regions in a multiregional trial. *Drug Information Journal, 42,* 139–147.

196. Kerry, S., & Bland, J. (2001). Unequal cluster sizes for trials in English and Welsh general practice: implications for sample size calculations. *Statistics in Medicine, 20,* 377–390.

197. Keuls, M. (1952). The use of the 'Studentized range' in connection with an analysis of variance. *Euphytica, 1,* 112–122.

198. Kieser, M. (2018). *Fallzahlplanung in der medizinischen Forschung.* Wiesbaden: Springer.

199. Kieser, M., & Friede, T. (2000). Re-calculating the sample size in internal pilot study designs with control of the type I error rate. *Statistics in Medicine, 19,* 901–911.

200. Kieser, M., & Friede, T. (2000b). Blinded sample size re-estimation in multi-armed clinical trials. *Drug Information Journal, 34,* 455–460.

201. Kieser, M., & Friede, T. (2003). Simple procedures for blinded sample size adjustment that do not affect the type I error rate. *Statistics in Medicine, 22,* 3571–3581.

202. Kieser, M., & Friede, T. (2007). Planning and analysis of three-arm non-inferiority trials with binary endpoints. *Statistics in Medicine, 26,* 253–273.

203. Kieser, M., Friede, T., & Gondan, M. (2013). Assessment of statistical significance and clinical relevance. *Statistics in Medicine, 32,* 1707–1719.

204. Kieser, M., Kirchner, M., Dölger, E., & Götte, H. (2018). Optimal planning of phase II/III programs for clinical trials with multiple endpoints. *Pharmaceutical Statistics, 17,* 437–457.

205. Kieser, M., & Wassmer, G. (1996). On the use of the upper confidence limit for the variance from a pilot sample for sample size determination. *Biometrical Journal, 38,* 941–949.

206. King, M. T. (1996). The interpretation of scores from the EORTC quality of life questionnaire QLQ-C30. *Quality of Life Research, 5,* 555–567.

207. Kirby, S., Burke, J., Chuang-Stein, C., & Sin, C. (2012). Discounting phase 2 results when planning phase 3 clinical trials. *Pharmaceutical Statistics, 11,* 373–385.

208. Kirchner, M., Kieser, M., Götte, H., & Schüler, A. (2016). Utility-based optimization of phase II/III programs. *Statistics in Medicine, 35,* 305–316.

209. Ko, F.-S., Tsou, H.-H., Liu, J.-P., & Hsiao, C.-F. (2010). Sample size determination for a specific region in a multiregional trial. *Journal of Biopharmaceutical Statistics, 20,* 870–885.

210. Koch, A., & Röhmel, J. (2004). Hypothesis testing in the "gold standard" design for proving the efficacy of an experimental treatment relative to placebo and a reference. *Journal of Biopharmaceutical Statistics, 14,* 315–325.

211. Kolassa, J. E. (1995). A comparison of size and power calculations for the Wilcoxon statistic for ordered categorical data. *Statistics in Medicine, 14,* 1577–1581.

212. Kombrink, K., Munk, A., & Friede, T. (2013). Design and semiparametric analysis of non-inferiority trials with active and placebo control for censored time-to-event data. *Statistics in Medicine, 32,* 3055–3066.

213. Kuck, K.-H., Brugada, J., Fürnkranz, A., Metzner, A., Ouyang, F., Chun, J., et al. (2016). Cryoballoon or radiofrequency ablation for paroxysmal atrial fibrillation. *New England Journal of Medicine, 374,* 2235–2245.

214. Lachin, J. M. (1977). Sample size determinations for r × c comparative trials. *Biometrics, 33,* 315–324.

215. Lachin, J. M. (1981). Introduction to sample size determination and power analysis for clinical trials. *Controlled Clinical Trials, 2,* 93–113.

216. Lachin, J. M. (2011). Power and sample size evaluation for the Cochran-Mantel-Haenszel mean score (Wilcoxon rank sum) test and the Cochran-Armitage test for trend. *Statistics in Medicine, 30,* 3057–3066.

217. Lake, S., Kammann, E., Klar, N., & Betensky, R. (2002). Sample size re-estimation in cluster randomization trials. *Statistics in Medicine, 21,* 1337–1350.

218. Landau, S., & Stahl, D. (2013). Sample size and power calculations for medical studies by simulation when closed form expressions are not available. *Statistical Methods in Medical Research, 22,* 324–345.

219. Larkin, J., Del Vecchio, M., Ascierto, P. A., Krajsova, I., Schachter, J., Neyns, B., et al. (2014). Vemurafenib in patients with BRAF^{V600} mutated metastatic melanoma: an open-label, multicentre, safety study. *Lancet Oncology, 15,* 436–444.

220. Lassen, M. R., Gallus, A., Raskob, G. E., Pineo, G., Chen, D., Ramirez, L. M., et al. (2010). Apixaban versus enoxaparin for thromboprophylaxis after hip replacement. *New England Journal of Medicine, 363,* 2487–2498.

221. Latouche, A., & Porcher, R. (2007). Sample size calculations in the presence of competing risks. *Statistics in Medicine, 26,* 5370–5380.

222. Lee, M.-K., Song, H.-H., Kang, S.-H., & Ahn, C. (2002). The determination of sample sizes in the comparison of two multinomial proportions from ordered categories. *Biometrical Journal, 44,* 395–409.

223. Lee, P. H., & Tse, A. C. Y. (2017). The quality of the reported sample size calculations in randomized controlled trials indexed in PubMed. *European Journal of Internal Medicine, 40,* 16–21.

224. Lehmacher, W., & Wassmer, G. (1999). Adaptive sample size calculations in group sequential trials. *Biometrics, 55,* 1286–1290.

225. Levin, G. P., Emerson, S. C., & Emerson, S. S. (2013). Authors' response to 'Adaptive clinical trial designs with pre-specified rules for modifying the sample size: A different perspective'. *Statistics in Medicine, 32,* 1280–1282.

226. Li, Z., Chuang-Stein, C., & Hoseyni, C. (2007). The probability of observing negative subgroup results when the treatment effect is positive and homogeneous across all subgroups. *Drug Information Journal, 41,* 47–56.

227. Lin, D. Y., & Wei, L. J. (1989). The robust inference for the Cox proportional hazards model. *Journal of the American Statistical Association, 84,* 1074–1078.

228. Lin, D. Y., & Ying, Z. (1994). Semiparametric analysis of the additive risk model. *Biometrika, 81,* 61–71.

229. Linde, K., Berner, M. M., & Kriston, L. (2008). St. John's wort for depression. *Cochrane Database of Systematic Reviews, 4,* CD000448.

230. Little, R. J. A. (1993). Pattern-mixture models for multivariate incomplete data. *Journal of the American Statistical Association, 88,* 125–134.

231. Little, R. J. A. (1994). A class of pattern-mixture models for normal incomplete data. *Biometrika, 81,* 471–483.

232. Liu, F. (2013). A revisit to sample size and power calculation for testing odds ratio in two independent binomials. *Biometrics, 69,* 530–536.

233. Liu, S., Chu, C., & Rong, A. (2018). Weighted log-rank test for time-to-event data in immunotherapy trials with random delayed treatment effect and cure rate. *Pharmaceutical Statistics, 17,* 541–554.

234. Liu, W. (1996). On some single-stage, step-down and step-up procedures for comparing three normal means. *Computational Statistics & Data Analysis, 21,* 215–227.

235. Liu-Seifert, H., Siemers, E., Selzler, K., et al., & Mohs, R. for the Alzheimer's Disease Neuroimaging Initiative. (2016). Correlation between cognition and function across the spectrum of Alzheimer's disease. *The Journal of Prevention of Alzheimer's Disease, 3,* 138–144.

236. Lu, K. (2016). Distribution of the two-sample t-test statistic following blinded sample size re-estimation. *Pharmaceutical Statistics, 15,* 208–215.

237. Luik, A., Radzewitz, A., Kieser, M., Walter, M., Bramlage, P., Hörmann, P., et al. (2015). Cryoballon versus open irrigated radiofrequency ablation in patients with paroxysmal atrial fibrillation: The prospective, randomised, controlled, non-inferiority FreezeAF study. *Circulation, 132,* 1311–1319.

238. Luik, A., Merkel, M., Hoeren, D., Riexinger, T., Kieser, M., & Schmitt, C. (2010). Rationale and design of the FreezeAF trial: A randomized controlled noninferiority trial comparing isolation of the pulmonary veins with the cryoballoon catheter versus open irrigated radiofrequency isolation in patients with paroxysmal atrial fibrillation. *American Heart Journal, 159,* 555–560.

239. Luo, X., Huang, B., & Quan, H. (2019). Design and monitoring of survival trials based on restricted mean survival times. *Clinical Trials, 16,* 616–625.

240. Luo, X., Mao, X., Chen, X., Qiu, J., Bai, S., & Quan, H. (2019). Design and monitoring of survival trials in complex scenarios. *Statistics in Medicine, 38,* 192–209.

241. Lydersen, S., Fagerland, M. W., & Laake, P. (2009). Recommended tests for association in 2×2 tables. *Statistics in Medicine, 28,* 1159–1175.

242. Manatunga, A. K., Hudgens, M. G., & Chen, S. (2001). Sample size estimation in cluster randomised studies with varying cluster size. *Biometrical Journal, 43,* 75–86.

243. Mann, H. B., & Whitney, D. R. (1947). On a test of whether one of two random variables is stochastically larger than the other. *Annals of Mathematical Statistics, 18,* 50–60.

244. Manns, M. P., McHutchison, J. G., Gordon, S. C., et al. (2001). Peginterferon alfa-2b plus ribavirin compared with interferon alfa-2b plus ribavirin for initial treatment of chronic hepatitis C: A randomised trial. *Lancet, 358,* 958–965.

245. Maringwa, J. T., Quinten, C., King, M., Ringash, J., Osoba, D., Coens, C., Martinelli, F., Vercauteren, J., Cleeland, C. S., Flechtner, H., Gotay, C., Greimel, E., Taphoorn, M. J., Reeve, B. B., Schmucker-Von Koch, J., Weis, J., Smit, E. F., van Meerbeeck, J. P., Bottomley, A., &

EORTC PROBE project and the Lung Cancer Group. (2011). Minimal important differences for interpreting health-related quality of life scores from the EORTC QLQ-C30 in lung cancer patients participating in randomized clinical trials. *Supportive Care in Cancer, 19*, 1753–1760.

246. Maruo, K., Tada, K., Ishii, R., & Gosho, M. (2018). An efficient procedure for calculating sample size through statistical simulations. *Statistics in Biopharmaceutical Research, 10*, 1–8.

247. Matsouka, R. A., & Betensky, R. (2015). Power and sample size calculations for the Wilcoxon-Mann-Whitney test in the presence of death-censored observations. *Statistics in Medicine, 34*, 406–431.

248. Maurer, W., Hothorn, L., & Lehmacher, W. (1995). Multiple comparisons in drug clinical trials and preclinical assays: A-priori ordered hypotheses. In J. Vollmar (Ed.), *Biometrie in der chemisch-pharmazeutischen Industrie* (Vol. 6, pp. 3–21). Stuttgart: Gustav Fischer Verlag.

249. Mauri, L., & D'Agostino, R. B. (2017). Challenges in the design and interpretation of noninferiority trials. *New England Journal of Medicine, 377*, 1357–1367.

250. Mayer, C., Perevozskaya, I., Leonov, S., Dragalin, V., Pritchett, Y., Bedding, A., et al. (2019). Simulation practices for adaptive trial designs in drug and device development. *Statistics in Biopharmaceutical Research, 11*, 325–335.

251. MacPherson, H., Thomas, K., Walters, S., & Fitter, M. (2001). The York acupuncture safety study: prospective survey of 34,000 treatments by traditional acupuncturists. *British Medical Journal, 323*, 486–487.

252. McCullagh, P. (1980). Regression models for ordinal data. *Journal of the Royal Statistical Society, Series B, 43*, 109–142.

253. McDaniel, L. S., Yu, M., & Chappell, R. (2016). Sample size under the additive hazards model. *Clinical Trials, 13*, 188–198.

254. McKeown, A., Gewandter, J. S., McDermott, M. P., Pawlowski, J. R., Poli, J. J., Rothstein, D., et al. (2015). Reporting of sample size calculations in analgesic clinical trials: ACTTION systematic review. *The Journal of Pain, 16*, 199–206.

255. Mehta, C. R. (2013). Adaptive clinical trial designs with pre-specified rules for modifying the sample size: A different perspective. *Statistics in Medicine, 32*, 1276–1279.

256. Mehta, C., & Liu, L. (2016). An objective re-evaluation of adaptive sample size re-estimation: commentary on 'Twenty-five years of confirmatory adaptive designs'. *Statistics in Medicine, 35*, 350–358.

257. Mehta, C. R., Liu, L., & Theurer, C. (2019). An adaptive population enrichment phase III trial of TRC105 and pazopanib alone in patients with advanced angiosarcoma (TAPPAS trial). *Annals of Oncology, 30*, 103–108.

258. Mehta, C. R., & Pocock, S. J. (2011). Adaptive increase in sample size when interim results are promising: A practical guide with examples. *Statistics in Medicine, 30*, 3267–3284.

259. Mielke, M., Munk, A., & Schacht, A. (2008). The assessment of non-inferiority in a gold standard design with censored, exponentially distributed endpoints. *Statistics in Medicine, 27*, 5093–5110.

260. Miller, F. (2005). Variance estimation in clinical studies with interim sample size reestimation. *Biometrics, 61*, 355–361.

261. Millum, J., & Grady, C. (2013). The ethics of placebo-controlled trials: Methodological justifications. *Contemporary Clinical Trials, 36*, 510–514.

262. Ministry of Health, Labour and Welfare of Japan. (2007). *Basic principles on global clinical trials*. https://www.pmda.go.jp/files/000153265.pdf. Accessed: October 13, 2020.

263. Montgomery, S. A. (1994). Clinically relevant effect sizes in depression. *European Neuropsychopharmacology, 4*, 283–284.

264. Morris, T. P., White, I. R., & Crowther, M. J. (2019). Using simulation studies to evaluate statistical methods. *Statistics in Medicine, 38*, 2074–2102.

265. Müller, V., Hoffmann, B., Albay, Z., & Gerlach, F. M. (2012). The Frankfurt patient safety matrix—A safety culture instrument. *Zeitschrift für Allgemeinmedizin, 87*, 499–508.

266. Müller, H.-H., & Schäfer, H. (2001). Adaptive group sequential designs for clinical trials: Combining advantages of adaptive and of group sequential approaches. *Biometrics, 57*, 886–891.

267. Mütze, T., Munk, A., & Friede, T. (2016). Design and analysis of three-arm trials with negative binomially distributed endpoints. *Statistics in Medicine, 35,* 505–521.

268. Munzel, U., & Hauschke, D. (2003). A nonparametric test for proving non inferiority in clinical trials with ordered categorical data. *Pharmaceutical Statistics, 2,* 31–37.

269. Nam, J.-M. (1987). A simple approximation for calculating sample sizes for detecting linear trend in proportions. *Biometrics, 43,* 701–705.

270. National Institutes of Health/National Cancer Institute. (2017). Common terminology criteria for adverse events (CTCAE), version 5.0. https://ctep.cancer.gov/protocoldevelopment/electr onic_applications/docs/ctcae_v5_quick_reference_8.5x11.pdf. Accessed: October 13, 2020.

271. Newman, D. (1939). The distribution of the range in samples from a normal population, expressed in terms of an independent estimate of standard deviation. *Biometrika, 31,* 20–30.

272. Ng, H. K. T., & Tang, M.-L. (2005). Testing the equality of two Poisson means using the rate ratio. *Statistics in Medicine, 24,* 955–965.

273. Noether, G. E. (1987). Sample size determination for some common nonparametric tests. *Journal of the American Statistical Association, 82,* 645–647.

274. Nunn, A., Bath, P. M., & Gray, L. J. (2016). Analysis of the modified rankin scale in randomized controlled trials of acute ischaemic stroke: A systematic review. *Stroke Research and Treatment, 2016,* 7 pages.

275. O'Hagan, A., Stevens, J. W., & Campbell, M. J. (2005). Assurance in clinical trial design. *Pharmaceutical Statistics, 4,* 187–201.

276. Offen, W., Chuang-Stein, C., Dmitrienko, A., Littman, G., Maca, J., Meyerson, L., et al. (2007). Multiple co-primary endpoints: medical and statistical solutions—A report from the multiple endpoints expert team of the pharmaceutical research and manufacturers of America. *Drug Information Journal, 41,* 31–46.

277. Osoba, D., Rodrigues, G., Myles, J., Zee, B., & Pater, J. (1998). Interpreting the significance of changes in health-related quality-of-life scores. *Journal of Clinical Oncology, 16,* 139–144.

278. Othus, M., Barlogie, B., Leblanc, M. L., & Crowley, J. J. (2012). Cure model as a useful statistical tool for analyzing survival. *Clinical Cancer Research, 18,* 3731–3736.

279. Palmer, R., Enderby, P., Cooper, C., Latimer, N., Julious, S., Paterson, G., et al. (2012). Computer therapy compared with usual care for people with long-standing aphasia poststroke. *Stroke, 43,* 1904–1911.

280. Peckham, E., Brabyn, S., Cook, L., Devlin, T., Dumville, J., & Torgerson, D. J. (2015). The use of unequal randomisation ratios in clinical trials: An update. *Contemporary Clinical Trials, 45,* 113–122.

281. Peto, R., & Peto, J. (1972). Asymptotically efficient rank invariant test procedures. *Journal of the Royal Statistical Society Series A, 135,* 185–207.

282. Piaggio, G., Elbourne, D. R., Pocock, S. J., Evans, S. J., & Altman, D. G. (2012). Reporting of non-inferiority and equivalence randomized trials: Extension of the CONSORT 2010 statement. *JAMA, 308,* 2594–2604.

283. Pigeot, I., Schäfer, J., Röhmel, J., & Hauschke, D. (2003). Assessing non-inferiority of a new treatment in a three-arm clinical trial including a placebo. *Statistics in Medicine, 22,* 883–899.

284. Pilz, M., Kunzmann, K., Herrmann, C., Rauch, G., & Kieser, M. (2019). A variational approach to optimal two-stage designs. *Statistics in Medicine, 38,* 4159–4171.

285. Pilz, M., Kunzmann, K., Herrmann, C., Rauch, G., & Kieser, M. (2019b). *Optimizing adaptive two-stage designs.* Paper presented at the 40th Annual Conference of the International Society for Clinical Biostatistics, Utrecht, 14–18 July 2019.

286. Pilz, M., Kilian, S., & Kieser, M. (2020). A note on the shape of sample size functions of optimal adaptive two-stage designs. *Communications in Statistics—Theory and Methods.* https://doi.org/10.1080/03610926.2020.1776875.

287. Pobiruchin, M., & Kieser, M. (2013). Sample size calculation and blinded sample size recalculation in clinical trials where the treatment effect is measured by the relative risk. *Communications in Statistics—Simulation and Computation, 42,* 1643–1653.

288. Pocock, S. (1977). Group sequential methods in the design and analysis of clinical trials. *Biometrika, 64,* 191–199.

289. Posch, M., & Bauer, P. (1999). Adaptive two stage designs and the conditional error function. *Biometrical Journal, 41,* 689–696.

290. Posch, M., Bauer, P., & Brannath, W. (2003). Issues in designing flexible trials. *Statistics in Medicine, 22,* 953–969.

291. Posch, M., Maurer, W., & Bretz, F. (2011). Type I error rate control in adaptive designs for confirmatory clinical trials with treatment selection at interim. *Pharmaceutical Statistics, 10,* 96–104.

292. Posch, M., Timmesfeld, N., König, F., & Müller, H.-H. (2004). Conditional rejection probabilities of Student's t-test and design adaptation. *Biometrical Journal, 46,* 389–403.

293. Posten, H. O. (1978). The robustness of the two-sample *t*-test over the Pearson system. *Journal of Statistical Computation and Simulation, 6,* 295–311.

294. Posten, H. O., Yeh, H. C., & Owen, D. B. (1982). Robustness of the two-sample *t*-test under violations of the homogeneity of variance assumption. *Communication in Statistics—Theory and Methods, 11,* 109–126.

295. Posten, H. O. (1984). Robustness of the two-sample t-test. In D. Rasch & M. L. Tiku (Eds.), *Robustness of statistical methods and nonparametric statistics* (pp. 92–97). Dordrecht: Springer.

296. Erdmann, S., Kirchner, M., & Kieser, M. (2020). Optimal designs for phase II/III drug development programs including rules for discounting of phase II results. *BMC Medical Research Methodology, 20,* 253.

297. Preussler, S., Kirchner, M., Götte, H., & Kieser, M. (2020). Optimal designs for multi-arm phase II/III drug development programs. *Statistics in Biopharmaceutical Research.* https://doi.org/10.1080/19466315.2019.1702092.

298. Preussler, S., Kieser, M., & Kirchner, M. (2019). Optimal sample size allocation and go/no-go decision rules for phase II/III programs where several phase III trials are performed. *Biometrical Journal, 61,* 357–378.

299. Proschan, M. A. (2005). Two-stage sample size re-estimation based on a nuisance parameter: A review. *Journal of Biopharmaceutical Statistics, 15,* 559–574.

300. Proschan, M., Glimm, E., & Posch, M. (2014). Connections between permutation and t-tests: Relevance to adaptive methods. *Statistics in Medicine, 33,* 4734–4742.

301. Proschan, M. A., & Hunsberger, S. A. (1995). Designed extension of studies based on conditional power. *Biometrics, 51,* 1315–1324.

302. Proschan, M. A., Liu, Q., & Hunsberger, S. (2003). Practical midcourse sample size modification in clinical trials. *Controlled Clinical Trials, 24,* 4–15.

303. Proschan, M. A., & Wittes, J. (2000). An improved double sampling procedure based on the variance. *Biometrics, 56,* 1183–1187.

304. Quan, H., Zhao, P.-L., Zhang, J., Roessner, M., & Aizawa, K. (2010). Sample size considerations for Japanese patients in a multi-regional trial based on MHLW guidance. *Pharmaceutical Statistics, 9,* 100–112.

305. Rahbari, N. N., Lordick, F., Fink, C., Bork, U., Stange, A., Jäger, D., et al. (2012). Resection of the primary tumour versus no resection prior to systemic therapy in patients with colon cancer and synchronous unresectable metastases (UICC stage IV): SYNCHRONOUS—A randomised controlled multicenter trial (ISRCTN30964555). *BMC Cancer, 12,* 142.

306. Rankin, J. (1957). Cerebral vascular accidents in patients over the age of 60: II. Prognosis. *Scottish Medical Journal, 2,* 200–215.

307. Rauch, G., Hafermann, L., Mansmann, U., & Pigeot, I. (2020). Comprehensive survey among statistical members of medical ethics committees in Germany on their personal impression of completeness and correctness of biostatistical aspects of submitted study protocols. *BMJ Open, 10,* e032864.

308. Rauch, G., Schüler, S., & Kieser, M. (2018). *Planning and analyzing clinical trials with composite endpoints.* Cham: Springer Nature.

309. Roebruck, P., & Kühn, A. (1995). Comparison of tests and sample size formulae for proving therapeutic equivalence based on the difference of binomial probabilities. *Statistics in Medicine, 14,* 1583–1594.

310. Röhmel, J., & Mansmann, U. (1999). Unconditional non-asymptotic one-sided tests for independent binomial proportions when the interest lies in showing non-inferiority and/or superiority. *Biometrical Journal, 2,* 149–170.

311. Röhmel, J. (2005). Problems with existing procedures to calculate exact unconditional p-values for non-inferiority/superiority and confidence intervals for two binomials and how to resolve them. *Biometrical Journal, 47,* 37–47.

312. Rosen, W. G., Mohs, R. C., & Davis, K. L. (1984). A new rating scale for Alzheimer's disease. *American Journal of Psychiatry, 141,* 1356–1364.

313. Royston, P., & Parmar, M. K. B. (2013). Restricted mean survival time: An alternative to the hazard ratio for the design and analysis of randomized trials with time-to-event outcome. *BMC Medical Research Methodology, 13,* 152.

314. Rubin, D. B. (1996). Multiple imputation after 18+ years. *Journal of the American Statistical Association, 91,* 473–489.

315. Rufibach, K., Burger, H. U., & Abt, M. (2016). Bayesian predictive power: Choice of prior and some recommendations for its use as probability of success in drug development. *Pharmaceutical Statistics, 15,* 438–446.

316. Rutterford, C., Copas, A., & Eldridge, S. (2015). Methods for sample size determination in cluster randomized trials. *International Journal of Epidemiology, 44,* 1051–1067.

317. Sanchez-Kam, M., Gallo, P., Loewy, J., Menon, S., Antonijevic, Z., Christensen, J., et al. (2014). A practical guide to data monitoring committees in adaptive trials. *Therapeutic Innovation & Regulatory Science, 48,* 316–326.

318. Sander, A., Rauch, G., & Kieser, M. (2017). Blinded sample size recalculation in clinical trials with binary composite endpoints. *Journal of Biopharmaceutical Statistics, 27,* 705–715.

319. Sandvik, L., Erikssen, J., Mowinckel, P., & Rodland, E. A. (1996). A method for determining the size of internal pilot studies. *Statistics in Medicine, 15,* 1587–1590.

320. Schechtman, E. (2002). Odds ratio, relative risk, absolute risk reduction, and the number needed to treat—Which of these should we use? *Value in Health, 5,* 431–436.

321. Schmidtmann, J., Konstantinides, S., & Binder, H. (2018). Power of the Wilcoxon-Mann-Whitney test for non-inferiority in the presence of death censored observations. *Biometrical Journal, 61,* 1187–1200.

322. Schoenfeld, D. A. (1981). The asymptotic properties of nonparametric tests for comparing survival distributions. *Biometrika, 68,* 316–319.

323. Schoenfeld, D. A. (1983). Sample-size formula for the proportional-hazards regression model. *Biometrics, 39,* 499–503.

324. Schouten, H. J. A. (1999). Sample size formula with a continuous outcome for unequal group sizes and unequal variances. *Statistics in Medicine, 18,* 87–91.

325. Schulz, K. F., Altman, D. G., & Moher, D. (2010). CONSORT 2010 statement: Updated guidelines for reporting parallel group randomized trials. *Annals of Internal Medicine, 152,* 1–8.

326. Scosyrev, E., & Glimm, E. (2019). Power analysis for multivariable Cox regression models. *Statistics in Medicine, 38,* 88–99.

327. Senn, S., & Bretz, F. (2007). Power and sample size when multiple endpoints are considered. *Pharmaceutical Statistics, 6,* 161–170.

328. Sertkaya, A., Birkenbach, A., Berlind, A., & Eyraud, J. (2014). *Examination of clinical trial costs and barriers for drug development.* https://aspe.hhs.gov/report/examination-clinical-trial-costs-and-barriers-drug-development. Accessed: October 13, 2020.

329. Shan, G., & Ma, C. (2014). A comment on sample size calculation for analysis of covariance in parallel arm studies. *Journal of Biometrics and Biostatistics, 5,* 184.

330. Shieh, G. (2017). Power and sample size calculations for contrast analysis in ANCOVA. *Multivariate Behavioral Research, 52,* 1–11.

331. Shiffman, M. L., Suter, F., Bacon, B. R., Nelson, D., Harley, H., Solá, R., et al. (2007). Peginterferon alfa-2a and ribavirin for 16 or 24 weeks in HCV genotype 2 or 3. *The New England Journal of Medicine, 357,* 124–134.

332. Shih, W. (1997). Sample size and power calculations for periodontal and other studies with clustered samples using the method of generalized estimating equations. *Biometrical Journal, 39,* 899–908.

333. Shih, W. J. (2009). Two-stage sample size reassessment using perturbed unblinding. *Statistics in Biopharmaceutical Research, 1,* 74–80.

334. Shih, W. J., & Zhao, P.-L. (1997). Design for sample size re-estimation with interim data for double-blind clinical trials with binary outcomes. *Statistics in Medicine, 16,* 1913–1923.

335. Shih, Y., & Lee, J.-H. (2018). Sample size calculations for group randomized trials with unequal group sizes through Monte Carlo simulations. *Statistical Methods in Medical Research, 27,* 2569–2580.

336. Shun, Z., Yuan, W., Brady, W. E., & Hsu, H. (2001). Type I error in sample size re-estimations based on observed treatment difference. *Statistics in Medicine, 20,* 497–513.

337. Sinclair, J. C., & Bracken, M. B. (1994). Clinically useful measures of effect in binary analyses of randomized trials. *Journal of Clinical Epidemiology, 47,* 881–889.

338. Singer, J. (1999). Letter to the editor: A method for determining the size of internal pilot studies. *Statistics in Medicine, 18,* 1151–1153.

339. Siqueira, A. L., Todd, S., & Whitehead, A. (2015). Sample size considerations in active-control non-inferiority trials with binary data based on the odds ratio. *Statistical Methods in Medical Research, 24,* 453–461.

340. Sones, W., Julious, S. A., Rothwell, J. C., Ramsay, C. R., Hampson, L. V., Emsley, R., et al. (2018). Choosing the target difference and undertaking and reporting the sample size calculation for a randomised controlled trial—the development of the DELTA^2 guidance. *Trials, 19,* 542.

341. Sozu, T., Sugimoto, T., & Hamasaki, T. (2011). Sample size determination in superiority clinical trials with multiple co-primary correlated endpoints. *Journal of Biopharmaceutical Statistics, 21,* 650–668.

342. Sozu, T., Sugimoto, T., Hamasaki, T., & Evans, S. R. (2015). *Sample size determination in clinical trials with multiple endpoints.* New York: Springer.

343. Speich, B. (2019). Adequate reporting of the sample size calculation in surgical randomized controlled trials. *Surgery, 167,* 812–814.

344. Spiegelhalter, D. J., Abrams, K. R., & Myles, J. P. (2004). *Bayesian approaches to clinical trials and health-care evaluation* (1st ed.). Chichester: Wiley.

345. Spiegelhalter, D. J., Freedman, L. S., & Parmar, M. K. B. (1994). Bayesian approaches to randomized trials (with discussion). *Journal of the Royal Statistical Society Series A, 157,* 357–416.

346. Spiegelhalter, D. J., Reedman, L. S., & Blackburn, P. R. (1986). Monitoring clinical trials—Conditional power or predictive power. *Controlled Clinical Trials, 7,* 8–17.

347. Stroup, W. W. (2013). *Generalized linear mixed models: Modern concepts, methods and applications.* Boca Raton: CRC Press.

348. Stein, C. (1945). A two-sample test for a linear hypothesis whose power is independent of the variance. *Annals of Mathematical Statistics, 16,* 243–258.

349. Stucke, K., & Kieser, M. (2012). A general approach for sample size calculation for the three-arm 'gold standard' non-inferiority design. *Statistics in Medicine, 31,* 3579–3596.

350. Su, P.-F. (2017). Power and sample size calculation for the additive hazard model. *Journal of Biopharmaceutical Statistics, 27,* 571–583.

351. Sugimoto, T., Sozu, T., & Hamasaki, T. (2012). A convenient formula for sample size calculations in clinical trials with multiple co-primary continuous endpoints. *Pharmaceutical Statistics, 11,* 118–128.

352. Sullivan, L. M., & D'Agostino, R. B. (1996). Robustness and power of analysis of covariance applied to data distorted from normality by floor effects: Homogeneous regression slopes. *Statistics in Medicine, 15,* 477–496.

353. Sullivan, L. M., & D'Agostino, R. B. (2002). Robustness and power of analysis of covariance applied to data distorted from normality by floor effects: Non-homogeneous regression slopes. *Journal of Statistical Computation and Simulation, 72,* 141–165.

354. Sullivan, L. M., & D'Agostino, R. B. (2003). Robustness and power of analysis of covariance applied to ordinal scaled data as arising in randomized controlled trials. *Statistics in Medicine, 22*, 1317–1334.

355. Sun, W., Grosser, S., & Tsong, Y. (2017). Ratio of means vs. difference of means as measures of superiority, noninferiority, and average bioequivalence. *Journal of Biopharmaceutical Statistics, 27*, 338–355.

356. Tam, W., Lo, K., & Woo, B. (2020). Reporting sample size calculations for randomized controlled trials published in nursing journals: A cross-sectional study. *International Journal of Nursing Studies, 102*. https://doi.org/10.1016/j.ijnurstu.2019.103450.

357. Tamhane, A. C. (1987). An optimal procedure for partitioning a set of normal populations with respect to a control. *The Indian Journal of Statistics, Series A, 49*, 335–346.

358. Tang, Y. (2018). Exact and approximate power and sample size calculations for analysis of covariance in randomized clinical trials with or without stratification. *Statistics in Biopharmaceutical Research, 10*, 274–286.

359. Tang, Y., & Fitzpatrick, R. (2019). Sample size calculation for the Andersen-Gill model comparing rates of recurrent events. *Statistics in Medicine, 38*, 4819–4827.

360. Tavernier, E., & Giraudeau, B. (2015). Sample size calculation: inaccurate a priori assumptions for nuisance parameters can greatly affect the power of a randomized controlled trial. *PloS ONE, 10*, e0132578.

361. Teerenstra, S., Eldridge, S., Graff, M., de Hoop, E., & Borm, G. F. (2012). A simple sample size formula for analysis of covariance in cluster randomized trials. *Statistics in Medicine, 31*, 2169–2178.

362. Therneau, T. M., & Grambsch, P. M. (2000). *Modeling survival data: Extending the Cox model*. New York: Springer.

363. Thompson, D. M., Fernald, D. H., & Mold, J. W. (2012). Intraclass correlation coefficients typical of cluster-randomized studies: Estimates from the Robert Wood prescription health projects. *Annals of Family Medicine, 10*, 235–240.

364. Timmesfeld, N., Schäfer, H., & Müller, H.-H. (2007). Increasing the sample size during clinical trials with t-distributed test statistics without inflating the type I error rate. *Statistics in Medicine, 26*, 2449–2464.

365. Todd, S., Valdés-Márquez, E., & West, J. (2012). A practical comparison of blinded methods for sample size reviews in survival data clinical trials. *Pharmaceutical Statistics, 11*, 141–148.

366. Tofte, S., Graeber, M., Cherill, R., Omoto, M., Thurston, M., & Hanifin, J. M. (1998). Eczema area and severity index (EASI): A new tool to evaluate atopic dermatitis. *Journal of the European Academy of Dermatology and Venereology, 11*, S197.

367. Tukey, J. W. (1953). *The problem of multiple comparisons*. Mimeographed monograph.

368. Turner, E. L., Li, F., Gallis, J. A., Prague, M., & Murray, D. M. (2017). Review of recent methodological developments in group-randomized trials: Part 1—Design. *American Journal of Public Health, 107*, 907–915.

369. Turner, E. L., Prague, M., Gallis, J. A., Li, F., & Murray, D. M. (2017b). Review of recent methodological developments in group-randomized trials: Part 2—Analysis. *American Journal of Public Health, 107*, 1078–1086.

370. Uesaka, H. (2009). Sample size allocation to regions in a multiregional trial. *Journal of Biopharmaceutical Statistics, 19*, 580–594.

371. Van Breukelen, G. J. P., & Candel, M. J. J. M. (2012). Efficient design of cluster randomized and multicenter trials with unknown intraclass correlation. *Statistical Methods in Medical Research, 24*, 540–556.

372. Van der Linden, P., De Villé, A., Hofer, A., Heschl, M., & Gombotz, H. (2013). Six percent hydroxyethyl starch 130/0.4 (Voluven®) versus 5% human serum albumin for volume replacement therapy during elective open-heart surgery in pediatric patients. *Anesthesiology, 119*, 1296–1309.

373. Van Elteren, P. H. (1960). On the combination of independent two sample tests of Wilcoxon. *Bulletin of the Institute of International Statistics, 37*, 351–361.

374. Varga, Z., Tsang, Y. C., & Singer, J. (2017). A simple procedure to estimate the optimal sample size in case of conjunctive coprimary endpoints. *Biometrical Journal, 59,* 626–635.

375. Vickers, A. J. (2003). Underpowering in randomized trials reporting a sample size calculation. *Journal of Clinical Epidemiology, 56,* 717–720.

376. Vollandt, R., & Horn, M. (1997). Evaluation of Noether's method of sample size determination for the Wilcoxon-Mann-Whitney test. *Biometrical Journal, 39,* 823–829.

377. Vourćh, M., Feuillet, F., Mahe, P.-J., Sebille, V., Asehnoune, K., & The BACLOREA trial group. (2016). Baclofen to prevent agitation in alcohol-addicted patients in the ICU: Study protocol for a randomized controlled trial. *Trials, 17,* 415–421.

378. Wachtlin, D., & Kieser, M. (2013). Blinded sample size recalculation in longitudinal clinical trials using generalized estimating equations. *Therapeutic Innovation & Regulatory Science, 47,* 460–467.

379. Waksman, J. A. (2007). Assessment of the Gould-Shih procedure for sample size re-estimation. *Pharmaceutical Statistics, 6,* 53–65.

380. Wang, H., Chow, S.-C., & Li, G. (2002). On sample size calculation based on odds ratio in clinical trials. *Journal of Biopharmaceutical Statistics, 12,* 471–483.

381. Wang, H., Chen, B., & Chow, S.-C. (2003). Sample size determination based on rank tests in clinical trials. *Journal of Biopharmaceutical Statistics, 13,* 735–751.

382. Wang, Y., Fu, H., Kulkarni, P., & Kaiser, C. (2013). Evaluating and utilizing probability of study success in clinical development. *Clinical Trials, 10,* 407–413.

383. Wassmer, G., & Brannath, W. (2016). *Group sequential and confirmatory adaptive designs in clinical trials.* New York: Springer.

384. Wellek, S. (2010). *Testing statistical hypotheses of equivalence and noninferiority* (2nd ed.). Boca Raton: CRC Press.

385. Wellek, S. (2015). Nearly exact sample size calculation for powerful non-randomized tests for differences between binomial proportions. *Statistica Neerlandica, 69,* 358–373.

386. Wellek, S., & Ziegler, P. (2016). *MIDN: Nearly exact sample size calculation for exact powerful nonrandomized tests for differences between binomial proportions.* R package version 1.0. https://CRAN.R-project.org/package=MIDN. Accessed: October 13, 2020.

387. Welsch, R. E. (1977). Stepwise multiple comparison procedures. *Journal of the American Statistical Association, 72,* 566–575.

388. Welte, T., Metzenauer, P., & Hartmann, U. (2008). Once versus twice daily formoterol via Novolizer® for patients with moderate to severe COPD—A double-blind, randomized, controlled trial. *Pulmonary Pharmacology & Therapeutics, 21,* 4–13.

389. Westfall, P. H., & Young, S. S. (1993). *Resampling-based multiple testing: Examples and methods for P-Value adjustment.* New York: Wiley.

390. Westfall, P. H., Tobias, R. D., & Wolfinger, R. D. (2011). *Multiple comparisons and multiple tests using SAS®* (2nd ed.). Cary, NC: SAS Institute.

391. Whitehead, J. (1993). Sample size calculations for ordered categorical data. *Statistics in Medicine, 12,* 2257–2271.

392. Whitehead, J. (1997). *The design and analysis of sequential trials* (2nd ed.). New York: Wiley.

393. Whittemore, A. S. (1978). Collapsibility of multidimensional contingency tables. *Journal of the Royal Statistical Society: Series B (Methodological), 40,* 328–340.

394. Williams, J. B., & Kobak, K. A. (2008). Development and reliability of a structured interview guide for the Montgomery Asberg depression rating scale (SIGMA). *British Journal of Psychiatry, 192,* 52–58.

395. Wilson, E. B. (1927). Probable inference, the law of succession, and statistical inference. *Journal of the American Statistical Association, 22,* 209–212.

396. Wittes, J. (2002). Sample size calculations for randomized controlled trials. *Epidemiologic Reviews, 24,* 39–53.

397. Wittes, J., & Brittain, E. (1990). The role of internal pilot studies in increasing the efficacy of clinical trials. *Statistics in Medicine, 9,* 65–72.

398. Wittes, J., Schabenberger, O., Zucker, D., Brittain, E., & Proschan, M. (1999). Internal pilot studies I: type I error rate of the I t-test. *Statistics in Medicine, 18,* 3481–3491.

399. Wollenberg, A., Howell, M. D., Guttman-Yassky, E., Silverberg, J. I., Kell, C., Ranade, K., et al. (2019). Treatment of atopic dermatitis with tralokinumab, an anti-IL-13 mAb. *Journal of Allergy and Clinical Immunology, 143,* 135–141.

400. Wollenberg, A., Howell, M.D., Guttman-Yassky, E., Silverberg, J.I., Kell, C., Ranade, K., Moate, R., & van der Merwe, R. (2019b). Treatment of atopic dermatitis with tralokinumab, an anti-IL-13 mAb—Supplementary Appendix. *Journal of Allergy and Clinical Immunology.* https://www.sciencedirect.com/science/article/pii/S0091674918308509#appsec1. Accessed: October 13, 2020.

401. Wu, J., & Wei, J. (2020). Cancer immunotherapy trial design with delayed treatment effect. *Pharmaceutical Statistics, 19,* 202–213.

402. Wüst, K., & Kieser, M. (2003). Blinded sample size recalculation for normally distributed outcomes using long- and short-term data. *Biometrical Journal, 45,* 915–930.

403. Wüst, K., & Kieser, M. (2005). Including long- and short-term data in blinded sample size recalculation for binary endpoints. *Computational Statistics and Data Analysis, 34,* 835–855.

404. Xing, B., & Ganju, J. (2005). A method to estimate the variance of an endpoint from an on-going blinded trial. *Statistics in Medicine, 24,* 1807–1814.

405. Xiong, X., & Wu, J. (2017). A novel sample size formula for the weighted log-rank test under the proportional hazards cure model. *Pharmaceutical Statistics, 16,* 87–94.

406. Xu, Z., Zhen, B., Park, Y., & Zhu, B. (2016). Designing therapeutic cancer vaccine trials with delayed treatment effect. *Statistics in Medicine, 36,* 592–605.

407. Zhang, F., Miyaoka, E., Huang, F., & Tanaka, Y. (2015). Test statistics and confidence intervals to established noninferiority between treatments with ordinal categorical data. *Journal of Biopharmaceutical Statistics, 25,* 921–938.

408. Zhang, X., Zhang, Y., Xiaofei, Y., Guo, X., Zhang, T., & He, J. (2016). Overview of phase IV clinical trials for postmarketing drug safety surveillance: A status report from the ClinicalTrials.gov registry. *BMJ Open, 6,* e010643. https://doi.org/10.1136/bmjopen-2015-010643.

409. Zhao, Y. D. (2006). Sample size estimation for the van Elteren test—A stratified Wilcoxon-Mann-Whitney test. *Statistics in Medicine, 25,* 2675–2687.

410. Zhao, Y. D., Qu, Y., & Rahardja, D. (2006). Power approximation for the van Elteren test based on location-scale family of distributions. *Journal of Biopharmaceutical Statistics, 16,* 1–13.

411. Zhao, Y. D., Rahardja, D., & Mei, Y. (2008). Sample size calculation for the van Elteren test adjusting for ties. *Journal of Biopharmaceutical Statistics, 18,* 1112–1119.

412. Zhou, M., & Kundu, S. (2016). Some practical considerations in three-arm non-inferiority trial design. *Pharmaceutical Statistics, 15,* 550–559.

413. Zimmermann, G., Kieser, M., & Bathke, A.C. (2020). Sample size calculation and blinded recalculation for analysis of covariance models with multiple random covariates. Submitted to *Journal of Biopharmaceutical Statistics, 30,* 143–159.

414. Zucker, D. M., Wittes, J. T., Schabenberger, O., & Brittain, E. (1999). Internal pilot studies II: Comparison of various procedures. *Statistics in Medicine, 18,* 3493–3509.

Index

© Springer Nature Switzerland AG 2020
M. Kieser, *Methods and Applications of Sample Size Calculation and Recalculation
in Clinical Trials*, Springer Series in Pharmaceutical Statistics,
https://doi.org/10.1007/978-3-030-49528-2

Printed in the United States
by Baker & Taylor Publisher Services